Miguel de Asúa
Science and Catholicism in Argentina (1750–1960)

Religion and Society

Edited by
Gustavo Benavides, Frank J. Korom,
Karen Ruffle and Kocku von Stuckrad

Volume 89

Miguel de Asúa

Science and Catholicism in Argentina (1750–1960)

A Study on Scientific Culture, Religion, and Secularisation in Latin America

DE GRUYTER

ISBN 978-3-11-135276-3
e-ISBN (PDF) 978-3-11-048877-7
e-ISBN (EPUB) 978-3-11-048749-7
ISSN 1437-5370

Library of Congress Control Number: 2022930555

Bibliographic information published by the Deutsche Nationalbibliothek
The Deutsche Nationalbibliothek lists this publication in the Deutsche Nationalbibliografie;
detailed bibliographic data are available on the Internet at http://dnb.dnb.de.

© 2023 Walter de Gruyter GmbH, Berlin/Boston
This volume is text- and page-identical with the hardback published in 2022.
Printing and binding: CPI books GmbH, Leck

www.degruyter.com

For Ignacio and Javier

Acknowledgements

This work was written with the support of the National Council for Scientific and Technological Research (Argentina). The research conducive to this book began a few years ago, in part as a follow-up of the issues explored in *Science in the Vanished Arcadia* (Leiden, 2014). But the main questions discussed here are a result of a perceived contrast between the mainstream scholarship of historical studies in science and religion in the anglophone world and my own experience of how that relationship developed in a society that underwent a different pattern of secularisation.

I am in great debt with family, old friends, and colleagues who along these years were kind enough to carry in their luggage many, many books which otherwise would not have arrived safely. They are (in alphabetical order): Ignacio de Asúa, Juan Garralda, Esteban Greif, Jorge Mercado, José Luis Narvaja SJ, Pablo Ubierna, and Carlos Wahren. I heartily thank them for their unselfish collaboration. Damasia Becú, Analía Busala, Father Rafael Braun (†), Daniel Miño SJ, Alejandro Palomo, Pablo Penchaszadeh, Msgr Luis Rivas, and again Pablo Ubierna generously helped at one point or another with bibliography, facilitated access to archives and hard-to-find documents, and/or were willing to discuss specific points of the argument deployed in this book. My thanks to all of them. Ubierna called my attention to several typos. Eduardo Míguez read carefully the first chapters and contributed insightful comments on points of Argentine political and economic history. All errors and interpretations are of course mine.

In the Bibliography there is a list of the archives consulted for this work. For their helpful assistance, I remain thankful to the librarians of the Ancient Periodicals Division of Biblioteca Nacional, Biblioteca Prebisch, the Library of Academia Nacional de la Historia (Mariana Lagar and Ana Oneto), the Library of the former Schools of Philosophy and Theology at Colegio Máximo in San Miguel (Gerardo Losada, Sergio Boada, Fernando Santamaría), the Library of Instituto de Investigaciones Históricas Emilio Ravignani (University of Buenos Aires), the Library of the School of Medicine (University of Buenos Aires), the Library of Museo Nacional de Ciencias Naturales (Ms Marta del Priore), the Library of Asociación Argentina Amigos de la Astronomía, the Library of the Instituto de Cultura Religiosa Superior, the Biblioteca del Maestro, the Library of San José School (Mr Oscar Palmyero), the Library of the extinct Observatorio de San Miguel, the Library of Congress (Ms Malea Walker, Serial and Government Publications Division), and the Library of the School of Theology and Central Library of Universidad Católica Argentina.

Some chapters contain revised versions or fragments of previously published material, listed below. The corresponding permissions for reproduction have been required and granted in each of the cases so that all of them are reproduced with permission. I remain obliged to the helpful critiques of many of the referees.

- "Jesuit Science in the Missions of Paraguay and Río de la Plata," in *Latin American Perspectives on Science and Religion*, edited by Ignacio Silva (London: Pickering & Chatto, 2014), 71–84, reproduced with permission of INFORMA UK LIMITED through PLSclear (chapter 1).
- "'Los phisicos modernos quasi todos son copernicanos.' Copernicanism and its Discontents in Colonial Río de la Plata," *Journal for the History of Astronomy* 48, no. 2 (2017): 160–70, © 2017 SAGE Publications (chapter 2).
- "Traces on a Muddy Shore. Science and Religion in Colonial and Early Independent Río de la Plata." *Annals of Science* 78, no. 2 (2021): 197–220, partially reproduced by permission of the publisher Taylor & Francis Ltd (chapter 2).
- "Draper, the 'Conflict Thesis,' and Secularising Politics in Late Nineteenth-Century Argentina," *Journal of Religious History* 43, no. 3 (2019): 305–27, © 2019, John Wiley & Sons Ltd (chapter 4).
- "Darwin among the Pagans: Secularisation and the Reception of the Theory of Evolution in Buenos Aires," *Science and Christian Belief* 31, no. 1 (2019): 4–25, reproduced by permission generously granted by the editor (chapters 4 and 5).
- "Science, Catholicism and Politics in Argentina (1910–1935)," *British Journal for the History of Science* 53, no. 2 (2020): 139–158, © 2020 The British Society for the History of Science, reproduced with permission (chapter 6).
- "Science and Integral Catholicism in Interwar Argentina," *Church History and Religious Culture* 99 (2019): 485–503, reproduced with permission graciously granted by the editor (chapter 7).
- "Argentina Catholic Democratic Scientists and the Projects of a Research University (1932–1959)," *Catholic Historical Review* 106, no. 1 (2020): 107–32, reproduced by the generous permission of the editor and of the director of Catholic University of America Press (chapter 8).
- "Three Centuries of Scientific Culture and Catholicism in Argentina: A Case Study of Long Trends," in *Rethinking History, Science, and Religion: An Exploration of Conflict and the Complexity Principle*, edited by Bernard Lightman, 37–49, © 2019. Reprinted by permission of The University of Pittsburgh Press (conclusion).
- "Science, Catholicism, and the French (Latin) Pattern of Secularization. The Case of Argentina (1820–1958)," *Studies in Religion/Sciences Religieuses* 50,

no. 2 (2021): 237–57, © 2021 SAGE Publications (introduction and conclusion).

In a few cases, previous versions of the papers had been presented as communications or lectures in meetings, such as the Seventh Latin-American Congress on Science and Religion (Pontifical Catholic University of Rio de Janeiro, October 2012); the seminar on "Science and Religion: Exploring the Complexity Thesis" at the 25th International Congress of History of Science and Technology (Rio de Janeiro, July 2017); the meeting on Ecclesiology and Society organised by IMHI-CIHU (Conicet) and the Centre d'études en Sciences Sociales du Religieux (CéSor – CNRS/PSL-École des Hautes Études en Sciences Sociales) in Buenos Aires, October 2017; and the 17th International Meeting on Jesuit Missions (Unisinos, São Leopoldo, Rio Grande do Sul, October 2018). I would like to thanks Ignacio Silva, Bernie Lightman, Pablo Ubierna, and Eliane Deckmann Fleck, respectively for their kind invitations to participate in these events. I should also express my gratitude to the students in the various courses and seminars on the history of science and religion I taught during these years (at the Jesuit Colegio Máximo in San Miguel; at the Graduate Program on Philosophy of Science in the Department of Philosophy of Universidad Austral; and at the Department of History of Universidad Nacional de Tres de Febrero). I thank Pablo Figueroa SJ, Claudia Vanney, and Samuel Amaral, respectively for their kind invitations to teach these seminars.

I would also like to recognise the efficient help, expertise, and ongoing support of the editors of De Gruyter: Sophie Wagenhofer (Senior Acquisitions Editor), Aaron Sanborn-Overby (Acquisitions Editor), Katrin Mittmann (Content Editor) and Anne Hirschelmann (Production Editor). At different stages of the making of the book (which took longer than expected) they helped me in many practical ways. I also wish to thank Gustavo Benavides, Frank J. Korom, Karen Ruffle, and Kocku von Stuckrad for their welcome of the book in De Gruyter's series Religion and Society. The anonymous reader of the manuscript provided a crucial insight which helped to sharpen the structure of the argument for which I remain very thankful.

Finally, my wife Natividad and our sons Ignacio and Javier provided the emotional and logistic support and genial atmosphere that made the writing of this book possible. They tolerated countless inconveniences and helped me to face the many hurdles which had to be overcome to complete the job. I can only express to them my deepest and warmest gratitude.

Contents

Acknowledgements —— VII

Introduction —— 1
    Science and Religion: Conflict and Complexity —— 4
    Secularisation and Science —— 8
    Brief Outlook of the Book —— 16

Chapter 1
Jesuit Science —— 19
    Natural Histories —— 25
    Herbal Medicine and Medicine —— 31
    Astronomy —— 35
    Cartography —— 40
    Jesuit Missionary Science in Río de la Plata —— 43

Chapter 2
Catholic Enlightenment. Science and Religion in Colonial and Early Independent Río de la Plata —— 50
    Historical Background —— 52
    Late Scholasticism: Modern Science and the Authority of the Church —— 56
    Biblical Geology —— 62
    A Clergyman's Natural Theology —— 67
    Non-Clerical Natural Theology —— 70
    Conclusions —— 75

Chapter 3
Religious and Secular Spaces in Post-Independence Río de la Plata (1820–1827). From Natural Theology to Secular Science —— 80
    Ecclesiastical Reform —— 82
    From Convent to Cemetery and Agricultural Station —— 86
    The Secularisation of Hospitals —— 89
    The Convent of Santo Domingo —— 91
    Natural History among the Clergy —— 96
    The University of Buenos Aires and *idéologie* —— 99
    Conclusions —— 106

Chapter 4
The "Conflict Thesis," Darwin, and Secularising Politics in Late Nineteenth-Century Argentina — 110

Historical Background — 111
Debate over Draper in the Liberal Press — 119
The Conflict Thesis in the Parliamentary Debates of the 1870s — 124
The 1879 Parliamentary Debate over Science and Religion — 125
Estrada — 128
Darwin and the Conflict Thesis on Stage — 132
Catholic Answers — 140
Excursus on "Madness and Crime" — 143
The Great Debate — 146
Sequels — 151
Conclusions — 155

Chapter 5
Science and Secularism: Transitional Times — 158

Positivism, Religion, and the Unknowable — 159
Spiritualism, Theosophy, and Science — 164
The Bankruptcy of Science — 171
Shifting Meanings — 175
The Uncomfortable Truth — 180
Soul versus Brain — 189
Conclusions — 193

Chapter 6
Science, Catholicism, and Politics in Argentina, 1910–1935 — 195

Church and Politics in Transition — 196
Catholics and Science — 200
Gallardo — 206
Biologist and Naturalist — 208
Science, Religion, and Evolution — 211
The Bishop Astronomer — 213
Observatories — 216
The Rise of Catholic Scientists — 218
Conclusions — 219

Chapter 7
Science and Integral Catholicism in Interwar Argentina — 221

Historical Background — 222

Catholics and Einstein —— 224
Catholics, University Reform, and Hans Driesch —— 228
Evolution and Belief —— 232
Maritain at the Crossroads —— 234
Against Draper —— 236
Catholics and Eugenics —— 238
Conclusions —— 244

Chapter 8
Argentine Catholic Democratic Scientists and their Projects of a Research University (1932–1959) —— 247
Lewis: An Early Proposal for "Free" Universities —— 248
Braun Menéndez: Research Institutes as the Basis of the University —— 252
Durelli —— 257
The Catholic Institute of Sciences —— 260
The First Catholic Universities —— 261
The Other Side of the Coin —— 267
Science and Religion —— 270
Conclusions —— 271

Conclusions —— 275
Religious and Secular Space —— 276
Science and Secularisation as Social Differentiation —— 278
Secularisation as Appropriation of Religious Rituals and Symbols —— 281
Catholic Scientists —— 284
A Last Word —— 287

Abbreviations used in this book —— 291

Bibliography —— 293
Manuscripts —— 293
Newspapers —— 293
Government Series —— 294
Printed Works —— 294

Index of Names —— 349
Index of Places —— 357
Index of Subjects —— 359

Introduction

A few years ago, when I told a distinguished colleague that I was working on a book on science and religion in Argentina, he replied rather shocked: "How is that possible? What has science to do with religion?" Not many months passed before I told a moral theologian and long-time editor of a Catholic magazine the same thing. This time the answer was: "Catholic bishops in Argentina never had anything to do with science, they just ignored it." These anecdotes from both sides of the religious divide are fairly representative of a general climate of opinion. But contrary to what is currently held, science and religion have interacted in different and complex ways since the eighteenth century in the territory of the Argentine Republic. In the following chapters, I shall examine these interactions at close range. It seems undeniable that here as in other Catholic societies the Enlightenment's cult of progress, positivism, and other ideologies which enlisted the authority of science contributed to the transition from the Old Regime to modernity and to the social process of secularisation.[1] But this is different from saying that it was the triumphal march of science as a rational inquiry into nature which turned people's minds away from the dark recesses of religious ignorance; it is also a far cry from seeing "religion as inchoate and backward superstition or as an apathetic froth obscuring the surface of the real until blown away by revolution."[2] The assertion that the growth of scientific knowledge per se "caused" the demise of the religious is a core legitimizing element in the grand narratives of secularisation. I expect to show that at least in Argentina things were the other way around: it was the particular dynamics of the process of secularisation which contributed to shaping the relationships between science and religion.

Colin Russell has affirmed that in Catholic cultures "the polarity between sacred and secular was often much sharper than in Britain and the United States, with the result that progressive science-based ideologies were more frequently in explicit contention with conservative political and ecclesiastical forces."[3] The

---

[1] For science and ideology, see Eric C. Martin, "Science and Ideology," in *Internet Encyclopedia of Philosophy*, revised in 2020, https://iep.utm.edu/sci-ideo/.
[2] David Martin, "Does the Advance of Science Mean Secularisation?," *Science and Christian Belief* 19, no. 1 (2007): 12. For the notion of master-narrative, see Allan Megill, "Recounting the Past: 'Description,' Explanation and Narrative in Historiography," *American Historical Review* 94, no. 3 (1989): 627–53.
[3] Colin Russell, "The Conflict of Science and Religion," in *Science and Religion. A Historical Introduction*, ed. Gary B. Ferngren (Baltimore: Johns Hopkins University Press, 2002), 6.

large majority of historical studies on science and religion deal with anglophone countries. This long-term study (from 1750 to 1960) of science and religion in Argentina, a society which underwent what David Martin has characterised as the "French" pattern of secularisation, confirms that the study of Romance language societies with a Catholic majority may help to raise new questions and to sharpen the focus of old ones. Recent efforts have expanded the scope of studies on science and religion to creeds other than Christianity and to non-Western cultures, but the amount of historical work on Latin America is still small.[4] Argentina is one of the countries of the southern cone of South America (the others are Uruguay and Chile), which offers a suitable case study: a settler society with a massive southern European immigration, an elite drenched in French literary culture, a non-trivial scientific tradition of manageable dimensions, and a capital city, Buenos Aires, integrated into the Atlantic economy and open to many currents of European thought.[5] Grace Davie declared that "the forms and processes of religious life in any given place can only be fully understood within *a long-term* and *relatively specific* historical perspective" (my emphasis).[6] Recent work underlining the diversity and the local character of the answers to challenges related to science and religion recommend the adoption of a circumscribed approach in these studies.[7] Situating the inquiry within the framework of a particular national history will highlight the relevance of political factors in the local debates between science and religion. A *longue durée* historical approach best reveals the interplay between the slow groundswell of secularisation and the changing surface interactions between science and religion. This approach should attend to the relationships among political and religious protagonists, large movements of ideas, and the interplay of scientific, ideological, and religious discourse.[8]

---

[4] John H. Brooke and Ronald L. Numbers, eds., *Science and Religion around the World* (Oxford: Oxford University Press, 2011). In Ignacio Silva, ed., *Latin American Perspectives on Science and Religion* (London: Pickering and Chatto, 2014) the number of historical essays is a minority.
[5] Jeremy Adelman, *Republic of Capital: Buenos Aires and the Legal Transformation of the Atlantic World* (Stanford: Stanford University Press, 1999).
[6] Grace Davie, *The Sociology of Religion* (London: Sage, 2007), 91.
[7] David N. Livingstone, *Putting Science in Its Place. Geographies of Scientific Knowledge* (Chicago: The University of Chicago Press, 2003); idem, "Science, Region, and Religion: The Reception of Darwinism in Princeton, Belfast, and Edinburgh," in *Disseminating Darwinism: The Role of Place, Race, Religion, and Gender*, ed. Ronald L. Numbers and John Stenhouse (Cambridge: Cambridge University Press, 1999), 7–38.
[8] David N. Livingstone, "Science, Religion, and the Cartographies of Complexity," *Historically Speaking* 8 (2007): 15–16.

Since conceptual definitions are a focus of debate in the field, I should begin by stipulating the meanings of the terms of discussion.[9] "Religion" is here understood as institutional, organised religion, in this case the Roman Catholic Church in Argentina, its hierarchy, clergy, and lay members. Of course, there were important changes in the institutions and the social perceptions of the sacred between the era of the Iberian colonial church and the forms of Catholicism in the different stages of the conformation of Argentina as a national state. Protestant Churches tended not to engage publicly with scientific issues. Jews have been increasingly active in the sciences since the postwar era, which is the *ad quem* of this study.[10] A strict notion of science that could encompass the whole period under consideration is out of the question. I shall take "science" to mean a methodologically organised and empirically oriented inquiry of the natural world, including theories, practices, institutions, and social images. Also, "science" denotes the natural *and* exact sciences (including math), as is common usage in Romance languages. It should be noticed that in a survey like the present one, we shall deal with the transference of scientific know-how and the reception of scientific theories and their local interpretation and recreation.[11]

---

9 See Peter Harrison, *The Territories of Science and Religion* (Chicago: University of Chicago Press, 2015); idem, "The Modern Invention of 'Science-and-Religion': What Follows?," *Zygon* 51, no. 3 (2016): 742–47.

10 For minority religions, see Daniel P. Monti, *Presencia del protestantismo en el Río de la Plata durante el siglo XIX* (Buenos Aires: La Aurora, 1969); Haim Avni, *Argentina and the Jews: A History of Jewish Immigration* (Tuscaloosa: University of Alabama Press, 1991); José Míguez Bonino, *Faces of Latin American Protestantism*, trans. Eugene L. Stockwell (Grand Rapids: William B. Eermans, 1997); Arnoldo Canclini, *400 años de protestantismo argentino: historia de la presencia evangélica en la Argentina* (Buenos Aires: Facultad Internacional de Educación Teológica, 2004); Claudia Häfner, *Heimischwerdung am La Plata. Von der Deutschen Evangelischen La Plata Synode zur Iglesia Evangélica del Río de la Plata* (Münster: LIT Verlag, 2008); Susana Bianchi, *Historia de las religiones en la Argentina: Las minorías religiosas*, 2nd ed. (Buenos Aires: Sudamericana, 2009); Raanan Rein, *Argentine Jews or Jewish Argentines? Essays on Ethnicity, Identity, and Diaspora* (Leiden: Brill, 2010); Paula Seiguer, *"Jamás he estado en casa." La Iglesia Anglicana y los ingleses en la Argentina* (Buenos Aires: Biblos, 2017).

11 For overviews of the question of imperial diffusion of science, which is only collaterally relevant to our inquiry, see Lewis Pyenson, "Science and Imperialism," in *Companion to the History of Modern Science*, ed. R. C. Olby, G. N. Cantor, J. R. R. Christie and M. J. S. Hodge (London: Routledge, 1990), 920–33 and Pratik Chakrabarti and Michael Workboys, "Science and Imperialism since 1870," in *Modern Science in National, Transnational and Global Contexts*, ed. Hugh R. Slotten, Ronald L. Numbers and David N. Livingstone, vol. 8, *The Cambridge History of Science* (Cambridge: Cambridge University Press, 2020), 9–31.

Science and Religion: Conflict and Complexity

The expression "conflict thesis" denotes the belief that science and religion have been fighting each other in the course of Western history on the grounds of an intrinsic incompatibility. Since the 1970s, it became customary in anglophone history of science to attribute the first articulated formulation of the conflict thesis to two books written in the last decades of the nineteenth century: John W. Draper's *History of the Conflict between Religion and Science* (1875) and Andrew Dixon White's *A History of the Warfare of Science with Theology on Christendom* (1895).[12] Recently, James Ungureanu has argued that Draper and White, writing within the tradition of liberal Protestantism, "did not in fact posit an endemic and irrevocable conflict between 'science and religion' … [on the contrary, they] hoped their narratives would actually preserve religious belief."[13] Lawrence Principe has pointed out that Draper's point of depart was his own religion, which was anti-Trinitarian, deterministic, pseudo-Averroistic, and "opposed to Christianity—and to Roman Catholic Christianity in particular."[14] Richard Schaefer has shown how Andrew White argued in favour of a religion purified from dogma by science.[15] One aspect of Draper's and White's works seems to lie beyond the scope of this revisionist scholarship: their anti-Catholicism. Science, according to Draper, "has never subjected any one to mental torment, physical torture, least of all to death, for the purpose of upholding or promoting her ideas … But in the Vatican—we have only to recall the Inquisition—the hands that are now raised in appeals to the Most Merciful are crimsoned. They have been steeped in blood!"[16] White's indictment was milder in tone: for more than a thousand years, the French Catholic Church "kept experimental science in servitude…humiliated Buffon in natural science [and threw] its weight against

---

[12] Perhaps the earliest exposition of the conflict thesis can be found in David C. Goodman, *The "Conflict Thesis" and Cosmology*, unit 1, *Science and Belief: from Copernicus to Darwin* (Milton Keynes: The Open University Press, 1974), 30–49; see also James R. Moore, *The Post-Darwinian Controversies. A Study of the Protestant Struggle to Come to Terms with Darwin in Great Britain and America, 1870–1900* (Cambridge: Cambridge University Press, 1979), 19–49.

[13] James C. Ungureanu, *Science, Religion, and the Protestant Tradition: Retracing the Origins of Conflict* (Pittsburgh: University of Pittsburgh Press, 2019), 13.

[14] Lawrence M. Principe, "The Warfare Thesis," in *The Warfare between Science and Religion. The Idea That Wouldn't Die*, ed. Jeff Hardin, Ronald Numbers and Ronald A. Benzley (Baltimore: Johns Hopkins University Press, 2018), 19.

[15] Richard Schaefer, "Andrew Dickson White and the History of a Religious Future," *Zygon* 50 (2015): 7–27.

[16] John W. Draper, *History of the Conflict between Religion and Science* (New York: Appleton, 1875), ix.

Newton in the physical sciences."[17] As we shall see in chapter 4, Draper was enthusiastically read by liberals and Freemasons in Buenos Aires for it supplied them with anticlerical rhetorical ammunition. But Draper's narrative coexisted and at times overlapped with the Enlightenment narrative that took Turgot's notion of progress as its point of departure. In Catholic countries such as France, Italy, and those of Iberian America the conflict narrative took a different turn than in English-speaking societies. I have argued elsewhere that Draper and White played only subsidiary roles in the kind of history of science made famous by George Sarton and transferred from Europe to Argentina by Aldo Mieli in the 1930s; in this historiographic tradition, the conflict thesis was given a supersessionist slant: science was not conceived as a frontal enemy of religion in an ever recurring battle, but as a substitute of an outdated theological worldview in a history of progress, in character with Condorcet's *Esquisse d'un tableau historique des progrès de l'esprit humain* (1795, posthumous) and Comte's famous three-stage scheme.[18] In his well-known work about the Christian roots of modernity, Löwith declared that the culmination in Comte of the principle of progress formulated by Condorcet and the law of the three stages by Turgot and Saint-Simon worked "the decisive transformation of the theology of history into a philosophy of history as inaugurated by Voltaire."[19] It is difficult to overestimate the influence that this genealogy of progress given a scientific twist had on the elite which raised the banner of secularism in Argentina.

In a restricted sense, the "conflict thesis" is a historiographic construct.[20] But the notion of an intrinsic confrontation between science and religion is also a deeply entrenched social belief with ideological and emotional overtones attached to it so that no scholarly work—no matter how well argued—will do much to alter it. It has been claimed that the force of "a perceived affinity" between non-religion and science "is cultural and historic, rather than intellectual

---

17 Andrew D. White, *A History of the Warfare of Science with Theology in Christendom*, vol. 1 (New York: Appleton, 1897), 408–409.
18 Miguel de Asúa, "The 'Conflict Thesis' and Positivist History of Science. A View from the Periphery," *Zygon. Journal of Religion and Science* 53, no. 4 (2018): 1131–48.
19 Karl Löwith, *Meaning in History* (Chicago: The University of Chicago Press, 1949), 90–91.
20 For detailed historiographical discussions, see Geoffrey Cantor, "What Shall We Do with the Conflict Thesis?," in *Science and Religion. New Historical Perspectives*, ed. Thomas Dixon, Geoffrey Cantor and Stephen Pumfrey (Cambridge: Cambridge University Press, 2010), 283–98; Russell, "The Conflict of Science and Religion;" Principe, "The Warfare Thesis;" Ungureanu, *Science, Religion, and the Protestant Tradition*, 3–21; and Ronald L. Numbers, "Revisiting the Battlefields of Science and Religion. The Warfare Thesis Today," in *Rethinking History, Science, and Religion. An Exploration of Conflict and the Complexity Principle*, ed. Bernard Lightman (Pittsburgh: University of Pittsburgh Press, 2019), 183–90.

and epistemic."²¹ Certainly, the belief that science and religion are born enemies goes far beyond the esoteric discussions of scholars; it is an important postulate of different anti-religious worldviews, ranging from the radical Enlightenment through Marxism in its many guises and disguises, to the New Atheism. Numbers affirmed that "the warfare thesis, though now repudiated by almost all professional historians, retains a powerful presence among ideological extremists on both the unbelieving left and fundamentalist right."²² In this broad sense, the "warfare" between science and religion can be seen as a guiding fiction staging the power struggle between the values and epistemological authority of either of them. As put by Peter Harrison, debates between religion and science "are often proxies for more deep-seated ideological or, in its broadest sense, 'theological' battles."²³ The long-term history of the relationships between science and religion in Argentina shows the sprawling ideological roots of the conflict view and how it served the political agenda of secularist factions.

In 1986, the late David Lindberg and Ronald L. Numbers published the revised papers of a 1981 conference on science and religion.²⁴ Soon after this groundbreaking work—a collective manifesto of the vitality and potential of a burgeoning area—John Hedley Brooke, then at the University of Lancaster, published his *Science and Religion*, a historical tour de force whose first historiographical chapter might be the most quoted item in the specialised literature of the last three decades.²⁵ In it, Brooke argued against any single overarching interpretation of the relationships between Western science and Christian religion (be it "conflict" of "harmony") and called for a recognition of the complexity involved in each of the cases. Brooke's work can be seen as an implicit historical criticism of Ian Barbour's four "ways of relating science and religion" (conflict, independence, dialogue, and integration) understood as a universally valid typology.²⁶ A recent volume edited by Bernard Lightman is indicative of the

---

21 Lois Lee, "Feeling Rational: Affinity and Affinity Narrations in British Science-Non-Religion Relations," in *Science, Belief and Society: International Perspectives on Religion, Non-Religion and the Public Understanding of Science*, ed. Stephen H. Jones, Tom Kaden and Rebecca Catto (Bristol: Bristol University Press, 2019), 174.
22 Numbers, "Revisiting the Battlefields," 190.
23 Harrison, *The Territories of Science and Religion*, 197.
24 David C. Lindberg and Ronald L. Numbers, eds., *God and Nature. Historical Essays on the Encounter between Christianity and Science* (Berkeley: University of California Press, 1986).
25 John H. Brooke, *Science and Religion. Some Historical Perspectives* (Cambridge: Cambridge University Press, 1991), 2–15.
26 Ian Barbour, *Religion in an Age of Science: The Gifford Lectures 1990–1991*, vol. 1 (San Francisco: HarperSanFrancisco, 1990), 3–30. For a criticism, see John H. Brooke and Geoffrey Cantor, *Reconstructing Nature. The Engagement of Science and Religion* (Edinburgh: T&T Clark, 2000),

wide array of historiographical questions related to the notion of Brooke's "complexity." The editor contends that more than a thesis, "complexity is a heuristic principle that should guide our research so that we are sensitive to how different contexts shape past understandings of science, religion, and their dynamic interface."[27] Brooke himself declares that "complexity is a historical reality, not a thesis, and that instead of being placed alongside other theses, its primary role is to function as critique."[28] As far as complexity implies a heightened sensitivity to local historical scenarios, it could be seen as the historiographic condition of possibility of historical inquiries like the one in this book. Tags such as harmony, conflict, and indifference are used here on occasion as regulative notions and not as literal descriptions of processes. If it is to be applied to the kind of historical material we are dealing with, complexity should include the political aspects of the case studies.[29] This statement will be amply illustrated in the following pages.

It has been observed that an overemphasis on complexity could result in a confusion of disiecta membra. In a longing for comprehensive patterns, Numbers has provocatively claimed that "the turn to complexity has left most people yawning or, worse yet, unconvinced ... [so that] without abandoning the gospel of complexity and retreating to uncomplicated master-narratives we can—and, I believe, should—search for mid-scale patterns."[30] One of the patterns he proposes is secularisation, for which he draws upon surveys of religious beliefs among scientists in the United States. In this book, I shall use the notion of secularisation in a broader and I think more complex sense, as the Ariadna's thread to not get lost in a labyrinth of historical microstudies.

275–76 and Geoffrey Cantor and Chris Kenny, "Barbour's Fourfold Way," *Zygon* 36, no. 4 (2001): 765–81. Ian Barbour's reply was "On Typologies for Relating Science and Religion," *Zygon* 37, no. 2 (2002): 345–59.

27 Bernard Lightman, "Introduction," in *Rethinking History, Science, and Religion*, 7.
28 John H. Brooke, "Afterword. The Instantiation of Historical Complexity," in *Rethinking History, Science, and Religion*, 235.
29 This concern has already been voiced by David N. Livingstone, "Which Science? Whose Religion?," in *Science and Religion around the World*, 282.
30 Ronald L. Numbers, "Simplifying Complexity: Patterns in the History of Science and Religion," in *Science and Religion. New Historical Perspectives*, 264.

Secularisation and Science

In a recent paper, Harrison highlighted three issues concerning the relations among secularisation, the conflict thesis, and the notion of historical complexity: (a) the "embedment" of the conflict thesis within secularisation stories; (b) the increasing secularisation of scientific discourse over the last 150 years, and (c) the very much debated question of the "causes" of secularisation in the West. In reference to the latter point, he highlights Taylor's position, who "locates the engine for changes in belief not in the corrosive powers of science, but in transformations that were internal to Western Christianity."[31] In the present inquiry, I shall attend to three interconnected issues in relation to science, religion, and secularisation: (a) the changing character of the relations between science and religion (Catholicism) in a society that underwent a "French" pattern of secularisation; (b) how the crucial phases of secularisation might have borne on the interactions between science and religion; and (c) the larger issue of science as an "engine" of secularisation in this particular society. I shall show that these three questions boil down to the following statement: the successive modes of relationships between science and religion were to a considerable extent predicated upon the dynamics of secularisation. If science had a role in secularisation, it was due neither to its epistemic power nor as a substitute for religious belief, but to its being an element in the rhetoric of science-based ideologies of secularism or as part of the broad technological transformations of modernity.

José Casanova has sorted out three "moments" in the theory of secularisation: (1) the Weberian differentiation between the secular spheres (the state, economy, and science) and the religious sphere; (2) the decline of religious beliefs and practice; (3) privatisation (and eventually marginalisation) of religion, which he sees not as a structural trend but as a historical option and is at the core of his argument about the public role of religions in the contemporary world.[32] In an analogous way, Taylor opens his *Secular Age* distinguishing between three types of secularity: (1) the separation of churches from political structures (the state), which entails the privatisation of religion; (2) "the falling off of religious belief and practice;" (3) the move from a society where belief in God is unchallenged and indeed, unproblematic, "to one in which it is understood to be one option among others."[33] I shall employ these familiar distinc-

---

[31] Peter Harrison, "Conflict, Complexity, and Secularization in the History of Science and Religion," in *Rethinking History, Science, and Religion*, 230.
[32] José Casanova, *Public Religions in the Modern World* (Chicago: Chicago University Press, 1994), 19–39, 211–15.
[33] Charles Taylor, *A Secular Age* (Cambridge, MA: Harvard University Press, 2007), 1–3.

tions as need be. In this study, "secularisation" is not given any a priori meaning; instead, the particular use and scope of the notion will be discussed in connection with the historical circumstances. Casanova defined secularism as "a whole range of modern secular worldviews and ideologies which may be consciously held and explicitly elaborated into philosophies of history and normative-ideological state projects, into projects of modernity and cultural programs."[34] Lee has distinguished secularisation (a shift from a "religion-governed jurisdiction" to one in which religion is secondary) from secularism, which she defines as the ideological expression of "the secular" (those phenomena in which religion is "a secondary concern, reference, point, or authority").[35] It has been claimed that lack of consensus in the meaning of secularism "should not be any surprise, given its multiform history and multipurpose potential".[36] I shall employ the term "secularism" in connection with the science-based ideologies that dominated Argentina's intellectual landscape between the second half of the nineteenth century and the early decades of the twentieth century. In most Romance language societies, secularisation, understood as the exclusion of the Catholic Church from the sphere of government and public institutions, took the form of laicism (laïcité, laicismo), in which the state is associated with unbelief more than with neutrality (more on this below).[37] The history of the relationships between science and religion in Catholic countries is also (and perhaps mostly) a history of the relationships between science and unbelief, as has been explored, for example, in anglophone societies by Lightman and in Europe by the contributors to a recent volume edited by Carolin Kosuch.[38] Here, the terms "secular" and cognates are used loosely as substitutes for laicismo and its family of

---

[34] José Casanova, "The Secular, Secularizations, Secularisms," in *Rethinking Secularism*, ed. Craig Calhoun, Mark Juergensmeyer and Jonathan VanAntwerpen (Oxford: Oxford University Press, 2011), 55. See also the discussion in Marian Burchardt, Monika Wohlrab-Sahr and Matthias Middell, "Multiple Secularities beyond the West: An Introduction," in *Multiple Secularities beyond the West. Religion and Modernity*, ed. M. Burchardt, M. Wohlrab-Sahr and M. Middell (Boston: De Gruyter, 2015), 1–15.

[35] Lois Lee, *Recognizing the Non-Religious. Reimagining the Secular* (Oxford: Oxford University Press, 2015), 46, 42, 39.

[36] Phil Zuckerman and John R. Shook, "Introduction: The Study of Secularism," in *The Oxford Handbook of Secularism*, ed. Phil Zuckerman and John R. Shook (New York: Oxford University Press, 2017), 6.

[37] See, for example, Jean Baubérot, *Histoire de la laïcité en France*, 3rd ed. (Paris: PUF, 2005), 3–4.

[38] Bernard Lightman, "Unbelief," in *Science and Religion around the World*, 252–77; idem, "The Victorians: Tyndall and Draper," in *The Warfare between Science and Religion: The Idea That Wouldn't Die*, 65–83; Carolin Kosuch, ed., *Freethinkers in Europe. National and Transnational Secularities, 1789–1920s* (Berlin: De Gruyter, 2020).

words, for which there are no precise English equivalents and are here understood as denoting "the absence of religion from public space and public affairs" or "a sociopolitical sphere freed from religious symbols and clerical control."[39]

In his *General Theory of Secularization*, the late David Martin proposed six different patterns. One of them is the French (Latin) pattern, which would apply to Baroque autocracies overthrown by revolutions with a secular ideology, in which religion is a political issue and "coherent and massive secularism confronts coherent and massive religiosity."[40] This seems also to be applicable in broad terms to Iberian American republics up to mid-twentieth century (Martin's "South American pattern" is clearly a photo of the second half of the twentieth century, which lies beyond the scope of this book). In his well-known article on multiple modernities, Shmuel Eisenstadt called the attention on "the enduring importance [in Latin America]" of reference points, above all in Europe— Spain, France, and England," which "were critical to the self-conception of the Latin American societies."[41] Martin himself has pointed out the importance of the "Parisian model" on "satellite" Latin American cities such as Rio de Janeiro, Buenos Aires, and Santiago de Chile and the role played by the local intelligentsias who adopted "an enlightened and radical anticlericalism" as over against what happened in the anglophone world. The Parisian model appealed to the elites, the bearers of secularisation, not to the masses, "who remained in an enchanted, animated universe made up of a mixture of Catholicism and spiritism."[42]

In the 2005 updating of his classic work, Martin distinguished between "those countries, mainly Protestant, where Enlightenment and religion overlapped and even fused, and those countries, mainly Catholic, where Enlightenment and religion clashed."[43] Taylor employs a similar scheme in his postulation of two "ideal types" of secularisation: (a) the Baroque Catholic or "paleo-Durkheimian" societies of continental Europe in which one church is identified with society and (b) the Protestant anglophone or "neo-Durkheimian" societies in which pertaining to any of several churches implies belonging to a diffuse "church"

---

[39] Grace Davie, "Religion and *laïcité*," in *Modern France: Society in Transition*, ed. Malcolm Cook and Grace Davie (London: Routledge, 1999), 144; Casanova, "The Secular, Secularizations, Secularisms," 57.
[40] David Martin, *A General Theory of Secularization* (New York: Harper, 1978), 16–17.
[41] Shmuel Einsenstadt, "Multiples Modernities," *Daedalus* 129, no. 1 (2000): 13–14.
[42] David Martin, *On Secularization. Towards a Revised Theory* (Aldershot: Ashgate, 2005), 69–72.
[43] Martin, *On Secularization*, 20.

which connects with the political identity of the nation.⁴⁴ Historians of Christianity and of secularism have analysed the complex histories underlying the neat categorisations of sociologists and philosophers and arrived at an analogous categorisation.⁴⁵

Although Argentina falls under Martin's French (Latin) pattern (Taylor's "paleo-Durkheimian" model), there are some distinguishing features which should be kept in mind:⁴⁶ (a) the Catholic church was transferred from Europe as an integral part of Spain's colonial rule; (b) political independence in Río de la Plata brought with it the breaking of relationships with the Holy See, supportive of Spain's imperial power (they were restored in 1859); (c) all the govern-

---

44 Taylor, *A Secular Age*, 454–56, 486–87.

45 See, for example, Owen Chadwick, *The Secularization of the European Mind in the Nineteenth Century* (Cambridge: Cambridge University Press, 1975); René Rémond, *Religion and Society in Modern Europe*, trans. Antonia Nevill (Oxford: Blackwell, 1999); Hugh McLeod, *Secularization in Western Europe, 1848–1914* (Basingstoke: Macmillan, 2000); Jacqueline Lalouette, *La république anticléricale, XIXᵉ–XXᵉ siècles* (Paris: Éditions du Seuil, 2002); Hugh McLeod and Werner Ustorf, eds., *The Decline of Christendom in Western Europe, 1750–2000* (Cambridge: Cambridge University Press, 2003); Manuel Borutta, *Antikatholizismus: Deutschland und Italien im Zeitalter der europäischen Kulturkämpfe* (Göttingen: Vandenhoeck & Ruprecht, 2010); Lisa Dittrich, *Antiklerikalismus in Europa: Öffentlichkeit und Säkularisierung in Frankreich, Spanien und Deutschland (1848–1914)* (Göttingen: Vandenhoeck & Ruprecht, 2014); Todd H. Weir, *Secularism and Religion in Nineteenth-Century Germany* (Cambridge: Cambridge University Press, 2014); Michael Rectenwald, *Nineteenth-Century British Secularism: Science, Religion, and Literature* (Basingstoke: Palgrave Macmillan, 2016).

46 Many studies of secularisation in Argentina have tried to fit the process in the mould of the famous French "two thresholds" (the claim is that the Latin American country did not proceed beyond the first "threshold"). This type of account, which so long as it remains descriptive is perfectly correct (the Third Republic *was* a model for the local elite), runs the risk of taking the French case as paradigmatic and Argentina as a failed version of it. It is not my intention to discuss here the general process of secularisation in Argentina, but I should remark that departing from Martin's global view, I take the French settlement of *laïcité* not as an *explanans* but as a regional particularity to be explained. See Jean Baubérot, "The Two Thresholds of Laïcization," in *Secularism and Its Critics*, ed. Rajeev Bhargava (Oxford: Oxford University Press, 1998), 94–136; Jean Baubérot, "Los umbrales de la laicización en la Europa latina y la recomposición de lo religioso en la modernidad tardía," in *La modernidad religiosa: Europa latina y América latina en perspectiva comparada*, ed. Jean-Pierre Bastian (Mexico: FCE, 2004), 94–110; Fortunato Mallimaci, "Laïcité de subsidiarité en Argentine: entre élargissement des droits sociaux et recherche mutuelle de légitimité," in *Laïcité, laïcités. Reconfigurations et nouveaux défis*, ed. Jean Baubérot, Micheline Milot and Philippe Portier (Paris: Éditions de la Maison des sciences de l'homme, 2014), 221–52; Roberto Di Stefano, "Por una historia de la secularización y de la laicidad en la Argentina," *Quinto Sol* 15, no. 1 (2011): 1–31; Roberto Di Stefano and José Zanca, eds., *Fronteras disputadas: religión, secularización y anticlericalismo en la Argentina (siglos XIX y XX)* (Buenos Aires: Imago Mundi, 2016).

ments after independence maintained the system of *patronato*, inherited from Spain (control of appointment of bishops, levy of tithes, granting of the *placet* to papal bulls and missives);[47] (d) since mid-nineteenth century and from a constitutional point of view, Catholicism has never been the state or established religion of the country; the Constitution of 1853 determined that the government would "support" the Roman Catholic church and provide for the conversion to Christianity of the native peoples; it required that the president should be a Catholic; freedom of worship was explicitly warranted.[48]

The story of science dispelling the dark shadows of religious ignorance and superstition—one of the most successful cultural myths of the Enlightenment—is currently associated with master-narratives of secularisation, either Casanova's "stadial progressive philosoph[ies] of history" or Taylor's "subtraction stories." Casanova formulates the stadial secularist philosophy of history as the identification of processes of modernisation with radical secularisation so that "to be secular ... means to leave religion behind, to emancipate oneself from religion, overcoming the nonrational forms of being, thinking, and feeling associated with religion."[49] Taylor's prime example of subtraction story is disenchantment, according to which "science gave us 'naturalistic' explanation of the world" and then "people began to look for alternatives to God."[50] He argues that what nurtured the sense that religion should be superseded by a more rational epistemological stance was not science itself but the worldview of an "impersonal order" in the universe reinforced by the "new [Early Modern] cosmic imagination" of Providential Deism. The attraction of materialism, a further step in this progression, would depend on moral considerations, on the accompanying feeling of "maturity, of courage...over against the childish fears and sentimentality." But once science supersedes religion, its own logic requires that the whole process be explained in terms of epistemological advantages.[51] In part building on Taylor's account and concentrating on the emergence of early modern natural philosophy, Peter Harrison argues that secularisation is not the result of the opposition between science and belief but "the indirect result of the *conditions of belief* that attended the successes of modern science" (my emphasis). He claims that science was involved in secularisation but "out of congenial interactions be-

---

[47] Robert McGeagh, "Catholicism and Sociopolitical Change in Argentina: 1943–1973" (PhD Diss., University of New Mexico, 1974), 15–24.
[48] *Constituciones argentinas. Compilación histórica y análisis doctrinario*, ed. Natalia Monti (Buenos Aires: Infojus, 2005), 135–37.
[49] Casanova, "The Secular, Secularizations, Secularisms," 67–68.
[50] Taylor, *A Secular Age*, 26.
[51] Ibid., 362–66.

tween religion and science rather than a growing separation and opposition."⁵² This paradoxical statement is in tune with Brad Gregory's account that sees secularisation as an unintended result of the Reformation.⁵³ Gregory himself declares that whereas modern science is widely thought to have falsified the notion of the world as God's creation, "it has not, it needs not, and this is not what happened historically."⁵⁴ He attributes to the decline of analogical thought and the rise of a univocal conception of divine action (a shift from transcendence to immance) the result that God was conceived as just another intramundane entity competing with the order of natural effects.⁵⁵

In his treatment of the question, John H. Brooke has categorically affirmed that "there is no one, universal prove of secularisation that can be ascribed to science or to any other factor."⁵⁶ He insists that it was not science but those worldviews based on science and mechanistic or evolutionistic images of the world which contributed to secularisation understood, in Taylor's terms, as a situation in which belief is one among many possible options. Against the view that scientific natural explanations undermined the belief in supernatural religious doctrines, Brooke argues that the distinction between natural and supernatural explanations was a consequence of secularisation (as demonstrated by Richard Boyle's explanation of natural phenomena as manifestations of God's power).⁵⁷ Brooke's "ironic pattern" is in line with a perspective that sees deism as a result of Christian culture, forcefully formulated by Michael Buckley.⁵⁸ Ba-

---

52 Peter Harrison, "Science and Secularization," in *Narratives of Secularization*, ed. Peter Harrison (London: Routledge, 2018), 56.
53 Brad S. Gregory, *The Unintended Reformation. How a Religious Revolution Secularized Society* (Cambridge, MA: Harvard University Press, 2012). Much of Peter Harrison's early work has to do with Reformation and early modern science.
54 Gregory, *The Unintended Reformation*, 384. See also William C. Placher, *The Domestication of Transcendence. How Modern Thinking about God Went Wrong* (Louisville, KY: Westminster John Knox Press, 1996).
55 It should be noticed that both Harrison and Gregory underline the transition from analogical to univocal interpretations of divine causation as a key issue in the origins of modern thought, but the former downplays the role of Scotist philosophy whereas the latter follows the more conventional interpretation.
56 John H. Brooke, "Science and Secularization," in *The Cambridge Companion to Science and Religion*, ed. Peter Harrison (Cambridge: Cambridge University Press, 2010), 114.
57 Ibid., 114–15. It should be pointed out that while these authors have rightfully discussed the question of science and secularisation departing from the crucial century of the Reformation and the Early Modern period, the present inquiry begins with the transition from the late Baroque to Catholic Enlightenment; much of it deals with questions posed by the Enlightenment and nineteenth-century thought.
58 Michael Buckley, *At the Origins of Modern Atheism* (New Haven: Yale University Press, 1987).

con's scientific utopia was a secularisation of millenarian expectations, the mechanistic worldview supported the argument of design, and the notion of natural laws arose from the idea of a divine legislator. It was the rationalisation of belief implied by natural theology which backfired.[59]

One of the strongpoints in Brooke's contribution is his mention of Mary Douglas's insight that the alleged incompatibility between science and religion ("a nineteenth-centuy relic") depends on the conception of the religious—if this is understood as "the subjective sense of awe and mystery we can hardly deny that the more modern science reveals, the more awe-inspiring (and so presumably sympathetic to religion) the universe appears."[60] Once again, the "zero-sum" epistemic view of the conflict between science and religion presupposes a conception of religion as belief in a set of doctrines, something along the lines of Smart's "doctrinal and philosophical dimension" of religion.[61] Since the 1960s, a large body of literature has been exploring the ways in which the notion of "religion" has been built upon a cultural basis, its definition "itself the historical product of discursive processes."[62] Wilfred Cantwell Smith asserted that "all through the eighteenth century and even beyond [the term Christianity] refers to an ideal, first transcendent, then intellectual."[63] It is the Enlightenment image of religion as a propositional attitude (the mental state of somebody toward a proposition) which underlies the conviction that religion and science compete for the same explanatory space. Condorcet put it tersely: "There does

---

[59] Brooke, "Science and Secularization," 119–20.
[60] Mary Douglas, "The Effects of Modernization on Religious Change," *Daedalus* 111, no. 1 (Winter 1982): 9.
[61] Ninian Smart, *The World's Religions*, 2nd ed. (Cambridge: Cambridge University Press, 1998), 17.
[62] Talal Asad, *Genealogies of Religion: Discipline and Reasons of Power in Christianity and Islam* (Baltimore: The Johns Hopkins University Press, 1993), 29. See the seminal work by Wilfred C. Smith, *The Meaning and End of Religion* (Toronto: Mentor Books, 1964) and also Jonathan Z. Smith, *Imagining Religion. From Babylon to Jonestown* (Chicago: The University of Chicago Press, 1982); Hans G. Kippenberg, *Discovering Religious History in the Modern Age*, trans. Barbara Harshav (Princeton: Princeton University Press, 2002) (the English version has the added benefit of an important "Introduction to the American edition"); Peter Harrison, *"Religion" and the Religions in the English Enlightenment* (Cambridge: Cambridge University Press, 2002); idem, *The Territories of Science and Religion*; Guy G. Stroumsa, *A New Science. The Discovery of Religion in the Age of Reason* (Cambridge, MA: Harvard University Press, 2010); Brent Nongbri, *Before Religion. A History of a Modern Concept* (New Haven: Yale University Press, 2013); Daniel L. Pals, *Nine Theories of Religion*, 3rd ed. (New York: Oxford University Press, 2015) (this list is just orientative).
[63] Smith, *The Meaning and End of Religion*, 70.

not exist any religious system, or supernatural extravagance, which is not founded on an ignorance of the laws of nature."[64]

In accordance with historians of science, sociologists of religion have been sceptical about the view that science has been the cause of secularisation in the West—and this applies even to the original supporters of the "secularisation thesis." Bryan Wilson affirmed that "religion and science can coexist as alternative orientations to the world" and denied that "the confrontation of science with religion...was *in itself* essentially harmful to religion" (my emphasis); "the real conflict," he said, resided with the proclivity of people to consider science "more reliable than religion." In the end, it was a question of the competing intellectual prestige of both spheres, with "the results of scientific endeavour" (i.e. technology) captivating the minds due to a shift in society's increasing response to pragmatic values and attitudes.[65] For Steve Bruce, at the time of assessing the balance between science and religion any zero-sum notion of knowledge should be set aside: "the decreasing plausibility of any body of ideas cannot be explained simply by the presence of some (to us) more plausible ones."[66] What favoured secularisation, he claims, was "the subtle impact of naturalistic ways of thinking" and the "underlying rationality" of science, which led to a "self-aggrandisement" which "make us less likely than our forebears to entertain the notion of a divine force external to ourselves."[67]

One of the first critics of secularisation theory, David Martin, pointed out that cross-national data showed no positive correlation between degree of scientific advance and diminishing religious influence. He argued that much of the power of the master-narrative of the Enlightenment resided in its conflating epistemological and moral planes when it combined the idea of "scientific truth" with that of "innocent virtue" as triumphing over religious error and vice.[68] Another critic of secularisation theory, Casanova, also dismisses any talk about the conflict between science and religion. In his discussion of science as one of the "carriers of secularisation," he affirms that what counted was "the new [scientific] method's claim to differentiated autonomy" and distinguishes between sci-

---

[64] Condorcet, *Outlines of a Historical View of the Progress of the Human Mind* (London: J. Johnson, 1795), 298.
[65] Bryan Wilson, *Religion in Secular Society. Fifty Years On*, ed. Steve Bruce (Oxford: Oxford University Press, 2016), 49–54.
[66] Steve Bruce, *Secularization. In Defence of an Unfashionable Theory* (Oxford: Oxford University Press, 2011), 43.
[67] Ibid., 47.
[68] Martin, "Does the Advance of Science Mean Secularisation?," 13.

ence and "the scientific worldview which claimed to have replaced religion the way a new scientific paradigm replaces an outmoded one."[69]

Rational choice theorists such as Rodney Stark and Roger Finke have appealed to surveys that show that in the United States scientists are no essentially more irreligious than the rest of the population.[70] The same use of surveys is prominent in the treatment of the question by historian of science Ronald Numbers, who calls attention to the high percentage of believers among scientists in the United States and maintains that what took place in the nineteenth century was not a growing atheism but a privatisation of religious views among scientists.[71]

Summing up, sociologists of religion, historians of science, philosophers, and historians of Christianity concur in the idea that it was not the theoretical content of scientific theories but scientific worldviews or ideologies which played a part, if any, in the secularisation of the West. The historical case analysed in this book bears out this claim and shows the full extent of its complexity.

Brief Outlook of the Book

The following episodes will show that it was the prevailing systems of thought and the particular configuration of the local Catholic Church which contributed to give shape to the relationships between science and religion as they developed in Argentina during the nineteenth and early twentieth century. The first chapter deals with the Jesuit missions of Paraguay and Río de la Plata, a setting in which science was at the service of the apostolic mission of the Society of Jesus. Chapter 2 explores Catholic Enlightenment in the region. This chapter analyses university teaching of philosophy and theology, Biblical geology, and natural theology as practiced by the clergy and lay persons; I shall argue that the variety of uses to which natural theology was put was a manifestation of the ambivalent character of Catholic Enlightenment. In Chapter 3, I shall look at the first disruption of the so far mostly harmonious relationships between science and religion. The ecclesiastical reform of 1822, as a result of which expropriated lands and buildings of religious orders were transformed

---

[69] Casanova, *Public Religions in the Modern World*, 212, 21 and 24.
[70] Rodney Stark and Roger Finke, *Acts of Faith. Exploring the Human Side of Religion* (Berkeley: University of California Press, 2000), 72–76.
[71] Ronald L. Numbers "Science, Secularization, and Privatization," in idem, *Science and Christianity in Pulpit and Pew* (New York: Oxford University Press, 2007), 129–36.

in scientific facilities, together with the arrival of Europan revolutionary emigrés acting as science teachers and the spread of French *idéologie* and Bentham's utilitarianism, helped to promote the idea of science as a secular enterprise. Chapters 4 and 5 correspond to periods in which science confronted religion. The liberal government of the 1880s pushed secularising laws which strongly modified the relationships between the state and the Catholic Church. The analysis of the reception in Buenos Aires of Darwinism and of Draper's book will show how evolution and the warfare thesis were used in the public sphere as rhetorical devices to support the government's secularising program. In this period, secularisation took the form of the differentiation of the religious sphere from the rest of society. The political and intellectual elite who led this process was also responsible for the introduction of modern science in the country geared to the ideal of economic and social progress. In chapter 5, I shall discuss how the liberal conservative and mostly positivist outlook of the 1880s gave way to various secularist philosophies and worldviews, ranging from spiritualism to materialism. In the first decades of the twentieth century, the secularist thrust passed from the liberal conservative elite to the socialists whose number increased exponentially with the arrival of millions of immigrants huddling in the port cities. Evolutionism (mostly Haeckel's) took more aggressive forms and there were attempts at introducing something like a "secular religion" around the figure of Argentina's most famous natural scientist at that time. Chapter 6 is about the emergence of prestigious Catholic scientists, the teaching of science in Catholic schools, and Catholic institutions of scientific research—a phenomenon that weakened considerably the prevalent idea of science as a secular pursuit. The "Catholic revival" of the 1930s was actualy a surge of integral Catholicism which paralleled a succession of authoritarian conservative governments with nationalistic leanings. A few Catholic scientists felt at home in this atmosphere; their reactionary political stance was not necessarily accompanied by backward scientific options (chapter 7). Chapter 8 is devoted to a group of prestigious physiologists who were at the same time liberal Catholics; their plans of founding a Catholic university based upon small centres of advanced scientific research did not suceed. In the mid-1950s, the clash between those supporting the creation of confessional (actually, Catholic) universities and the political factions who thought that higher education should be the exclusive preserve of the state was the last encounter between secularists and Catholics in the period under consideration. In any case, arguments dealing with science and secularism were absent from the heated public discussions over this issue. The conclusion recapitulates the chapters focusing on three moments in the history of science and secularisation in the country (as seizure of ecclesiastical property in the 1820s; as social differentiation in the 1880s; and as the promotion of a secular religion of science in the 1920s).

Chapter 1
Jesuit Science

Any understanding of learning in Río de la Plata during the colonial period demands a balanced view of the cultural role played by the Jesuits in the cities and the missions of Paraquaria. Alas, the layers of partisan interpretations accumulated over the centuries turn this into a task fraught with difficulties. Voltaire's acid mockery of sybarite Jesuits exploiting starving natives in *Candide* is well known: "The Fathers have everything and the people nothing; it is a masterwork of reason and justice," says Cacambo, Candide's valet. In his *Relation du Paraguai* (1777), he further compared the Guaraní missions to Lacedaemonia, echoing Diderot, who in the *Supplément au Voyage de Bougainville* (1772) had portrayed the Fathers as "cruel Spartans in black robes" (a judgement harsher than any passed by Bougainville himself).[1] While the luminaries of the French Enlightenment compared the Jesuit mission towns of historical Paraguay to an obscurantist militarised society, others saw the Jesuit reductions as a Christian commonality evocative of Plato's *Republic*, Thomas More's *Utopia* or Tommaso Campanella's City of the Sun: a community of good savages sharing their goods and living in rustic plenitude under the paternal supervision of benevolent priests.[2] Twentieth-century cultural productions, such as Fritz Hochwälder's *Das heilige Experiment* (1942) and Robert Bolt's script for Roland Joffé's British film *The Mission* (1986) draw much of their force from the historical ambiguities surrounding the "Jesuit Republic."[3]

During one century and a half, the Jesuits were the driven force in higher education, the sciences, and the arts in Río de la Plata. The inventory of their possessions, expropriated by the Spanish crown in 1767, is a witness to the material

---

[1] Voltaire, *Candide, où l'optimisme* ([Geneva], 1759), 108; idem, *Relation du Paraguai sous la domination des jesuites* (Basle, 1777), 16; Denis Diderot, "Supplément au Voyage de Bougainville," in idem, *Le neveu de Rameau et autres dialogues philosophiques* (Paris: Gallimard, 1972), 287; Louis Antoine de Bougainville, *Voyage autour du monde ... nouvelle edition augmentée*, vol. 1 (Neuchâtel: Imprimerie de la Société Typographique, 1772), 119–41. But notice that icons of the Enlightenment such as Montesquieu and Raynal had a rather favourable view of the Jesuit missions of Paraguay. (Unless a translator is mentioned, all translations from the originals are mine.)
[2] Girolamo Imbruglia, *The Jesuit Missions of Paraguay and a Cultural History of Utopia (1568–1789)* (Leiden: Brill, 2017).
[3] James S. Saeger, "*The Mission* and Historical Missions: Film and the Writing of History," *The Americas* 51, no. 3 (1995): 393–415; Frederick Hale, "Appropriating the Closure of Jesuit Missions: Fritz Hochwälder's *Das heilige Experiment*," *Acta Theologica* 28, no. 1 (2008): 58–75.

https://doi.org/10.1515/9783110488777-003

grandeur of their missions, colleges, and estancias; their spiritual heritage was equally immense.[4] Debates about the organisation and contents of learning in Córdoba and Buenos Aires after the expulsion of the Society of Jesus from the Viceroyalty of Peru reflected the positive or negative responses toward its cultural legacy. In 1610, the Jesuits had founded a college in Córdoba, which in little more than a decade was granted the power of awarding degrees by Pope Gregory XV; it would eventually grow into the University of Córdoba, the intellectual powerhouse of the Society in Río de la Plata. Shortly after, they established the University of Chuquisaca in Charcas, Upper Peru (present-day Bolivia). These tended to be conservative institutions which taught philosophy, theology, civil and canon law; original inquiry about the natural world (astronomy, natural history, *materia medica*, cartography) flourished in the freer atmosphere of the missions. As a consequence of the 1767 expulsion of the Jesuits from Spanish America, this world of "missionary science" would be irretrievably lost.

The ejection of the Jesuits from the Guaraní missions was the result of a long series of events. In the 1720s and 1730s, a protracted armed conflict erupted between the members of the Society of Jesus and the Spanish and creole landowners of Paraguay, who wished to release the Guaraní from the missions to use them in their own *encomiendas* (a system of forced labour); they also resented the Jesuits' extensive landholdings and their efficient and profitable production of *yerba mate* (Paraguayan tea) of superior quality. The 1750 Treaty of Madrid, signed between Spain and Portugal, assigned the lands east of the Uruguay River (what is now Rio Grande do Sul in Brazil) to Portugal. This implied that the seven Jesuit missions in that area had to be moved west of the river. The armed Guaraní militias resisted but in the end lost what came to be known as the "Guaraní wars": the rebellion was crushed by a conjoint Spanish and Portuguese force in February 1756 and its leader, José (Sepé) Tiaragú was executed. The failed Guaraní uprising triggered a spate of anti-Jesuit literature that exploited the theme of the rich Jesuit "kingdom" of Paraguay and its rebellious attitude against the crown.[5] The treaty was annulled in 1761.

---

[4] Francisco Javier Brabo, *Inventarios de los bienes hallados a la expulsión de los jesuitas y ocupación de sus temporalidades por decreto de Carlos III* (Madrid: Rivadeneyra, 1872).

[5] Ernesto J. A. Maeder, *Misiones del Paraguay. Construcción jesuítica de una sociedad cristiano-guaraní (1610–1768)* (Resistencia: Contexto, 2013), 243–69; Julia J. S. Sarreal, *The Guaraní and Their Missions. A Socioeconomic History* (Stanford: Stanford University Press, 2014), 97–122. For anti-Jesuit literature, see Pierre-Antoine Fabre, José Eduardo Franco, and Carlos Fiolhais, "The Dynamics of Anti-Jesuitism in the History of the Society of Jesus," *Jesuit Historiography Online*, last modified November 2018, https://referenceworks.brillonline.com/entries/jesuit-historiography-online/the-dynamics-of-anti-jesuitism-in-the-history-of-the-society-of-jesus-COM_192530.

Obeying Charles III's order of 27 February 1767, Governor Francisco de Paula Bucareli commanded the expulsion of the Jesuits from the cities of Río de la Plata and from the Guaraní and Chaco missions. By March 1768 all of them had been dispatched to Europe. Those who were Spanish or Spanish-American ended up in Faenza or the Papal States; the rest (most of them German-speaking Jesuits) returned to their countries of origin. It should be noticed that "the Fathers from Paraquaria—brutally and stupidly—were not allowed to take with them the written materials that would have assisted them when in exile they wrote their numerous works."[6] At that time, there had been 457 Jesuits in *Paraquaria*: 295 Spanish, 81 born in Río de la Plata, 53 Germans, 17 Italians, four English, two from Peru, two Portuguese, one Greek, one French, and one Flemish.[7] The large structure of Jesuit missionary work and learning, which it had taken a century and a half to build up, collapsed within a year. Their states, buildings, and libraries were from then on administered by the crown.

The Jesuits had been expelled from Portugal in 1759 and from France in 1764. In Spain, anti-Jesuit momentum came to a head with the "Esquilache riots," a movement of revolt triggered by unpopular laws which took place in early 1766 and for which the Jesuits were held responsible (current opinion is that they were not). The accusation was spelled out in the *Dictamen fiscal* (1766) authored by Pedro Rodríguez de Campomanes, a public prosecutor of the Council of Castile. The document was a detailed exposition of the alleged damage inflicted on the state by the members of the Society of Jesus; one of its nine chapters dealt with the Guaraní missions.[8] Campomanes and others contended "that the Jesuits had created in Paraguay a state within the state where the Indians lived in slav-

---

6 Magnus Mörner, "The Role of the Jesuits in the Transfer of Secular Baroque Culture to the Río de la Plata Region," in *The Jesuits. Cultures, Sciences, and the Arts, 1540–1773*, ed. John W. O'Malley, Gauvin A. Bailey, Steven J. Harris and T. Frank Kennedy (Toronto: University of Toronto Press, 1999), 312.
7 Guillermo Furlong, *Los jesuitas y la cultura rioplatense* (Buenos Aires: Biblos, 1994), 162. The *Historical Dictionary of the Society of Jesus* gives the same total number but a slightly different distribution according to country of origin; see João Baptista, "Expulsión de Hispanoamérica y Filipinas (1767–1770)," in *Diccionario Histórico de la Compañía de Jesús. Biográfico-temático*, ed. Charles E. O'Neill and Joaquín M. Domínguez (Rome: Institutum Historicum Societatis Iesu; Madrid: Universidad Pontificia de Comillas, 2001), 2:1353–59. See also Pablo Hernández, *El extrañamiento de los jesuitas del Río de la Plata* (Madrid: Victoriano Suárez, 1908); Maurice Whitehead, "On the Road to Suppression. The Jesuits and Their Expulsion from the Reductions of Paraguay," in *The Jesuit Suppression in Global Context*, ed. Jeffrey D. Burson and Jonathan Wright (Cambridge: Cambridge University Press, 2015), 83–99.
8 John Lynch, *Edad Moderna. Crisis y recuperación, 1598–1808*, vol. 5, *Historia de España* (Barcelona: Crítica, 2005), 489–94.

ery and the Crown had no power."⁹ The Guaraní wars and the subsequent international campaign of anti-Jesuitical books and pamphlets (much of which was instigated by the Portuguese minister Sebastião de Carvalho e Melo, Marquis of Pombal) were important factors contributing to the expulsion of the Jesuits from Spain.¹⁰

Magnus Mörner has argued that the ultimate cause of the expulsion should be sought in Charles III's regalist tendencies and his efforts to concentrate power.¹¹ The "Jansenist" position of some of Charles III's ministers such as Campomanes and Pedro Abarca de Bolea, Count of Aranda, also fostered the anti-Jesuit animus of the crown.¹² In Bourbon Spain, "Jansenist" (a name of abuse given to their opponents by the Jesuits) meant a person who envisaged a church dependent on the state and independent from Rome (something approximating Gallicanism); it had little to do with the theological questions of grace and predestination associated with Jansenius or with the moral stance and spirituality of seventeenth-century French Jansenists. At most, it expressed the desire for a more enlightened religion, free of superstition and of the most extreme forms of baroque popular piety. Jesuits defended the infallibility and supreme authority of the papacy over the universal Church (ultramontanism), the jurisdiction of the Church over against that of the state, and a style of religion impregnated by the values and spirit of the Counter-Reformation. It should be noticed that the more radicalised currents of the French Enlightenment did not cross the Pyrenees and if they did, their effect was marginal. During the reigns of Charles III and Charles IV Spain remained solidly Catholic; the conflicts that shattered the religious and learned world were internal dissensions within the Catholic

---

9 Sarreal, *The Guaraní and Their Missions*, 114.
10 Dale K. Van Kley, "Plots and Rumors of Plots: The Role of Conspiracy in the International Campaign against the Society of Jesus, 1758–1768," in *The Jesuit Suppression in Global Context*, 13–39; Niccolò Guasti, "From Expulsions to Canonical Suppression (1759–1773)," in *The Oxford Handbook of the Jesuits*, ed. Ines G. Županov (New York: Oxford University Press, 2019), 918–49.
11 Magnus Mörner, "Introduction," in idem, ed., *The Expulsion of the Jesuits from Latin America* (New York: Alfred A. Knopf, 1965), 3–30. This is also the opinion of Teófanes Egido: "Today...it can be affirmed that the expulsion of the Jesuits from Spain in 1767 was a triumph of regalist ideology." Idem, "La expulsión de los jesuitas de España," in *Historia de la Iglesia de España*, ed. Ricardo García-Villoslada, vol. 4 (Madrid: La Editorial Católica, 1979), 791.
12 Aranda was an admirer of Voltaire; Manuel de Roda, the minister of Grace and Justice, was a Jansenist. These and other high officials had been poor students living in pension houses (*manteistas*, for the kind of cloak they used). They had a grudged against those who had lodged in the rich university colleges (*colegiales*) and had been introduced into a Jesuit network of privilege. Lynch, *Edad Moderna*, 483.

Church.¹³ Something analogous could be said of the role played by Jansenism and Gallicanism in the suppression of the Society of Jesus by Clement XIV in 1773.¹⁴ Far from seeing the dissolution of the Jesuits as the consequence of the French Enlightenment, current opinion tends to consider it as a consequence of an opportunist "converge of convenience" among reformist and regalist rulers, Jansenist, and Gallican feeling.¹⁵ This complex state of affairs should be born in mind at the time of explaining the relationships between science and religion in colonial Río de la Plata.

The reductions (*reducciones*) formed the living heart of the complex of religious, economic and educational institutions which the Jesuits built up in Paraquaria. This religious province, configured in 1610, extended over present-day Argentina, Paraguay, Uruguay, part of Bolivia, and south-western Brazil.¹⁶ Over the course of 160 years, the Society of Jesus founded around one hundred missions. Eventually, many of them were destroyed, moved or merged. The core of the Jesuit Republic was constituted by the famous "33 towns," in which, at the peak of their population curve in 1732, lived more than 140,000 Guaraní, a people of roaming agriculturalists.¹⁷ The number of Jesuits was comparatively small. By 1692 there were 249 of them in the province, of which 73 were in the missions.¹⁸

---

13 Charles Noel, "Clerics and Crown in Bourbon Spain, 1700–1808: Jesuits, Jansenists, and Enlightened Reforms," in *Religion and Politics in Enlightened Europe*, ed. James E. Bradley and Dale K. Van Kley (Notre Dame: University of Notre Dame Press, 2001), 119–53; Andrea J. Smidt, "Bourbon Regalism and the Importation of Gallicanism: The Political Path for a State Religion in Eighteenth-Century Spain," *Anuario de Historia de la Iglesia* 19 (2010): 25–53.
14 Dale K. Van Kley, "Jansenism and the International Suppression of the Jesuits," in *Enlightenment, Awakening, and Revolution, 1660–1815*, ed. Stewart J. Brown and Timothy Tackett, vol. 7, *The Cambridge History of Christianity* (Cambridge: Cambridge University Press, 2006): 302–8.
15 Guasti, "From Expulsion to Canonical Suppression," 920.
16 For the history of the missions see Magnus Mörner, *The Political and Economic Activities of the Jesuits in the La Plata Region; the Hapsburg Era* (Stockholm: Victor Pettersons Bokindustri Aktiebolag, 1953); Philip Caraman, *The Lost Paradise, an Account of the Jesuits in Paraguay, 1607–1768* (London: Sidgwick & Jackson, 1975); Sarreal, *The Guaraní and Their Missions*; Maeder, *Misiones del Paraguay*; Guillermo Wilde, "The Missions of Paraguay: Rise, Expansion and Fall," in *A Companion to Early Modern Catholic Global Missions*, ed. Ronnie Po-chia Hsia (Leiden: Brill, 2017): 73–101.
17 Ernesto J. Maeder and A. S. Bolsi, "La población de las misiones guaraníes," *Estudios Paraguayos* 2, no. 1 (1974): 125. For the Guaraní, see for example Alfred Métraux, "The Guaraní," in *The Tropical Forest Tribes*, ed. Julian H. Stewart, vol. 3, *Handbook of South-American Indians* (Washington: Government Publishing Office, 1948), 69–94; Bartomeu Melià, *El guaraní conquistado y reducido. Ensayos de etnohistoria* (Asunción: Centro de Estudios Antropológicos, Universidad Católica, 1986).
18 Mörner, *The Political and Economic Activities of the Jesuits*, 98.

The main economic basis of the reductions were the cultivation and commercialization of yerba mate and cattle ranching; the colleges were supported by large estancias manned by African slaves (around 3,000 by 1767).[19]

The reductions had been designed to free the Guaraní from the system of *encomienda* which consisted of granting a Spaniard a number of aborigines who worked for him in conditions of servitude, in exchange for being supported and Christianized; the system lend itself to all kind of abuses. The mission towns allowed the Jesuit Fathers to segregate the natives from Spanish colonial society, thereby preserving them from what the Jesuits saw as the scandalous way of life of the settlers, and also enabled the Fathers' religious instruction of the natives. This paternalistic system, which worked beyond any initial expectations, prompted the emergence of a rich hybrid culture. The printing press of the missions published works in the Guaraní language; orchestras of Mocoví boys performed baroque concerts with great success in the capital city of Buenos Aires; and massive churches, whose ruins still loom out of the luxuriant vegetation of the jungle, were built by Guaraní artisans who interpreted creatively the designs and models of the Italian and German Jesuit architects.[20] Science was a feature of this cultural experience and reflected the imaginative effort to develop a missionary style marked by inculturation.[21] As a result, missionary discourse on the natural world was a flowing baroque mélange in which particular aspects of aboriginal knowledge of nature were resignified and included into a body of Western knowledge.[22] Guillermo Wilde has claimed that the art of the Guaraní in

---

19 Nicholas P. Cushner, *Jesuit Ranches and the Agrarian Development of Colonial Argentina, 1650–1767* (Albany: State University of New York Press, 1983).
20 For a panoramic view of culture in the reductions, see Guillermo Furlong, *Misiones y sus pueblos guaraníes* (Buenos Aires: Imprenta Balmes, 1962); idem, *Los jesuitas y la cultura rioplatense*; Gauvin Alexander Bailey, *Art on the Jesuit Missions in Asia and Latin America, 1542–1773* (Toronto: University of Toronto Press, 1999), 155–60; 169–73; T. Frank Kennedy, "Music and the Jesuit Mission in the New World," *Studies in the Spirituality of the Jesuits* 39, no. 3 (2007); and Guillermo Wilde, ed., *Saberes de la conversión. Jesuitas, indígenas e imperios coloniales en las fronteras de la cristiandad* (Buenos Aires: Editorial Sb, 2011).
21 Inculturation is here understood as the adaptation of the Christian message to non-Western cultures and the influence of these cultures on Christian teaching. See Joseph W. Olson, *Jesuit Inculturation in the New World. Experiments in Missions of 16th, 17th, and 18th Centuries* (Denver: Outskirts Press, 2008).
22 It is usual for Portuguese- and Spanish-speaking historians of science to refer to Iberoamerican Jesuit science as "baroque." See, for example, Carlos Ziller Camenietzki, "Baroque Science between the Old and the New World: Father Kircher and His Colleague Valentin Stansel (1621–1705)," in *Athanasius Kircher. The Last Man who Knew Everything*, ed. Paula Findlen (New York: Routledge, 2004), 311–28 and Juan Pimentel, "Baroque Natures: Juan E. Nieremberg, American

the missions incorporated "local motives into the hegemonic representations of Christianity, conveying a certain idea of continuity (and harmony) between nature and civility, between rainforest and mission town, between Christian and pagan worlds."[23] In a similar manner the Jesuit natural histories of Paraguay evoke the synthetic character of missionary art. Early modern European discourse on nature was permeable enough to absorb elements of local cultures embodied in the names of creatures and the folk taxonomies. It was in the Italian exile when former missionary Jesuits from Paraguay sought to come to grips with Enlightenment natural history, either building bridges towards it (as in José Sánchez Labrador's encyclopedia) or rejecting it (this was the case of José Jolís).

In the Jesuit reductions of the New World, science was at the service of religion and religion spurred the cultivation of science. The purpose of this chapter is to take a closer look at the various kinds of their interactions in a context of general harmony. With that aim in mind, I shall review the Jesuit achievements in four fields of knowledge: natural history, medical botany, astronomy, and cartography. I hope that this brief survey will highlight what seem to be two characteristic traits of Jesuit science in Paraguay and Río de la Plata: (1) that science was a practical enterprise subordinated to missionary activity, which favoured the cultivation of applied aspects of knowledge; and (2) as a result of the inculturated style of Jesuit evangelization, segments of native lore on nature were incorporated into a matrix of European science.

Natural Histories

The first systematic, all-encompassing image of the natural world of Ibero-America was moulded by Jesuit missionaries with the creation of a new literary genre: the natural and civil / religious history of the New World, whose "newness" these works strove to capture. The natural histories of the Jesuits condensed the vast natural and human panoramas of the recently discovered lands and codified, in what became a standard format, the multifarious experience of the wilderness and the encounters with so many different peoples. The first work of this kind was Joseph de Acosta's *Historia natural y moral de las Indias*, the complete version of which was published in 1590. It is significant that the two first books of

---

Wonders and Preterimperial Natural History," in *Science in the Spanish and Portuguese Empires, 1500–1800*, ed. Daniela Bleichmar et al. (Stanford: Stanford University Press, 2009), 93–114.
23 Guillermo Wilde, "The Political Dimensions of Space-Time Categories in the Jesuit Missions of Paraguay (Seventeenth and Eighteenth Centuries)," in *Space and Conversion in Global Perspective*, ed. Giuseppe Marcocci, A. Maldavsky, W. de Boer and I. Pavan (Leiden: Brill, 2014), 203.

this work, which deal exclusively with natural history, had been originally conceived as the first part of Acosta's *De procuranda indorum salute* (Salamanca, 1588), which expounds a program for the evangelization of Spanish America based on ideas developed by its author in the course of the first Jesuit provincial congregation in Peru, 1576. From the beginning, natural history was conceived of as the description of the theatre of the Society of Jesus' apostolic activities in the New World. Joseph de Acosta structured his *Natural and Moral History* within the framework of Aristotelian philosophy, as a study of the causes of natural phenomena. Later Jesuit natural histories, closer to a Plinian ideal, treated the geography, the plants, the beasts, and the aboriginal peoples of the land they described more systematically.[24]

It is possible to distinguish three types of Jesuit natural histories in Paraguay. The first type consist of those works written by the official chroniclers of the Society, such as the *Chorographic Description of the Great Chaco Gualamba* (Madrid, 1733) and the sections on natural history of the *History of the Conquest of Paraguay, Tucumán and Río de la Plata*, both by Pedro Lozano (1697–1752). To this group also belongs the *History of Paraguay*, written by José Guevara (1719–1806), Lozano's successor as official chronicler. These natural histories consist of a series of chapters on the geography, plants, animals, and peoples of the land.

Second, we have the recorded memories of Jesuits exiled in the Habsburg empire and England: Martin Dobrizhoffer's *Historia de abiponibus* (Vienna, 1784), Florian Paucke's *Hin und Her*, and Thomas Falkner's *Description of Patagonia*. Given the vivid descriptions of the peoples among which the missionaries lived, these works are usually read as ethnographic accounts. Falkner's *Description* does not follow the usual sequence from geography to ethnography; it is the most properly chorographic of the natural histories, combining geography and history with an emphasis on the Patagonian peoples.

A third kind of natural histories is that written in Italian exile as a riposte to the assumption of the inferiority of the nature of the New World when compared with the Old, proposed by Buffon, Cornelius de Pauw, William Robertson, and other authors. Works like the *Essay on the Natural History of the Great Chaco* written by José Jolís (whose first part was published in Italian in 1789) or the long article "On some Observations of American Natural History" (Milan, 1810) by the Spaniard of Dutch descent José María Termeyer (1737–1814), are versions of a family of books on different regions of Ibero-America authored by exiled Jes-

---

[24] See Luis Millones Figueroa and Domingo Ledesma, eds., *El saber de los jesuitas. Historias naturales y el nuevo mundo* (Frankfurt: Vervuert; Madrid: Iberoamericana, 2005) and the chapter on Jesuits in Miguel de Asúa and Roger K. French, *A New World of Animals. Early Modern Europeans and the Creatures of Iberian America* (Aldershot: Ashgate, 2006), 141–82.

uits in the course of the last two decades of the eighteenth century and the first of the nineteenth century. Impregnated with nostalgia for the lost homeland, these works were written in the context of the "Dispute of the New World."[25] Besides the famous *Ancient History of Mexico* (1780) by Francisco Javier Clavigero, we should recall the works of Juan de Velasco on Quito, and of Felipe Gómez de Vidaurre and Juan Ignacio Molina on Chile. While extolling the fauna and landscapes of the New World, this literature is heavy with proto-independentist feelings.[26] As for the Río de la Plata, we should add the massive encyclopaedia in three parts by the Spanish Jesuit José Sánchez Labrador (1717–1798): *Natural Paraguay*, *Cultivated Paraguay*, and *Catholic Paraguay*, which focused on natural history, agrarian economy, and the missions, respectively.[27]

The New World was conceived by the Jesuits as a stage where the kingdom of God enacted its drama of salvation and retribution. In his *History of the Conquest* Lozano introduces each of the chapters on natural history with phrases evocative of imagery from the Biblical creation narrative, emphasizing how humans can benefit from a variety of native creatures.[28] The trees are beneficial for they embellish the land; their fruits support the life of the dwellers, their wood fuels human needs, and their virtues help restore health.[29] When Lozano talks about animals, what he introduces first are the cattle. In Guevara's natural history, the order of treatment of the creatures follows more closely the sequence of

---

[25] Antonello Gerbi, *The Dispute of the New World; the History of a Polemic, 1750–1900*, rev., ed. and trans. Jeremy Moyle (Pittsburgh: University of Pittsburgh Press, 1973).

[26] Jorge Cañizares Esguerra, *How to Write the History of the New World: Histories, Epistemologies, and Identities in the Eighteenth-Century Atlantic World* (Stanford: Stanford University Press, 2001), 234–54.

[27] The manuscript of *El Paraguay Natural* comprehends four illustrated volumes. See Sánchez Labrador, *El Paraguay Natural*, Paraquaria 16–19, Institutum Historicum Societatis Iesu (IHSI), Rome. Parts of it have been edited: Aníbal Ruiz Moreno, ed., *La medicina en "El Paraguay natural"* (Tucumán: Universidad Nacional de Tucumán, 1946); Mariano Castex, ed., *Peces y aves del Paraguay Natural Ilustrado, 1767* (Buenos Aires: Compañía General Fabril, 1968); Eliane C. Deckmann Fleck, ed., *As artes de curar em um manuscrito jesuítico inédito do Setecentos. O Paraguay Natural Ilustrado do padre José Sánchez Labrador (1771–1776)* (São Leopoldo: Oikos-Editorial Unisinos, 2015); Esteban O. Lavilla and Guillermo Wilde, eds., *Los anfibios y reptiles del Paraguay Ilustrado de Joseph Sánchez Labrador (Rávena, 1776)* (Tucumán: Fundación Miguel Lillo, 2020).

[28] Lozano, Pedro, *Historia de la conquista del Paraguay, Río de la Plata y Tucumán*, ed. Andrés Lamas, vol. 1 (Buenos Aires: Imprenta Popular, 1873–75), 271.

[29] Ibid., 1:215.

creation in the Genesis narrative.³⁰ Guevara broaches geography, trees, plants and herbs, aquatic animals (including whales and other sea monsters), the birds, quadrupeds, serpents, and insects.

The perception of Paraguay as a paradise in the New World was a *topos* of literature about Guaraní missions. In *Il cristianesimo felice*, the learned Ludovico Muratori contemplates a "spectacle worthy of the eyes of Paradise."³¹ The comparison of the jungle to the Garden of Eden was reinforced by the disturbing presence of serpents. In *Spiritual Conquest* (Madrid, 1639) by Antonio Ruiz de Montoya (1585–1652), the only chapter devoted to animals amounts to a collection of tales about serpents, which Lozano enthusiastically reproduces in his own history.³² That snakes play so prominently a role in these Jesuit natural histories is attributable not only to the real fact that they were one of the dangers which anybody living in the missions had to face, but also to the symbolic certainty that serpents constitute the dangerous dimension of every Paradise worthy of that name.

For a Jesuit missionary, the New World was a geographical void sparsely inhabited by ferocious savages, which called for the civilising and Christianizing action of the Society of Jesus. In other words, a fallow land that could yield a rich crop of conversion if properly cultivated. The natural histories purported to give a circumstantiated account of these lands and these peoples; something of a "field-guide" for missionaries.³³ The idea of providing a substantiated chronicle of the theatre where action was to be deployed was natural for the Jesuits, who were familiar with the method of "composition: seeing the place," recommended by Ignatius of Loyola in his *Spiritual Exercises*, which consisted of visualizing with the imagination the surroundings where a spiritual drama took place.³⁴ In his *History of the Conquest of Paraguay*, Lozano justifies the geographical account he is about to begin, arguing that "a new world, not well known at the present time" needs to be described. This description, he goes on, is like "pointing out the theatre where the triumphs of faith and virtue of

---

30 Lozano's sequence of creatures is different from that in the narrative of creation of the book of Genesis, since he has cattle and quadrupeds created on earth *before* birds and fish are created in the element of water.
31 Ludovico Muratori, *Il cristianesimo felice nelle missioni de' Padri della Compagnia di Gesù nel Paraguai* (Venice: Giambattista Pasquali, 1752), 91.
32 Antonio Ruiz de Montoya, *Conquista Espiritual*, ed. Ernesto J. Maeder (Rosario: EDEHI, 1989), 49–53.
33 Anthony Pagden, *The Fall of Natural Man. The American Indian and the Origins of Comparative Ethnology* (Cambridge: Cambridge University Press, 1982), 149–51.
34 Ignace of Loyola, "The Spiritual Exercises," in idem, *Personal Writings* (London: Penguin, 2004), 294, 312 and 329 (§§ 47, 151 and 232).

those valiant knights [the Jesuits] against the forces of the abyss are to be represented."[35]

The natural histories of *Paraquaria* included the native names of plants and animals. If, as the ethnobiologists point out, naming the species implies a classification, then it is evident that when the Jesuits wove the local names of plants and animals into the text, fragments of folk taxonomies were incorporated as well.[36] In his *Description of the Great Chaco*, Lozano describes seven species of bees: *yamacuá*, or mongrel bee, *yalamacuá*, similar to the European bee; *aneacuá*, or little black bee, amongst others.[37] In fact, the Jesuit chronicler took these names from the *Vocabulary of the Lule and Tonocote Tongues*, written by his colleague Father Antonio Machoni (1672–1753).[38] We should recall here the enormous effort made by the Jesuits to master the native languages with a view to preaching, which resulted in vast amounts of grammars, dictionaries, catechisms, manuals for confession, and spiritual literature written in the native languages. Between 1580 and 1590, General Claudio Acquaviva established that all Jesuits that were preparing themselves for priesthood in Brazil, Peru, and Mexico, were to learn the native languages.[39] Naming the creatures of Paraguay in the local tongues was certainly a pragmatic solution, inasmuch as they had no European names, but the missionaries went further. Their natural histories are permeated by the native names of plants and animals, which they evidently wished to learn and use. The gesture of using the autochthonous names brings with it echoes of the Biblical story of Adam naming the creatures—the only difference being that this was a Guaraní paradise. In the natural histories written in the Italian exile, the use of native languages was deliberate and polemic. In their critique of the theory of degeneration proposed by Buffon (Georges-Louis Leclerc)—according to which animals of the New World were inferior to those of the Old World—Jolís and Termeyer censored the French naturalist for using erroneously or desultorily the Guaraní names. In the work of the Catalan Jolís at

---

[35] Lozano, *Historia de la conquista*, 1:7.
[36] Scott Atran, *Cognitive Foundations of Natural History* (Cambridge: Cambridge University Press; Paris: Éditions de la Maison des sciences de l'homme, 1990).
[37] Pedro Lozano, *Descripción Corográfica del Gran Chaco Gualamba*, ed. Radamés Altieri (Tucumán: Universidad Nacional de Tucumán, Instituto de Antropología, 1941), 46.
[38] Antonio Machoni, *Arte y vocabulario de la lengua lule y tonocote* (Madrid: Herederos de Juan García Infanzón, 1732).
[39] Sabine MacCormack, "Grammar and Virtue: The Formulation of a Cultural and Missionary Program by the Jesuits in Early Colonial Peru," in *The Jesuits II. Cultures, Sciences and the Arts, 1540–1773*, ed. John W. O'Malley, G. A. Bailey, S. J. Harris, and T. F. Kennedy (Toronto: University of Toronto Press, 2006), 576–601.

least, native names of beasts came to symbolize his lost South American adoptive homeland and his right to describe its creatures.

Neither Jolís nor Termeyer saw themselves as naturalists; they justified their works on natural history on the grounds that they had actually lived in the land they described, and thus were able to give a first-hand account of it. Jolís declares that his aim is "to present the readers and in particular the scholars, with an exact and faithful portrait of the practices and customs of the savages in contraposition to the damaging image presented by some authors [he is thinking of Buffon here] in their writings."[40] In order to legitimize such a project, he asserts his nine years of life among the natives and his three entries into the Deep Chaco. Against the savant Buffon, who had never set foot on America, Jolís brandishes his direct familiarity with the land and argues that scientific nomenclature is not necessary for his purposes, which can best be served by common local names. His epistemic authority was based on the principle of *autopsia* ("I was *there* and I saw it with my own eyes"), in other words, an expression of "local science."[41]

It has been pointed out that early modern natural history was a mixture of texts from classical antiquity, material of the medical tradition, and popular lore on nature.[42] Certainly, the conceptual "openness" of discourse on natural history preceding that of Buffon and Linnaeus facilitated the weaving into its fabric of the aboriginal names of plants and animals and the local modes of ordering the natural world; a cultural operation characteristic of the cultural syncretism of the missions.

It should be added that Jesuit writing experienced a transformation. In the early ones, such as those written by Ruiz de Montoya or Lozano, nature is rich in prodigies and marvels and the boundary between the profane and the divine is porous and unstable. In his *Paraguay natural*, written in the Italian exile, Sánchez Labrador displays a more enlightened attitude: (a) he criticises the use of explanations in terms of "occult qualities," characteristic of early modern Jesuit autors such as Eusebius Nieremberg and, at times, Athanasius Kircher;[43] (b) he

---

[40] José Jolís, *Ensayo sobre la historia natural del Gran Chaco*, trans. M. L. Acuña (Resistencia: Universidad Nacional del Nordeste, Facultad de Humanidades, Instituto de Historia, 1972), 42.
[41] Asúa and French, *A New World of Animals*, 84.
[42] Brian W. Ogilvie, *The Science of Describing. Natural History in Renaissance Europe* (Chicago: University of Chicago Press, 2006), 216–21.
[43] Building upon Benito Feijóo's enlightened criticism, Sánchez Labrador contends against explanations in terms of occult qualities, which "leave nature as occult as before among its draperies." Idem, *El Paraguay Natural*, Introduction, Paraquaria 16, v; cf. Miguel de Asúa, *Science in*

does not set himself against systematic taxonomy; and (c) at times, he attempts to engage in a dialogue with Buffon.

Herbal Medicine and Medicine

One of the major challenges faced by the Jesuits on the Paraguay missions was the devastating attacks of smallpox and measles, which troubled the towns at regular intervals. According to the chronicles, a severe outbreak could kill between 20,000 and 30,000 Guaraní. The smallpox outbreak of 1739–40 killed between 40 and 56 per cent of the population of several reductions.[44] In the towns, care of the body was indissolubly associated with care of the soul. The missionary from Tyrol Anton Sepp tells in his chronicle: "I was priest and Samaritan at the same time, cleaning not only the sores of the soul but also those of the body and easing the pains of the sick with the sacred oil of apostolic love and Christian fervour."[45] Faced with massive deaths, the Fathers cared first and foremost for the administration of the sacraments. In the 1635–37 outbreak of smallpox in the reduction of Jesús María, the Jesuits in charge gathered all the dying Guaraní in a single place in order to impart them a brief, simplified catechism that would allow them to be baptised.[46] The *litterae annuae* insist once and again that it was as a result of measures like this that the catechumens died after having received the sacraments. During the seventeenth century, the Fathers resorted to all kinds of religious cures, which included the use of consecrated fluids or objects, the touch of cards inscribed with prayers, relics, and other resources associated with the cult of saints. Contrarily, Sánchez Labrador, an enthusiastic reader of the enlightened Benedictine Benito Feijóo, provides a sketchy account of the epidemics of smallpox in the missions which is thoroughly medical. This author attributed the increased mortality among the Guaraní to their "corrupted humours" and their disregard for healing measures.[47] In any case, if in the

---

the *Vanished Arcadia. Knowledge of Nature in the Jesuit Missions of Paraguay and Río de la Plata* (Leiden: Brill, 2014), 71 and Asúa and French, *A New World of Animals*, 162–82.
44 Robert Jackson, "Mortality Crises in the Jesuit Missions of Paraguay, 1730–1740," *World History Review* 1, no. 2 (2004): 11.
45 Anton Sepp, *Continuación de las labores apostólicas*, ed. Werner Hoffmann (Buenos Aires: Eudeba, 1973), 149.
46 *Littera annua* for the years 1635–1637, in *Cartas anuas de la Provincia del Paraguay, Chile y Tucumán, de la Compañía de Jesús (1615–1637)*, ed. Carlos Leonhardt (Buenos Aires: Peuser, 1929), 594.
47 Sánchez Labrador, *El Paraguay natural*, pt. 1, bk. 3, chap. 9, Paraquaria 16, 536.

eighteenth century Jesuit medicine in Paraguay and Río de la Plata evolved toward a more empirical approach, divine causation of illness was always considered foremost.⁴⁸ Florian Paucke tells that after the 1760 smallpox outbreak in his mission of San Francisco Javier, he organized a thanksgiving mass for the saint, "for he had saved from the death most of the children in the village".⁴⁹

The colleges of Córdoba and Buenos Aires had pharmacies attached to them, which primarily served the religious community but also sold medication to the city inhabitants. They could be favourably compared to many of those the Society of Jesus had in Spain.⁵⁰ The number of Jesuits that acted as pharmacists, physicians, surgeons, and *infirmarii* in the missions was never high. In 1724, for example, there were two in the reductions along the Paraná River and only one in the towns of the Uruguay River. At the time of the expulsion, the number of lay brothers in these jobs was twelve (most of them were from Germany and other lands of the Habsburg Empire, as was tradition in the Society).⁵¹ In the missions, the Fathers had to make do with the most basic remedies of the kind used by travellers and soldiers. Martin Dobrizhoffer (1718–1791) mentions that the missionaries sought to replace "the want of physicians, surgeons and druggists by easily obtained remedies, by reading medical books and by other means."⁵² The literature on medical botany and medicine in *Paraquaria* is not negligible. The "herbal of the missions" was compiled by Pedro Montenegro, a lay brother born in Galicia (Spain) in 1663, who worked in the Madrid Hospital for many years and arrived in Buenos Aires by 1690. Montenegro's work was finished in

---

48 Eliane C. Deckmann Fleck, "'Da mística às luzes'—medicina experimental nas reduções jesuítico-guaranis da Província Jesuítica do Paraguai," *Revista Complutense de Historia de América* 32 (2006): 153–78.
49 Florian Paucke, *Hacia allá y para acá (una estada entre los indios mocovíes, 1749–1767)*, trans. Ernesto Wernicke, vol. 3, bk. 1 (Tucumán: Universidad Nacional de Tucumán; Buenos Aires: Institución Cultural Argentino-Germana, 1942–1944), 38.
50 See, for example, the inventory of the Jesuit pharmacy in Córdoba in Eliane C. Deckmann Fleck and Roberto Poletto, "Transcrição do *Inventário formado por Lorenzo Infante Boticário em la Ciudad de Córdoba de los bienes medicinales, Julio de 1772,*" *Antiguos Jesuitas en Iberoamérica* 1, no. 1 (2013): 162–247.
51 Thomas Falkner (1707–1784), Wenceslas Horsky (1723–1791), Peter Kornmayr (1691–1769), Norbert Ziulak (1716–1769), Sigismund Aperger (1678–1772), Rupert Dalhammer (1710–1780), Christian Mayr (1729–1773), Juan de la Cruz Montealegre (1739–1810), Joseph Jenig (1724–1770), Thomas William Brown (1745–1768), Esteban Font (1726–1772), and Thomas Heyrle (1697–1768). Asúa, *Science in the Vanished Arcadia*, 105–13. See also Renée Gicklhorn, *Missionsapotheker: Deutsche Pharmazeuten in Lateinamerika des 17. und 18. Jahrhunderts* (Stuttgart: Wissenschaftliche Verlagsgesellschaft, 1973).
52 Martin Dobrizhoffer, *An Account of the Abipones, an Equestrian People of Paraguay*, trans. Sarah Coleridge, vol. 2 (London: Murray, 1822), 256.

1710 and is usually referred to as *Materia medica misionera*, the title of one of the extant manuscripts.⁵³

Montenegro's aim was to write a work with practical advice on the recollection, preparation, and uses of medical herbs that could be beneficial in the mission towns. In the prologue, he humbly declares that after 12 years of being in America and given the lack of apothecaries, "he was forced to become an author." On this path, "he was moved more by charity and the wish to benefit his brothers than by ambition."⁵⁴ Montenegro's intellectual endeavour was framed within a religious worldview. In the prologue of his work, he proclaims that He who ultimately communicates the remedies to the human being is God, although this knowledge is transmitted through intermediaries i.e. the *auctoritates*. Moreover, the existence of so many plants with curative virtues in the New World was a gift of Divine Providence, which sought to compensate for the lack of physicians and pharmacists.⁵⁵

Montenegro's *materia medica* consists of textual accounts and pictures of around 150 plants. There is no discernible criterion of ordering (neither alphabetical, nor by disease, nor symptom). Most of the plants are flora of Paraguay, but there are a few chapters copied from Piso and Markgraf's *Historia naturalis Brasiliae* (Leiden, 1648). For each plant he gives a brief description and in many cases he distinguishes varieties of a single "species," which are obviously those used by the Guaraní. Actually, Montenegro's *Materia medica* is a textual witness to the articulation of Indian and Western knowledge; a blend of folk taxonomy and the categories of traditional European pharmacopoeia as found in Andrés Laguna's famous commentary on Dioscorides (Salamanca, 1563).

Montenegro mentions his native informers. The most assiduous and reliable was Clemente, a convert to Christianity who worked as a male nurse in the reductions. The Jesuit underlines that Clemente was a "good Christian."⁵⁶ Medical attention in the missions was in charge of Guaraní male nurses, the *curuzuya*,

---

53 For the manuscripts with Montenegro's *materia medica*, see Asúa, *Science in the Vanished Arcadia*, 117–23, to which should be added those described in Harald Thun, "El saber médico de los guaraníes y la medicina de los jesuitas. Transmisiones y transformaciones," in *Indigenous Knowledge as a Resource. Transmission, Reception, and Interaction of Knowledge between the Americas and Europe, 1492–1800*, ed. Laura Dierksmeier, Fabian Fechner and Kazuhisa Takeda (Tübingen: Tübingen University Press, 2021), 41–73. Two of them are edited: Pedro Montenegro, *Materia medica misionera* (Buenos Aires: Biblioteca Nacional, 1945) and Carmen Martín Martín and José L. Valverde, eds., *La farmacia en la América colonial: el arte de preparar medicamentos* (Granada: Universidad de Granada y Hermandad Farmacéutica Granadina, 1995).
54 Montenegro, *Materia medica*, 7.
55 Ibid., 5–10.
56 Ibid., 363.

who were trained by the Fathers. It is plausible that many of these were familiar with traditional Guaraní medicine. The Jesuit herbals include a good deal of material obviously derived from the knowledge and use of plants in folk medicine. As ethnobotanical investigations show, it is almost impossible to distinguish between therapeutic and magical uses in Guaraní medicine, much of which is shamanic.[57] In the context of the confrontation between the curative resources of the Fathers and those of the Guaraní shaman, it is logical that Montenegro insisted on the fact that Clemente was a Christian: in some way, this warranted that the reports of the *curuzuya* were free from the magical contents which the Jesuit considered diabolical.

A group of manuscripts with fragments of Montenegro's herbal circulated under the name of Sigismund Aperger.[58] Born in Innsbruck, Aperger arrived in the Río de la Plata in 1716.[59] Having no formal qualifications as an apothecary or physician, he could have learned his art from Heinrich Peschke, a famous pharmacist at the college in Córdoba. After being ordained in 1717, Aperger distinguished himself helping in the epidemic outbreak that arose that year in Córdoba and soon spread to the missions. He spent the rest of his life as pharmacist in the reductions and was the only Jesuit who remained after the expulsion in 1767 (he was too sick to travel). Aperger's *Tratado Breve de la Medicina*, is a practical handbook to be used in the missions, which draws on contemporary medical works.[60] Other works like this circulated in the missions, such as the *Libro de cirugía*, a manuscript with extracts from printed medical books which was described in the early twentieth century, was lost, and has recently been recovered.[61]

---

[57] See Héctor Keller, "Notas sobre medicina y magia entre los guaraníes de Misiones, Argentina. Un enfoque etnobotánico," *Suplemento Antropológico* 42, no. 2 (2007): 345–83; Alfred Métraux, "Le shamanisme chez les Indiens de l'Amérique du Sud tropicale," *Acta Americana* 2 (1944): 197–219, 320–41.
[58] Elsewhere, I showed that the manuscripts that circulated under the name of Aperger were actually copied from Montenegro's treatise. See Asúa, *Science in the Vanished Arcadia*, 123–26.
[59] Guillermo Furlong, "Un médico colonial: Segismundo Aperger," *Estudios* 54 (1936): 117–48.
[60] For Aperger's *Tratado breve* see Sabine Anagnostou, *Jesuiten in Spanisch-Amerika als Übermittler von heilkundlichem Wissen* (Stuttgart: Wissenschaftliche Verlagsgesellschaft, 2000), 331–411. The first part of Aperger's work is an herbal, which has been lost. This genre of writing was also practiced by Jesuits in all of Ibero-America. Perhaps the most famous example is the *Florilegio medicinal* (Mexico: Herederos de Juan Joseph Guillena Carrascoto, 1712), written in the second half of the seventeenth century in New Spain by the Moravian Jesuit Juan de Esteyneffer (Johannes Steinhöffer), a missionary in Sonora.
[61] Félix Garzón Maceda, *La medicina en Córdoba: apuntes para su historia*, vol. 1 (Buenos Aires: Rodríguez Giles, 1916–1917), 479–81; Eliane C. Deckmann Fleck and Franz Obermeier, "O libro

Thomas Falkner was an English surgeon who during many years acted as the only person with real medical training for the whole territory of Paraguay and the River Plate.[62] Falkner arrived in the Río de la Plata in 1730 as the surgeon of a slave ship. Since he was very sick and unable to return to England in his own vessel, he was left to be nursed by the Jesuits of the College in Buenos Aires. He converted to Catholicism and entered the Society in 1723. Falkner studied in Córdoba and explored the central region of present-day Argentina and northern Patagonia, founding reductions in the province of Buenos Aires for the nomad peoples of the Pampas. Eventually, he returned to Córdoba, where he stayed from 1756 to 1767. When later living at Spetchley Park, Worcestershire, England, where he was appointed chaplain to his patron Robert Berkeley, Falkner dictated to a scribe the text of *A Description of Patagonia* (Hereford, 1774), a book surrounded by many uncertainties with respect to its production and editing. The work includes a short list of plants with healing properties, references to two species of tea (*albahaca de campo* in Córdoba and *culén* in Chile), extensive narratives of his travels throughout the territory of present-day Argentina, and even some political opinions, like his suggestion that the Spanish territories of Patagonia were so large and so poorly defended, that it would be easy for a foreign expeditionary force to take hold of them. Falkner is also said to have written a long treatise on *American Distempers Cured by American Drugs*, but this work is lost.[63]

Astronomy

If there were ever a scientific discipline that could properly be called "Jesuit," it would be astronomy. Jesuits observed phenomena like comets and eclipses in the Southern Cone of South America from the seventeenth century onwards. Nicolò Mascardi, a missionary in Chile and Patagonia, was a representative of a group of Jesuits which carried out astronomical observations along and across the Americas. Mascardi, Eusebius Kino in New Spain, Jean-Raymond [Juan

---

de medicina, cirugía, e botica: um manuscrito anônimo de Matérica médica rioplatense da primeira metade do século XVIII," *Antíteses* 11, no. 21 (2018): 132–56.

62 C. W. Sutton (rev. Geoffrey Scott), "Falkner, Thomas," in *Oxford Dictionary of National Biography*, published online 23 September 2004, doi.org/10.1093/ref:odnb/9124; Miguel de Asúa, "Acerca de la biografía, obra y actividad médica de Thomas Falkner S.I. (1707–1784)," *Stromata* 62, no. 3–4 (2006): 227–54.

63 George Oliver, *Collections towards Illustrating the Biography of the Scotch, English and Irish Members of the Society of Jesus* (London: Charles Dolman, 1854), 89.

Ramón] Coninck in Peru, and Valentin Stansel in Bahia (Brazil) organised a network of astronomical observation that converged in Rome. Here, Athanasius Kircher, the Jesuit polymath and epigone of baroque learning, kept an active correspondence with these men while dispatching his books and model of science to the outermost recesses of the New World.[64] If the pattern of communication was centripetal during the seventeenth century, from periphery to the centre, the growth of the missions in the eighteenth century brought with it a more articulated and decentralized network into being, so that a Jesuit astronomer in the jungle of Paraguay, such as Buenaventura Suárez (1679–1750), was able to correspond with many Iberian-American and European colleagues.

Suárez was perhaps the most remarkable character among those who worked in the Paraguay missions as inquirers of nature.[65] He was almost entirely self-taught, for his formal university education went no further than one year of philosophy and another of theology at the University of Córdoba. Suárez spent most of his life in the reduction of San Cosme, but also lived for one- to two-year periods in other mission towns and also in Asunción—it is possible to trace his whereabouts from the records of his astronomical observations. Father Suárez had natural a gift not only for the exact sciences, but also for all kinds of craftwork. He was adept at metallurgy and set up a foundry where he fabricated bells and organs tubes. Among the instruments he made, there was a pendulum clock accurate to a second, quadrants for measuring the altitude of heavenly bodies, and several telescopes ranging from eight to 23 feet. For the lenses, he used rock crystal, which abounds in that region. It is highly plausible that the Guaraní artisans collaborated in significant ways with Suárez, for it is well known that many of them excelled at making instruments. Sepp mentions that they could smelt metal for bells, make chiming clocks similar to those fabricated in Augsburg, and musical instruments such as fagots, oboes, violins, harps, organs, lutes, and flutes.[66] The fabrication, possession, and use of astronomical instruments was not unusual among the Jesuits, as testified to in the case of the Swiss Martin Schmidt, his countryman Karl Rechberg, the Belgian Ignace

---

[64] For Mascardi's letters with Kircher, see Giuseppe Rosso, "Nicolò Mascardi missionario Gesuita esploratore del Cile e della Patagonia (1624–1674)," *Archivum Historicum Societatis Iesu* 19 (1950): 3–74; cf. the discussion in Andrés I. Prieto, *Missionary Scientist. Jesuit Science in Spanish South America, 1570–1810* (Nashville, TN: Vanderbilt University Press, 2011), 126–40.

[65] See Guillermo Furlong, "Buenaventura Suárez (1679–1750)," in idem, *Glorias santafecinas. Buenaventura Suárez, Francisco Javier Iturri, Cristóbal Altamirano. Estudios biobibliográficos* (Buenos Aires: Editorial Surgo, 1929), 79–140.

[66] Anton Sepp, *Jardin de Flores Paraquario*, ed. Werner Hoffmann (Buenos Aires: Eudeba, 1974), 179.

Chomé, and the Spanish José Quiroga—the most renowned cartographer of the Paraguay missions.⁶⁷ Sundials, used to rationalize time in terms of liturgical rhythms, were a usual fixture in the landscape of the reductions. Suárez calculated a "Perpetual table" for the use of sundials.⁶⁸

The Jesuit astronomer of *Paraquaria* concentrated on the observation of lunar and solar eclipses and the appearances and disappearances of Jupiter's satellites. Determining the time of the disappearance of the Jovian moons in the shadow of the planet was a standard method for the calculation of longitudes in the early eighteenth century (in the wake of imperial expansion, this question came to play a fundamental role in the relation between astronomy and world politics.) Suárez was able to reckon the longitude of each of the missions with respect to the meridian of the Isla de Fierro in the Atlantic (a standard meridian of reference in the Iberian world). In order to obtain the data he needed for comparison with other astronomical observers, he corresponded with the Jesuit astronomer Nicasius Grammatici established in Ingolstadt, who provided him with his own observations from Madrid and Amberg, with those from the Jesuit Ignaz Kögler in Beijing, and of Joseph-Nicolas Delisle, then in St. Petersburg. Suárez also interchanged information with Pedro de Peralta y Barnuevo, cosmographer of the vice-royalty of Peru and author of an astronomical almanac (*Conocimiento de los Tiempos*), edited in Lima between 1721 and 1743. In turn, Grammatici was the channel through which Suarez's data on the satellites of Jupiter reached the Swedish astronomer Pehr W. Wargentin, who used 43 observations made by the Jesuit missionary in a paper published in 1748 in the *Acta Societatis Regiae Scientiarum Upsaliensis*.⁶⁹

Like Peralta in Lima, Suárez published astronomical almanacs in the missions of Paraguay for many years. An offshoot of these reckonings was his lunar calendar, the *Lunario de un siglo*, which encompassed several editions: Lisbon (1748), Barcelona (1752), and Ambato, Ecuador (1759). The *Lunario* was based on the tables of the French astronomer Philippe de la Hire (1640–1718) and on Suárez's own observations. It predicted the phases of the moon, lunar and solar eclipses, and also gave information about the liturgical year (golden number, epact, dominical letters, date of Easter, and so on). In a letter dated 29 March 1730, Suárez answered a query from the Father procurator in *Paraquaria* about

---

67 Miguel de Asúa, *Science in the Vanished Arcadia*, 165–67; for Suárez's telescopes, see ibid., 226–28.
68 Furlong, "Buenaventura Suárez (1679–1750)," 133–35.
69 Pehr W. Wargentin, "Series observationum primi satellitis Jovis, ex quibus theoria motuum ejusdem satellitis est deducta," *Acta Societatis Regiae Scientiarum Upsaliensis* 3 (1748): 1–32, cf. Asúa, *Vanished Arcadia*, 228–31.

the calculation of the epact for that year, which indicates that the astronomer's work was not an amateur's pastime but answered in direct ways to the needs of the community.[70] In the prologue of the *Lunario* Suárez deems his work is useful for "Agriculture and Medicine."[71] In fact, the Jesuits used to give advice on agriculture to the Spanish settlers and the *Lunario* was a convenient device in this respect. The medical function of the *Lunario* emphasizes the belief in the relationship between the heavenly bodies, climate, and epidemics; this included the election of favourable times for bloodletting or herb gathering (Suárez was the author of a list of medicinal herbs).[72]

Many data obtained by Suárez found their way into the pages of the *Philosophical Transactions of the Royal Society*. Created in 1660, the Royal Society was not only one of the cauldrons of the new science, but a clearing house through which information about geography and the natural products of the New World (among other exotic lands) reached London.[73] Suarez's observations were communicated to this learned body by one of its members, Jacob de Castro Sarmento (1691–1762), a Jewish-Portuguese physician exiled in England, who had introduced Newtonianism in Portugal and had written a treatise on the Newtonian theory of tides, the *Theórica verdadeira das marés* (London, 1737). Castro Sarmento had escaped the Portuguese Inquisition and was established in London as an associate of the rabbi Hakhim Nieto, also a Newtonian physician; the former was a fellow of the Royal Society and of the London College of Physicians.[74] Suárez translated Castro Sarmento's work into Spanish (the manuscript has been lost).[75] This means that Suárez was interested in Newtonian literature, while in the University of Córdoba teaching of natural philosophy followed a more conservative pattern (I shall discuss this further in the next chapter). The teaching of mathematics was introduced rather late, when the sixteenth provincial congregation, which met at Córdoba in 1762, asked the Father General for the creation of a Chair of Mathematics in the University, well after Suárez had died.

---

70 Furlong, "Buenaventura Suárez (1679–1750)," 133–35. The epact is the difference between the solar and lunar year, used to calculate the date of Easter.
71 I have used the copy of the Lisbon 1748 edition held in Biblioteca Nacional (Buenos Aires), with marginal annotations by Suárez.
72 Furlong, "Buenaventura Suárez (1679–1750)," 137.
73 Phyllis Allen, "The Royal Society and Latin America as Reflected in the *Philosophical Transactions*, 1665–1730," *Isis* 37, no. 3–4 (1947): 132–38.
74 For Castro Sarmento, see Richard Barnett, "Dr Jacob de Castro Sarmento and Sephardim in Medical Practice in 18th-Century London," *Transactions of the Jewish Historical Society of England* 27 (1982): 84–114 and Matt Goldish, "Newtonian, *Converso*, and Deist: The Lives of Jacob (Henrique) de Castro Sarmento," *Science in Context* 10, no. 4 (1997): 651–75.
75 Furlong, "Buenaventura Suárez (1679–1750)," 139–40.

Two of the several reasons invoked by the Fathers in support of the creation of this chair are of interest to us.[76] To begin with, the Jesuits claimed that the teaching of mathematics was necessary for helping the missionaries find their way during their excursions through uncharted hills and jungles. Another reason was that knowledge of mathematics would be useful for establishing new reductions, for that discipline was the basis of architecture, the wood industry, and hydromechanics. It is evident that what the Jesuits congregated in Córdoba had in mind was the teaching of some kind of applied mathematics, in the form of astronomy, cartography, and mechanics.

Suarez's two papers in the *Philosophical Transactions* were published in 1748 and 1749.[77] The first paper describes a series of observations made from 1706 through 1730, and is in two parts; the first part deals with lunar and solar eclipses, and the following part with eclipses of the Jupiter satellites. The second paper describes lunar eclipses. These lunar eclipse observations were more sophisticated than any of his previous ones (the geographical accidents of the Moon are mentioned for the first time). The intermediary between Castro Sarmento and Suárez was Mateus Saraiva, a physician in Rio de Janeiro and fellow of the Royal Society.[78]

One of Suárez's most significant contributions was a table with the coordinates of all the mission towns of Paraguay. The *ad quem* of its first version is 1719, for it seems that a copy was sent to Germany that year.[79] In his *Paraguay natural*, Sánchez Labrador copied an analogous table attributed to Suárez, with the commentary that it corresponded to observations made between 1745 and 1747.[80] Perhaps on account of its practicality, this information was available in writings and maps that circulated within the boundaries of the Society of Jesus.

---

76 Guillermo Furlong, *Matemáticos argentinos durante la dominación hispánica* (Buenos Aires: Huarpes, 1945), 86–7.
77 Jacob de Castro Sarmento, "Observationes astronomicae variae factae in Paraquaria, Regione Americae Australis, ab anno 1706 [1700] ad annum 1730," *Philosophical Transactions of the Royal Society* 45 (1748): 667–74; [Jacob de Castro Sarmento], "Observationes aliquae astronomicae a Reverendo P. P. Suarez e S. J. in Paraquaria habitae, et per D. Suarez M. D. Cum Soc. Regali communicatae," *Philosophical Transactions of the Royal Society*, 46 (1749–1750): 8–10.
78 For a detailed discussion with reference to manuscript sources, see Asúa, *Vanished Arcadia*, 235–42.
79 Furlong, "Buenaventura Suárez (1679–1750)," 135.
80 José Sánchez Labrador, *El Paraguay natural*, pt. 1, bk. 3, chap. 5, Paraquaria 16, 473–75. Furlong claimed that the figures in both tables were different, see idem, "Buenaventura Suárez (1679–1750)," 136 (I was not able to see the first table).

Suarez's commitment to the land where he was born never blinded him with respect to the universal dimensions of the religious order to which he belonged. It is a telling symbol of this attitude that his *Lunario* brings an algorithm that allows the reader to use the work in different parts of the globe. In the extreme periphery, Suárez wrote a lunar calendar which he thought could be used in Siam, Warsaw or Athens (he listed the differences in longitude from the reduction of San Cosme to seventy cities of the world). Suárez's inconspicuous mission town joined in a dialogue of equals with large and famous cities—as Raphael proclaims at the beginning of More's *Utopia:* "the road to heaven is the same length from all directions."[81]

Cartography

Cartography is perhaps the area of Jesuit learning which better shows the connexion between science and missionary activity. Jesuit maps provide geographical information and at the same time give visual expression to the ideal of mission.[82] Furlong numbered 18 Jesuit maps of the River Plate region produced between 1647 and 1730 and 80 between 1730 and 1798.[83] Alexander von Humboldt himself declared that "the missionaries...were the only geographers of the most inland parts of the [American] continents."[84] Most Jesuits maps were modest affairs, rich in information but without aesthetic aspirations, drawn to illustrate the course of an expedition, to substantiate territorial claims, to chart the progress of the missionary enterprise, or to stimulate pious feelings. Mapping was a skill that came with missionary training. We have seen that together with their breviaries and Bibles, many Jesuits brought with them high-quality measuring instruments. The most famous cartographers of Río de la Plata, historical Paraguay, Chaco, and Patagonia were the Spaniards José Quiroga, José Cardiel, and Joaquín Camaño (1737–1820) (the latter worked in Italy after the expulsion). But many maps were anonymous. Prestigious European cartogra-

---

[81] Thomas More, *Utopia*, ed. Peter Marshall (New York: Simon and Schuster, 1988), 3.
[82] Luke Clossey, *Salvation and Globalization in the Early Jesuit Missions* (Cambridge: Cambridge University Press, 2008), 73.
[83] Guillermo Furlong, "Cartografía colonial," in *El momento histórico del Virreinato del Río de la Plata*, ed. Ricardo Levene, vol. 4, bk. 2, *Historia de la Nación Argentina*, 3rd ed. (Buenos Aires: El Ateneo, 1961), 193. See also Guillermo Furlong, *Cartografía jesuítica del Río de la Plata*, 2 vols. (Buenos Aires: Peuser, 1936).
[84] Alexander von Humboldt, *Personal Narrative of the Equinoctial Regions of America: During the Years 1799–1804*, trans. Thomasina Ross, vol. 2 (London: Henry Bohn, 1852–1853), 430.

phers made their maps of the southern cone of South America upon the information provided by Jesuit charts. The Dutch mapmaker and printer Willem Blaeu reproduced the map entitled "*Paraquaria* vulgo Paraguay cum adjacentibus" (c. 1647) in the twelfth volume of his *Le grand atlas, ou Cosmographie Blaviane* (Amsterdam, 1667);[85] the French Jean-Baptiste Bourgignon d'Anville, cartographer to the king, drew "Le Paraguay, où les RR. PP. de la Compagnie de Jesus ont répandu leurs Missions" based on Jesuit sources and edited in the *Lettres édifiantes et curieuses*.[86]

Many Jesuit maps represent the territory in a missionary key. In them, geography is at the service of religion. One of the principal preoccupations of Jesuit mapmakers was to convey information about the localisation of cities and towns, the routes of communication among them, and the lands of the native peoples they sought to Christianize. In this sense, most of the maps were thematic, insofar as geography was seen as a scenario for some or other particular event or process.[87] The map "Paraquariae Provinciae Societatis Jesu cum adjacentibus novissima descriptio," finished in 1722 and printed in 1726—perhaps the most complete of those made by the Jesuits until the middle of the eighteenth century—represents South America between latitude 10° to 36° south. Copies of this map were engraved in Rome by Joannes Petroschi in 1726, others by Matthaeus Seutter in Augsburg. Furlong has attributed this map to the lay brother Juan Francisco Dávila (or de Ávila) born in Buenos Aires, who employed ten years on its making.[88] In a letter written in September 1722, Brother Dávila explains that he made it as an answer to a request of the Provincial, because the Superior General needed to know the distances between the Jesuit establishments in order to arrange more effectively the transference of personnel in the country.[89] The highly tight organization of the Society of Jesus and the necessity of elevating frequent missionary reports (accompanied with maps) contributed to the consolidation of the Jesuit culture of mapping.[90] Furlong dubbed Quiroga's map of 1749, engraved by Ferdinando Franceschelli and published in Rome in 1753, as "per-

---

[85] Furlong, *Cartografía jesuítica*, 1:26–30.
[86] Jean-Baptiste Bourgignon d'Anville, "Le Paraguay, où les RR. PP. de la Compagnie de Jesus ont répandu leurs Missions" (map), *Lettres édifiantes et curieuses* 21 (1734): 278–79.
[87] Norman J. W. Thrower, *Maps and Civilization. Cartography in Culture and Society*, 3rd ed. (Chicago: The University of Chicago Press, 2008), 95.
[88] Furlong, *Cartografía jesuítica*, 1:44–49.
[89] Ibid., 1:47.
[90] David Buisseret, "Jesuit Cartography in Central and South America," in *Jesuit Encounters in the New World: Jesuit Chronicles, Geographers, Educators and Missionaries in the Americas (1549–1767)*, ed. Joseph A. Gagliano and Charles E. Ronan (Rome: Institutum Historicum Societatis Iesu, 1997), 161.

haps the most perfect map [of Paraguay and Rio de la Plata] made by the Jesuits in the 18th century."[91] It represents the territory of the Jesuit missions from Santa Cruz de la Sierra (Bolivia) down to Buenos Aires (latitude 22° to 36° south). Quiroga toured the missions and the central and northern part of present-day Argentina during three years while he worked on this chart. The map is circumscribed by two vertical columns and an upper and lower horizontal band with texts. In the left column there is a table which gives the latitude and longitude of the reductions and of the most important cities of *Paraquaria* (38 in total). The legend adverts that the lack of the coordinates of certain towns are due to the fact that the measurements had not yet been done. The right column and the horizontal band at the bottom contain short ethnographic descriptions of several native peoples, indicating their habitat, food, religions, and so on.[92]

Maps from Paraquaria were published in the *Neue Welt-Bott* and the *Lettres édifiantes*, the collection of letters and reports from Jesuit missionaries which served both as a source of information on exotic lands and as inspiring literature and propaganda media for the missions.[93] Maps were instruments of preaching and helped in confirming the believers in their faith. The map entitled "Part of Southern America where the zeal of the members of the Paraguay province of the Society of Jesus is deployed" (1760) has been attributed to José Cardiel.[94] This is a map of Jesuit martyrs. The chart represents the whole province of *Paraquaria*. In its upper margin there is a cartouche divided in two parts, each half containing the key to the iconic signs used in the map. The left half is "secular," its symbols represent the realm of earthly powers; the right one is "sacred," it accounts for the space of the religious power. The secular world is arranged hierarchically, with signs varying according to the size and importance of the city or settlement; Spanish cities that had been destroyed are indicated by a Maltese cross. On the semi-cartouche situated to the right, we find the symbols for different kinds of Jesuit mission towns and also places where the "barbarian gentiles" had killed Jesuit missionaries and destroyed the town; geographical spots where the missionaries who brought the light of faith to the pagan nations have been slain; places where Jesuits have been killed by the Portuguese. In the later cases the map has informative legends giving the name of the martyred Jesuit and the name of the people that caused his death. With its surface almost covered by leg-

---

[91] Guillermo Furlong, *El Padre José Quiroga* (Buenos Aires: Peuser, 1930), 72.
[92] Ibid., 75–79; cf. idem, *Cartografía jesuítica*, 1:71–75.
[93] For maps of Paraquaria published in the *Neue Welt-Bott*, see Vicente D. Sierra, *Los jesuitas germanos en la conquista espiritual de Hispano-América* (San Miguel: Facultades de Filosofía y Teología; Buenos Aires: Institución Cultural Argentino-Germana, 1944), 312.
[94] Furlong, *Cartografía jesuítica*, 1:97–98.

ends in an elegant handwritten lettering, the map bears a testimony of commitment and sacrifice. It was conceived to impress upon Catholic believers the very real way in which the pagan land had become consecrated by the blood of the Fathers. Jesuit maps with the plotting of places of martyrdoms were not unusual in Spanish America.[95] A similar message is proclaimed explicitly in Dávila's map of 1722 (engraved in 1726), dedicated to the Superior General Michelangelo Tamburini. The six distiches of the dedicatory describe "wide America" inhabited by "wild nude people" whose barbaric necks were progressively being subjugated by Christ's gentle yoke. The map, the verses proclaim, demonstrates how fallow the land would have remained were it not for the blood shed over it.[96]

The imaginary landscape of Jesuit cartography was a large geographical void sparely inhabited by savages, which called for the civilising and Christianizing action of the Society of Jesus.[97] As much attention was paid to the secular space—geographical accidents, secular cities—as to the sacred space, embodied in the reductions. Jesuit maps of Paraquaria represent the territory and its Christianization at the same time. Many of them include the names of the native peoples converted and convey as much ethnographical as geographical information. An implicit triumphalism is latent in these charts which exhibit the extent to which pagan territory became Christianized as a demonstration of the superior performance of the Jesuits over other missionary orders.

Jesuit Missionary Science in Río de la Plata

Missionary science was at the crossroads between the global and the local. David Livingstone has emphatically underlined its local character: "missionary science was quintessentially a spatial practice" in which "missionaries ... derived their authority from the places where they pursued their inquiries."[98] The attachment

---

95 In this connection, Buisseret mentions a map drawn by Father Kino to illustrate his story of the martyrdom of Father Saverio Saetta, a map of the Marañón by Father Fritz with five crosses marking the death of Jesuits at the hands of natives, and José Gumilla's map in *El Orinoco Ilustrado* (Madrid, 1741). See Buisseret, "Jesuit cartography," 118, 122 and 161.
96 Cited in Furlong, *Cartografía jesuítica*, 1:44–45.
97 See William M. Denevan, "The Pristine Myth. The Landscape of the Americas in 1492," *Annals of the Association of American Geographers* 82, no. 3 (1992): 369–85 and idem, "The 'Pristine Myth' Revisited," *Geographical Record* 101, no. 4 (2011): 576–91.
98 David N. Livingstone, "Scientific Inquiry and the Missionary Enterprise," in *Participating in the Knowledge Society. Researchers beyond the University Walls*, ed. Ruth Finnegan (Basingstoke: Palgrave Macmillan, 2015), 61, 52. For missionary science as practiced by nineteenth- and twentieth-century Protestant missionaries, see John Stenhouse, "Missionary Science," in *Modern Sci-*

of missionary science to a particular locality was articulated with a universal dimension related to the religious organisation, institution, or church that supported the missionary enterprise: "Nowhere is this remarkable relation between 'world' and 'local' Christianity clearer than in mission history."[99]

Science as practiced in the Jesuit missions is a splendid example of this dialectic between the global and the local, manifested in the circulation of personnel, texts, practices, and objects. As I have hinted at above, the geographical, geological, and astronomical data sent from the missions to Europe were the empirical foundation upon which Athanasius Kircher at the *Collegium romanum* and, more modestly, Eusebius Nieremberg at the *Colegio imperial* in Madrid built their learned monuments of baroque learning.[100] As a rule, the centripetal-centrifugal dynamics between missionary outposts and Rome was complemented by exchanges among the points of the network of Jesuit missions covering the world.[101] Analysing the movements of Jesuits and their scientific production over the "geography of knowledge," Steven Harris has claimed that the success of Jesuit science was grounded on the relation between the local and the global.[102] This applied not only to astronomy and what we call today the Earth sciences but also to natural history and medical botany as cultivated in the Jesuit missions.[103]

In the missions of historical Paraguay and Río de la Plata science seems to have been at the service of religion. Clossey has claimed that the successes of the Jesuits in many fields, from astronomy to botany, make "historians forget they were missionaries at all," but "the missionary sources make clear [the] subordi-

---

ence in National, Transnational, and Global Context, ed. Hugh R. Slatten, Ronald L. Numbers and David N. Livingstone, vol. 8, *The Cambridge History of Science* (Cambridge: Cambridge University Press, 2020), 90–107.

99 Clossey, *Salvation and Globalization*, 3. See also Thomas Banchoff and José Casanova, eds., *The Jesuits and Globalization. Historical Legacies and Contemporary Challenges* (Washington: Georgetown University Press, 2016).

100 Findlen, ed., *Athanasius Kircher*; John E. Fletcher, *A Study of the Life and Works of Athanasius Kircher, 'Germanus Incredibilis'* (Leiden: Brill, 2011); Millones Figueroa and Ledezma, *El saber de los jesuitas*; Pimentel, "Baroque Natures."

101 Steven J. Harris, "Confession-Building, Long-Distance Networks, and the Organization of Jesuit Science," *Early Science and Medicine* 1, no. 3 (1996): 287–318.

102 Steven Harris, "Mapping Jesuit Science: the Role of Travel in the Geography of Knowledge," in *The Jesuits. Cultures, Sciences, and the Arts*, 229. For surveys of global Jesuit science, see Steven Harris, "Jesuit Scientific Activity in the Overseas Missions, 1540–1773," *Isis* 96, no. 1 (2005): 71–79; Sheila J. Rabin, "Early Modern Jesuit Science. A Historiographical Essay," *Journal of Jesuit Studies* 1, no. 1 (2014): 88–104.

103 Miguel de Asúa, "Natural History in the Jesuits Missions," in *The Oxford Handbook of the Jesuits*, 708–36.

nation of science to religion."[104] Rubiés also partakes of the idea that the motivations of the Jesuit missionaries, "when undertaking ethnological research were often religious and apologetic."[105] Harris has argued that it is the Jesuits' "apostolic spirituality" which explains the legitimation of science within the Society.[106]

The practice of Jesuit science in the missions of *Paraquaria* supports this interpretation: the strategies deployed by the missionaries were largely practical and geared to the support of the infrastructure and daily functioning of the mission towns. An objection to this view has been raised by Mordechai Feingold, who finds "the common perception of Jesuit activity as motivated by religious concerns problematic" and challenges "the received view that they [Jesuit savants] ought to be considered first and foremost Jesuits."[107] His analysis concerns Jesuit savants and science teachers active in early modern Europe and England, which might account in part for his conclusions. The missions of South and North America were a different thing from the Jesuit colleges in France and the Holy Roman Empire. In the reductions of Paraguay, everything—including science—was aimed at the salvation of souls. However, all this said, Feingold's approach brings a welcome complexification to the idea of the primacy of the religious. While it is evident that, as a general statement, scientific knowledge in the missions was instrumental to a religious goal, when we consider the careers of individual Jesuits we enter the realm of personal motives and things get more hazy: it is plausible that some of the scientific pursuits of Jesuits such as Suárez, Falkner or Termeyer were driven by scientific curiosity as much as by their commitment to their religious mission. In fine, scientific endeavours might have resulted from an unstable balance between missionary practices, individual talents, and circumstantial political requirements.[108]

When cultivating disciplines like astronomy, medical botany, natural history, and cartography the missionaries were following a disciplinary pattern that by the eighteenth century had become established in the Old Society. These were

---

104 Clossey, *Salvation and Globalization in the Early Jesuit Missions*, 246–47.
105 Joan-Pau Rubiés, "The Jesuits and Enlightenment," in *The Oxford Handbook of the Jesuits*, 871.
106 Steven J. Harris, "Transposing the Merton Thesis: Apostolic Spirituality and the Establishment of the Jesuit Scientific Tradition," *Science in Context* 3, no. 1 (1989): 29–65.
107 Mordechai Feingold, "Jesuits: Savants," in *Jesuit Science and the Republic of Letters*, ed. Mordechai Feingold (Cambridge, MA: The MIT Press, 2003), 2, 7.
108 This is also the view of Antonella Romano, "Les jésuites entre apostolat missionnaire et activité scientifique (XVIe–XVIIIe siècles)," *Archivum Historicum Societatis Iesu* 74, no. 147 (2005): 213–36.

the fields in which the Jesuits had excelled in Europe and elsewhere and where their contributions should be looked for. But these were also the disciplines that could help to accomplish the missionary goals in the most direct and efficient manner possible. One can perceive a no-nonsense approach to learning in the writings of the Jesuits in Paraguay. Notwithstanding few exceptions, it seems possible to conclude that, as a whole, their scientific activities were clearly tailored for the ultimate apostolic aims of the Society: natural history was conceived as a description of the theatre where the drama of the conversion to Christianity of the native peoples was being enacted; astronomy was a means to set the geographical coordinates of the mission towns, and medical botany was obviously concerned with the health of the missionaries and their catechumens. In this context, "harmony" between science and religion should be construed in the sense that the overarching goal of scientific activity in the missions was to contribute to the fulfilment of what the Jesuits understood as their mission.

Before its dissolution in 1773, the Old Society deployed a variety of strategies in order to adapt the Christian message to the lands and peoples among whom the Jesuits lived and preached. The Jesuits' "'way of proceeding' was often characterised by flexible accommodation to local circumstances that combined a commitment to the Gospel with an affirmation of the positive characteristics of the cultures they encountered."[109] Ronnie Po-chia Hsia distinguishes two models of early modern Catholic missions: (a) in the New World, Christianization followed conquest and colonisation and thus had strong political connotations tied to Spanish imperialism; (b) in the Asiatic colonies of Portugal, a sea and commercial empire, Christian outposts depended of the good will of the local rulers and the settlement and running of missions had to be negotiated.[110] The missionary styles associated with these models found their expression in the missionary programs of Joseph de Acosta and Matteo Ricci, respectively.[111] I have argued elsewhere that the role played by science in the missions was different in each case. In a writing culture with a vast tradition of learning such as China, science was a *constitutive* factor of missionary activity insofar as the prestige of Western scientific knowledge served to legitimize the religious authority of the Jesuit missionaries. In the case of the oral cultures of Iberian America, science was *accessory* to the missionary enterprise providing material and symbolic

---

109 Thomas Banchoff and José Casanova, "The Jesuits and Globalization," in *The Jesuits and Globalization*, 1.
110 Ronnie Po-chia Hsia, "Introduction. Catholic Global Missions and the Expansion of Europe," in *A Companion to Early Modern Catholic Global Missions*, 1–4.
111 Ana C. Hosne, *The Jesuit Missions to China and Peru, 1570–1610* (Abingdon: Routledge, 2013), 71–96.

logistic support to the missions.[112] While in China science was part of a huge operation of *captatio benevolentiae* used to demonstrate the supposed superiority of European civilisation in order to attract the cultured elite, among the sedentary agriculturalists and hunter-gatherers of the southern cone of South America missionary science was characterised for its ability to absorb elements of native cultures into the matrix of European discourse, as well as for its pragmatic bent oriented toward the stability and progress of the mission towns.[113]

The expulsion of the Society of Jesus from South America sounded the death knell for Jesuit science in Río de la Plata. The natural histories of the missionaries were succeeded by the surveys of the natural resources of the New World performed by the large Spanish scientific expeditions launched by the Bourbons in the second half of the eighteenth century.[114] This was "Enlightenment science," a different thing from Jesuit missionary inquiries about nature.[115] While for the Jesuits natural history was an instrument of Christianization, for the enlightened officers of the crown, who catalogued and described the natural products of the colonies in order to better exploit its resources, science was conceived as a tool of empire. Félix de Azara should be included in this company.[116] A military engineer and one of the most accomplished representatives of Bourbon imperial science, Azara arrived in Río de la Plata in 1781 as part of the commission for the demarcation of limits between Spain and Portugal as a result of the Treatise of 1777. He remained twenty years in the region, in the course of which he became the foremost local authority in the fields of geography, topography, engineering, and nat-

---

[112] Miguel de Asúa, "La 'ciencia' en las misiones jesuitas como encrucijada de saber global y culturas locales," in *A Experiência da Missão Jesuítica na Primeira Modernidade*, ed. Luiz F. Medeiros Rodrigues and Maria C. Bohn Martins (São Leopoldo: Oikos, 2019): 180–200.
[113] Nicolas Standaert, *Handbook of Christianity in China*, vol. 1, *635–1800* (Leiden: Brill, 2001), 311.
[114] Arthur R. Steele, *Flowers for the King. The Expedition of Ruiz and Pavón and the Flora of Peru* (Durham, NC: Duke University Press, 1964); Iris W. Engstrand, *Spanish Scientists in the New World. The Eighteenth-Century Expeditions* (Seattle: University of Washington Press, 1981).
[115] Juan Pimentel, "The Iberian Vision: Science and Empire in the Framework of Universal Monarchy, 1500–1800," *Osiris*, 2nd series, 15 (2000): 17–30; Antonio Lafuente, "Enlightenment in an Imperial Context: Local Science in the Late-Eighteenth-Century Hispanic World," *Osiris*, 2nd series, 15 (2000): 155–73.
[116] Helen Cowie, "A Creole in Paris and a Spaniard in Paraguay: Geographies of Natural History in the Hispanic World (1750–1808)," *Journal of Latin American Geography* 10, no. 1 (2011): 175–97; Marcelo F. Figueroa, "Félix de Azara and the Birds of Paraguay. Making Inventories and Taxonomies at the Boundaries of the Spanish Empire, 1784–1802," in *Global Scientific Practice in an Age of Revolutions, 1750–1850*, ed. Patrick Manning and Daniel Rood (Pittsburgh: University of Pittsburgh Press, 2016): 147–62.

ural history (a self-taught naturalist, his books on birds and quadrupeds, in Spanish and French versions and written in the mould of Buffon's *Histoire naturelle*, soon became reference works for the regional fauna).[117] In like manner as his brother José Nicolás, who as a Spanish diplomat in Rome had been involved in the suppression of the Society of Jesus by Clement XIV in 1773, Félix was a notorious anti-Jesuit. This stance is particularly evident in his *Voyages*, a work on Paraguay and Río de la Plata in which he paints a bleak vision of the Jesuit missions.[118] Azara became a bone of contention between Catholic and liberal historians. The nineteenth-centuy liberal Catholic intellectual Félix Frías, after citing Buffon and Macaulay praising the Jesuit missions, described Azara as "an employee of the king, *his master*, who was aware that, *in matters concerning the Jesuits*, freedom of thought was a treasonable offence" (emphasis in the original).[119] Bartolomé Mitre, one of the founders of liberal historiography in Argentina, saw Azara as "a modest Humboldt of these parts of America."[120] Juan María Gutiérrez, an influential nineteenth-century literary figure, rector of the University of Buenos Aires (1861–1873) and one of the main architects of the idea of science as an engine of secularism in Argentina, claimed that "the facts of nature had been absurdly studied and described by the chroniclers of the Order [sic] of Jesus... He [Azara] has thrown light upon that night of absurdities."[121] Gutiérrez made an exception for Aperger and Buenaventura Suárez, "a hero of astronomical science," but argued that the Society did nothing to encourage their scientific activities.[122] While liberal and anticlerical writers contrasted Enlightenment science with

---

117 Barbara Beddall, "'Un naturalista original': Don Félix de Azara, 1746–1821," *Journal of the History of Biology* 8, no. 1 (1975): 15–66; idem, "Isolated Spanish Genius: Myth or Reality: Félix de Azara and the Birds of Paraguay," *Journal of the History of Biology* 16, no. 2 (1983): 225–58; Thomas Glick and David M. Quinlan (1975), "Félix de Azara: The Myth of the Isolated Genius in Spanish Science," *Journal of the History of Biology* 8, no. 1 (1975): 67–83.
118 Félix de Azara, *Voyages dans l'Amérique méridionale*, ed. C. Walckenaer, 2 vols. (Paris: Dentu, 1809).
119 Félix Frías, "Los derechos de los frailes," in idem, *Escritos y discursos*, ed. Pedro Goyena, vol. 3 (Buenos Aires: Librería de Mayo, 1884), 124.
120 Bartolomé Mitre, Letter to Diego Barros Arana, 20 October 1975, in *Correspondencia literaria, años 1859–1881*, vol. 20, *Archivo del General Mitre* (Buenos Aires: Biblioteca de La Nación, 1912), 75.
121 Juan María Gutiérrez, "Don Félix de Azara. Su mérito, sus servicios, su juicio sobre las misiones del Paraná y del Uruguay," *Revista de Buenos Aires* 18 (1869): 175–76.
122 Gutiérrez, "Don Félix de Azara," 188. For other views on Azara, see Harald Thun, "Félix de Azara, los jesuitas y el guaraní," in *Século das Luzes. Portugal e Espanha, o Brasil e a Região do Rio da Prata*, ed. Werner Thielemann (Frankfurt a. M.: Ibero-Amerikanisches Institut, 2006), 475–502.

Jesuit obscurantism, Catholics did the opposite. The next chapter argues that the ambivalent nature of Catholic Enlightenment in Río de la Plata does not easily fit into these categories.

Chapter 2
Catholic Enlightenment. Science and Religion in Colonial and Early Independent Río de la Plata

In the first decade of the nineteenth century, the University of Córdoba decided to purchase a cabinet of electrical instruments, a splendid collection owned by a rich landowner from Buenos Aires, Martín José de Altolaguirre, accountant of the local Royal Tribunal of Accounts. Altolaguirre, an agricultural innovator who had introduced the cultivation of flax and hemp in Río de la Plata, had inherited the instruments from his father Martín (also a high officer in charge of the finances of the viceroyalty), who had acquired them in Europe.[1] Three artisans of Buenos Aires had appraised the instruments in 9,372 pesos.[2] These were the kind of electrical apparatus made famous by the Abbé Nollet.[3] In 1798, Altolaguirre put forward his father's scientific instruments for sale; after prolonged haggling, a deal was closed in 1801 for half the original price and the collection was purchased by the Franciscan Pedro José Sullivan, then rector of the College of Montserrat, attached to the University of Córdoba.[4] Despite a crown's decree transferring the University from the Franciscans to the secular clergy, by the beginning of the nineteenth century the institution was still in the hands of the friars: both factions of the clergy, religious and secular, became enmeshed in a protracted conflict over the control of the university and the power and prestige attached to it. Several members of the *cabildo* (municipal body of government) of Córdoba partial to the disgruntled secular priests seized the opportunity to confront the friars by boycotting the transaction. A third party in the conflict were the enlightened functionaries of the crown, who fostered the diffusion of the new philosophy in the climate of educational reforms encouraged by the Bourbons in all their overseas territories. The legal documents produced in the course of the transaction testify to the views on natural theology of the parties involved. In his report from 8 February 1802, Pablo de Cires, the city attorney of the city of Córdoba, used the occasion to proclaim his enlightened viewpoints rallying

---

[1] Marcos Estrada, "La casa de Altolaguirre," *Genealogía* 12 (1957): 139–51.
[2] See the inventory of the instruments in Zenón Bustos, ed., *Anales de la Universidad Nacional de Córdoba. Segundo período*, vol. 3 (Córdoba: Casa editora Domenici and Tipografía La Industrial, 1901–1910), 305–29.
[3] Lewis Pyenson and Jean-François Gauvin, eds., *The Art of Teaching Physics. The Eighteenth-Century Demonstration Apparatus of Jean Antoine Nollet* (Montreal: Septentrion, 2002).
[4] Miguel de Asúa, *La ciencia de Mayo. La cultura científica en el Río de la Plata, 1800–1820* (Buenos Aires: FCE, 2010), 78–80.

against the "fantastic Aristotelian hallucinations." He also extolls the use of such machines as the microscope, the telescope, and the hygrometer for the investigation of nature. It was "the author of nature," he goes on, who had caused in due time the appearance on Earth of minds no lesser than those of the Greeks, which could invent the devices required to reveal the hidden beauties of creation.[5] Convened to deliberate on that same month, the members of the *cabildo* of Córdoba had refused to grant their approval for the transaction on the grounds that the founder of the university had envisaged that the graduates should be "doctors in theology" and not natural philosophers occupied with the discovery of new phenomena. One of them argued that the teaching of speculative physics was more useful to theology students than experimental physics, for scholastic natural philosophy fostered habits of argumentation.[6] In his statement defending the transaction, Father Sullivan resorted to traditional natural theological arguments. The Franciscan declares that the advances made possible by experimental machines and new inventions facilitated "the road to God inasmuch as they allowed for a better understanding of the wonderful workings of creation."[7] In February 1803, the public prosecutor José Márquez de la Plata approved the acquisition of the equipment and went to great lengths to explain why experimental physics was relevant to theology. In the first place, it contributed to the better demonstration of the existence and attributes of the Supreme Being, thus helping rebuke the claims of deists and materialists. Secondly, natural philosophy could be of use in distinguishing between false and true miracles. Finally, experimental philosophy aided in purifying the "sound and holy" religion from popular superstitions and false beliefs. Márquez de la Plata had in mind the formation of a clergy that could "enlighten the people" and "instruct them in all that is useful to the service of God, the king, and the community." This is the ideal of the enlightened parish priest, not only a religious guide and dispenser of sacraments, but also a teacher trained to lead his flock away from superstitions, mistakes, and "errors against philosophy, morals, politics, and religion."[8]

This story reveals how on occasion of the purchase of an electric cabinet by the university for the purpose of instruction, members of the clergy and Spanish colonial officers in Río de la Plata raised issues related to natural theology and

---

[5] Bustos, *Anales*, 3:332–35 (8 February 1802).
[6] Ibid., 3:340–41 (26 February 1802). This very same argument had been used by the University of Salamanca when rejecting the program of enlightened reform pushed forward by Charles III's ministers.
[7] Ibid., 3:348 (28 September 1802).
[8] Ibid., 3:352–61 (25 February 1803).

discussed the proper relationships between natural philosophy and religion. After the expulsion of the Jesuits, discourse and practices concerning the empirical investigation of the natural world in the southernmost settlements of the Spanish empire were as a rule in harmony with a religious view of the universe framed within the limits of Catholic orthodoxy. But within that framework, there was a spectrum of positions on the treatment of natural philosophy vis-à-vis religious authority and the handling of natural theology, including those who held deist views but did not expressed them openly in order to avoid problems with censorship. In a word, during the late eighteenth and early nineteenth centuries the solid functional integration between science and religion of Jesuit times began to show some fissures and reaccomodations.

The aim of this chapter is to get a picture of the relationships between science and religion in this period from late colonial to early independent times (1770–1815). After going through some historical background I shall cursorily review the limits imposed on the teaching of natural philosophy in Córdoba and Buenos Aires by authoritative church pronouncements. The next section discusses the handling of some aspects of Biblical geology by teachers of scholastic courses and the adoption of the Bible as an interpretive framework in the empirical study of the geological origin of Río de la Plata by the naturalist priest from Montevideo Dámaso Larrañaga. In the fourth section, I shall peruse the uses of natural theology in Larrañaga's *Diary of natural history* and in speeches by enlightened crown bureaucrats and supporters of independence, which will make possible to identify a stream of natural theological arguments, ranging from the demonstration of God from his works to the reflection on the workings of Providence in the fabric of nature. The conclusion will rearrange these fragments into a coherent image of the relationships between natural philosophy and religion in late eighteenth-century Río de la Plata.

Historical Background

During much of the colonial period, Buenos Aires was a forgotten outpost of the Spanish empire on the southern coast of Río de la Plata. Much of its economy depended on the shipping of the silver extracted in Potosí (in present-day Bolivia) to the metropolis; other settlements along the long inland route connecting the mines with the port, such as Córdoba, thrived on their role as intermediaries.⁹ As a result of the exhaustion of the ore supply and a mild liberalization of commerce by the crown, Buenos Aires became more fully integrated into the Atlantic trade system: the creation in 1776 of the Viceroyalty of Río de la Plata (present-day Argentina, Bolivia, Paraguay, and Uruguay) represented an official recognition of the growing importance of the region.¹⁰ This new political unit was part of the package of reforms pushed forward by the Bourbon monarchs in their efforts to increase the profits extracted from the colonies.¹¹ Much of the ideological undergirding of this programme should be sought in the principles of rational administration fostered by the Spanish Enlightenment. Unlike what was the case in France, Enlightenment in Spain was not essentially anti-religious. For the most, its advocates did not disown their traditional faith. Instead, they cultivated those aspects of the new philosophy that did not collide with Christian doctrines: the ideal of progress, physiocracy, and many elements of the new scientific worldview.¹² Versions of this Iberian Catholic Enlightenment were reproduced in the overseas territories and, after 1767, in the Italian states by expelled Jesuits, many of whom came from Río de la Plata.¹³

Estimations of the population of Buenos Aires in 1810 approach 40,000 inhabitants, whose livelihoods depended on commerce, bureaucracy, the church, and the military; little less than one third of the population were of African descent, mostly slaves.¹⁴ At the time of independence (a process starting in 1810), pastoral products amounted at most to 20% of exports, and the rest was silver. The fall of silver mining and improved commercial conditions after independence eventually shifted the basis of its economy so that in the end cattle products

---

9 David Rock, *Argentina, 1516–1987. From Spanish Colonization to Alfonsín* (Berkeley: University of California Press, 1987), 39–78.
10 Adelman, *Republic of Capital*, 19–108.
11 See Adrian J. Pearce, *The Origins of Bourbon Reform in Spanish South America, 1700–1763* (New York: Palgrave Macmillan, 2014); Allan J. Kuethe and Kenneth J. Andrien, *The Spanish Atlantic World in the Eighteenth Century: War and the Bourbon Reforms, 1713–1796* (Cambridge: Cambridge University Press, 2014).
12 Jean Sarrailh, *L'Espagne éclairée de la deuxième moitié du XVIIIᵉ siècle* (Paris: Imprimerie nationale, 1954); Richard Herr, *The Eighteenth-Century Revolution in Spain* (Princeton: Princeton University Press, 1958); Gabriel B. Paquette, *Enlightenment, Governance, and Reform in Spain and Its Empire, 1759–1808* (London: Palgrave Macmillan, 2008).
13 Andrea J. Smidt, "'Luces por la Fe': The Cause of Catholic Enlightenment in 18ᵗʰ-Century Spain," in *A Companion to the Catholic Enlightenment in Europe*, ed. Ulrich L. Lehner and Michael O'Neill Printy (Leiden: Brill, 2010), 403–52; Ulrich L. Lehner, *The Catholic Enlightenment* (Oxford: Oxford University Press, 2016).
14 César A. García Belsunce, *Buenos Aires y su gente, 1810–1830* (Buenos Aires: Emecé, 1976), 62; George R. Andrews, *The Afro-Argentines of Buenos Aires, 1800–1900* (Madison, WI: University of Wisconsin Press, 1980), 4.

became the main item in exports; the increase of investment and production took place only in the 1840s, giving place to a large expansion of the economy.[15] The invasion of Spain by Napoleon in May 1810 prompted the *criollos* (Spaniards born in Río de la Plata) to set up a local autonomous government; the formal Declaration of independence took place six years later. As a result of the revolutionary wars and the internecine conflicts that followed, the Viceroyalty of Río de la Plata was fragmented into distinct political units: the United Provinces of Río de la Plata (later Argentina and Uruguay) and the future republics of Bolivia and Paraguay.[16]

Separation from Spain brought along with it the interruption of diplomatic communications with the Holy See but the new independent governments maintained the system of *patronato*, which had shaped the relationships between the Spanish crown and the Catholic Church.[17] Apart from the secular clergy, there were many female and male religious orders in Buenos Aires, e.g. Dominicans, Franciscans, Recollects (a branch of the Friars Minor), Mercedarians, and Bethlehemites. The society was homogeneously Catholic; Protestants could only be found within the foreign communities.[18] In the period under consideration, the Inquisition based in Lima (Peru) did not play a significant role in Río de la Plata.[19] Prohibited books circulated among the clerical and lay elite; between 1782 and 1810, the Lima Inquisition conducted only two inquiries in Buenos

---

15 Susan M. Socolow, *The Merchants of Buenos Aires, 1778–1810* (Cambridge: Cambridge University Press, 1978); Carlos Newman and Barry Poulson, "Purely Animal: Pastoral Production and Early Argentine Economic Growth, 1825–1865," *Explorations in Economic History* 35, no. 3 (1998): 325–45; Samuel Amaral, *The Rise of Capitalism in the Pampas. The Estancias of Buenos Aires, 1785–1870* (Cambridge: Cambridge University Press, 1998); Eduardo Míguez, "Reforma y primitivismo. Tierra y fiscalidad en el Río de la Plata, de la colonia a la independencia," in *Cambio institucional y fiscalidad. Mundo hispánico, 1760–1850*, ed. Michel Bertrand and Zacarías Moutoukias (Madrid: Casa de Velázquez, 2018), 287–303. I owe this remark to Eduardo Míguez.
16 John Lynch, *The Spanish American Revolutions, 1808–1826*, 2nd ed. (New York: Norton, 1986), 38–127; Tulio Halperin Donghi, *Revolución y Guerra. Formación de una elite dirigente en Argentina*, 3rd ed. (Buenos Aires: Siglo XXI, 2014).
17 Eugene Shiels, *King and Church: The Rise and Fall of the Patronato Real* (Chicago: Loyola University Press, 1961); Austen Ivereigh, *Catholicism and Politics in Argentina, 1810–1960* (New York: St. Martin's Press, 1995), 39–43.
18 Bianchi, *Historia de las religiones en la Argentina*, 17–39.
19 For an earlier period, see Ana Schaposchnik, *The Lima Inquisition: The Plight of Crypto-Jews in Seventeenth-Century Peru* (Madison: University of Wisconsin Press, 2015). See also José Toribio Medina, *La Inquisición en las Provincias del Plata* (Santiago de Chile: Imprenta Elzeviriana, 1899), 259–68.

Aires and Mendoza and the penalties were lenient.[20] In what was still a stratified baroque society, public life had a basically religious tone.[21] The surviving late baroque colonial churches in Córdoba, Salta, and Buenos Aires—product of the craftmanship of Italian and German Jesuit architects—are a testimony to a kind of religiosity informed by the style and mentality of the Counter-reformation.[22] Against this background, some signs of an "enlightened piety" could be discerned.[23]

Despite appearances of dullness and conformity, the intellectual and political turmoil of the independence movement would reveal that a number of clergy had been trying with more or less success to come to terms with the new currents of thought. During the revolution, the Spanish prelates took the side of Spain while the large majority of the lower clergy sided with the new regime: of the 29 deputies who signed the Declaration of independence, eleven were priests.[24] Many of the priests of the generation who supported the Revolution of 1810 in Río de la Plata held Gallican or "Jansenist" positions—the latter should be understood "not as a doctrine on grace, but as a current of thought bent on removing the attributions of Roman primacy supported by men zealous for the study of history, ancient discipline, episcopal rights, and enemies of scholasticism."[25] Chilean historian Mario Góngora has underlined the connection between Spanish regalism and enlightened Gallicanism in Spanish America and claimed that the notion of the divine right of kings and the independence of the secular power was given a republican twist, insofar as "the doctrine [by itself] did not exclude any form of government."[26] The disconnection of the episcopal sees from Rome was functional to episcopalist positions. Church historian Américo Tonda af-

---

[20] Ernesto J. Maeder, "Libros, bibliotecas, control de lecturas e imprentas rioplatenses en los siglos XVI al XVIII," *Teología* 77 (2001): 5–24.
[21] Tulio Halperin Donghi, *Revolución y guerra*, 78–80.
[22] Gauvin Alexander Bailey, *The Spiritual Rococo. Decor and Divinity from the Salons of Paris to the Missions of Patagonia* (Abingdon: Routledge, 2016), 250–70.
[23] Jaime Peire, *El taller de los espejos. Iglesia e imaginario, 1765–1815* (Buenos Aires: Claridad, 2000), 180–86.
[24] Nancy Calvo, Roberto Di Stefano, and Klaus Gallo, eds., *Los curas de la revolución* (Buenos Aires: Emecé, 2002).
[25] Américo A. Tonda, *La eclesiología de los doctores Gorriti, Zavaleta y Agüero* (Rosario: Universidad Católica Argentina, Facultad de Derecho y Ciencias Sociales del Rosario, n. d.), 6.
[26] Mario Góngora, "Estudios sobre el Galicanismo y la 'Ilustración católica' en América española," *Revista Chilena de Historia y Geografía* no. 125 (1957): 120. See also Rómulo D. Carbia, *La Revolución de Mayo y la Iglesia* (Buenos Aires: Huarpes, 1945) for Bourbon regalism and the relationships between state and Church in Río de la Plata.

firmed that in Río de la Plata political "Jansenism" constituted "the common patrimony of the majority of secular clergy."[27]

Late Scholasticism: Modern Science and the Authority of the Church

The introduction of modern science into the universities of Spanish America has long been a topic of active research.[28] In the Viceroyalty of Río de la Plata, the schools of Córdoba and Charcas (Chuquisaca, present-day Bolivia) were the only centres of higher learning.[29] The former remained in the hands of the Society of Jesus until the year of their expulsion. The course of studies in Córdoba spanned three years of philosophy and four of theology. Law was introduced in 1790. After the expulsion of the Jesuits in 1767, the University was transferred to the secular clergy by royal decree. The viceroy and the local authorities ignored the order and handed it on to the Franciscans established in the region since the mid-sixteenth century; it was finally entrusted to the secular clergy by a new royal decree (1800) put into effect in 1808.[30] The 1813 plan of studies drafted by the secular priest Gregorio Funes introduced an elementary course of mathematics and recommended teaching physics with "the help of experiences" to which purpose a person acquainted with the use of instruments was to be employed by the school in order to profit from the aforementioned cabinet of physics purchased by the university.[31] In his plan of studies for the seminaries, the patriot priest Juan Ignacio Gorriti encouraged the study of "those parts of physics which should be necessary for the exercise of the ministry in order to the common good...such as the use of the compass." He considered the discussion of the causes of gravity as "an unending and useless dispute" and claimed that what

---

27 Tonda, *La eclesiología*, 63.
28 See, for example, John Tate Lanning, *Academic Culture in the Spanish Colonies* (London: Oxford University Press, 1940); idem, "Tradition and Enlightenment in the Spanish Colonial Universities," *Cahiers d'histoire mondiale* 10 (1967): 705–21; Walter B. Redmont, *Bibliography of the Philosophy in the Iberian Colonies of America* (The Hague: Martinus Nijhoff, 1972); Antonio Lafuente, Alberto Elena and M. L. Ortega, eds., *Mundialización de la ciencia y cultura nacional* (Madrid: Universidad Autónoma-Doce Calles, 1993).
29 For the University of Chuquisaca (Charcas), see Clément Thibaud, *La Academia Carolina y la independencia de América. Los abogados de Chuquisaca (1776–1809)* (Sucre: Editorial Charcas, 2010).
30 Juan M. Garro, *Bosquejo histórico de la Universidad de Córdoba* (Buenos Aires: Biedma, 1882).
31 Gregorio Funes, *Plan de estudios para la Universidad de Córdoba* (Córdoba: Imprenta de la Universidad, 1832).

mattered was "knowing the laws of gravity," for this was something positive which could be used "to obtain safe and useful results."[32] The famous anti-Jesuit rector of San Carlos, Father Baltasar Maciel, claimed that he "could not understand which of the principles of modern physics" were in opposition to Catholic dogma, when this was "perfectly well explained in any system contrary to that of Aristotle."[33]

Soon after been elected into office as governor of Buenos Aires in 1772, Juan José de Vértiz—a prototype of the enlightened colonial functionary—created the College of San Carlos with chairs of grammar (Latin) and philosophy (another chair of philosophy and two of theology were later added).[34] The courses taught at San Carlos were in charge of secular priests; this kind of course was also offered in the convents of the mendicant orders of the city.[35] Buenos Aires lacked a university (it was created in 1821) but this void was compensated by the existence of professional schools: the school of medicine of the Protomedicato (the institution regulating medical practice) created by the Irish physician Michael O'Gorman, which came into function in 1808; the Nautical School (1799–1806), founded by Manuel Belgrano; and the Academy of Mathematics, created in 1810 by the revolutionary government for the training of military officers.[36]

The character of the teaching of natural philosophy in colonial Río de la Plata and the degree and tempo of the introduction of modern experimental science in the courses in Córdoba and Buenos Aires have been widely discussed.[37] As made clear in the opening paragraph of the course on natural philosophy taught by the Jesuit Benito Riva in Córdoba c. 1762–1764, followers of the "inno-

---

[32] Juan Ignacio Gorriti, *Reflecciones sobre las causas morales de las convulsiones interiores en los nuevos estados americanos* (Valparaíso: Imprenta del Mercurio, 1836), 190, 187.
[33] José Carlos Chiaramonte, *La Ilustración en el Río de la Plata* (Buenos Aires: Sudamericana, 2007), 124–27; see also Juan Probst, *Juan Baltazar Maziel. El maestro de la generación de Mayo* (Buenos Aires: Instituto de Didáctica de la Facultad de Filosofía y Letras de la Universidad de Buenos Aires, 1946).
[34] Juan Probst, "Introducción," in *La enseñanza durante la época colonial (1771–1810)*, vol. 18, *Documentos para la Historia Argentina*, ed. Instituto de Investigaciones Históricas, Facultad de Filosofía y Letras, Universidad de Buenos Aires (Buenos Aires: Peuser, 1924), cxxvii–cli.
[35] Celina Lértora Mendoza, "La enseñanza elemental y universitaria," in *Nueva Historia de la Nación Argentina*, ed. Academia Nacional de la Historia, vol. 3 (Buenos Aires: Planeta, 1999), 369–402.
[36] Asúa, *La ciencia de Mayo*, 19–48.
[37] Guillermo Furlong, *Nacimiento y desarrollo de la filosofía en el Río de la Plata, 1536–1810* (Buenos Aires: Kraft, 1952); Celina Lértora Mendoza, *La enseñanza de la filosofía en tiempos de la colonia* (Buenos Aires: Fecic, 1979); José Carlos Chiaramonte, *La Ilustración en el Río de la Plata*.

vata Philosophia" were to be preferred to the Peripatetic philosophers.[38] But despite declarations to the contrary, the overall spirit was conservative. The reception of Newton's *Opticks* and *Principia* amounted to the inclusion of bits and pieces of experimental science into a basically scholastic framework, without discussion of the founding notions of Newtonian natural philosophy or of the mathematical worldview it implied.[39] The often repeated claim that singles out the Jesuit Thomas Falkner as Newton's "favourite disciple" is chronologically inconsistent.[40] By mid-eighteenth century, the person best qualified to grasp the technicalities of Newton's work was the astronomer Buenaventura Suárez, discussed in the previous chapter.

Contrarily to what took place elsewhere, the Jesuits seem to have held a slightly more favourable view of Copernicus than their successors the Franciscans.[41] Benito Riva treats briefly the Ptolemaic system, dispatches Tycho's in a pair of paragraphs, and discusses with some detail the Copernican system in the best light. In his arguments in favour of heliocentrism, he follows the enlightened Benedictine Feijóo, but unlike the latter, Riva rejects gravitational attraction: "what is unsatisfactory with this attractive virtue is that it is more occult that the occult qualities."[42] The Jesuit's arguments against heliocentrism (the venerable Peripatetic objections of Galileo's *Dialogue*) are taken from Tomás V. Tosca's *Compendium philosophicum*, which circulated widely in the Iberian world. Riva points out that the Copernicans read metaphorically those passages

---

38 Benito Riva, [*Cursus physicae*]. *De mundo et caelo*, fol. 1r, Fondo Antiguo de la Compañía de Jesús en Argentina (FACJA), Buenos Aires.
39 Celina Lértora Mendoza, "Introducción de las teorías newtonianas en el Río de la Plata," in *Mundialización de la ciencia*, 307–23; idem, "Nollet y la difusión de Newton en el Río de la Plata," in *The Spread of the Scientific Revolution in the European Periphery*, ed. Celina Lértora Mendoza, Efthymios Nicolaïdis and Jan Vandersmissen (Turnhout: Brepols, 2000), 123–36. The contrast with the reception of Newton in other centres of Spanish America is striking, see for example Luis C. Arboleda and Diana Soto Arango, "The Theories of Copernicus and Newton in the Viceroyship of Nueva Granada and the Audiencia de Charcas during the 18th Century," in *Universities and Science in the Early Modern Period*, ed. Mordechai Feingold and Víctor Navarro-Brotons (Dordrecht: Springer, 2006), 289–309.
40 Asúa, "Acerca de la biografía, obra y actividad médica de Thomas Falkner."
41 For a closer look at the reception of Copernicanism in Río de la Plata, see Miguel de Asúa, "'Los phisicos modernos quasi todos son copernicanos.' Copernicanism and its Discontents in Colonial Río de la Plata," *Journal for the History of Astronomy* 48 (2017): 160–79.
42 "Displicet quod hec virtus attractiva occultior est qualitatibus occultis." Riva, *Cursus*, fol. 13r. This is one of several arguments against gravity. Riva's discussion on Newton's natural philosophy ocuppies fols. 10r–15v.

of Scripture which imply a geocentric universe and remembers his students that it was admitted to treat heliocentrism as a hypothesis.[43]

The Franciscans were even more cautious than the Jesuits as regards the Copernican system. The 1786 pastoral exhortation on the Plan of studies for the order by Manuel María Truxillo, the Franciscan Commissar General for the Indies, extolled mathematicians and authors of handbooks and treatises of natural philosophy such as Pieter van Musschenbroek, the Franciscans Fortunato da Brescia (Brixia) and Lorenzo Altieri, the Oratorian Tosca, and the Minims Emmanuel Maignan and Jean Seguens, while he harshly dismissed Aristotelian philosophy.[44] Aftern a eulogy of astronomy, a science that helps to elevate our thoughts to the Supreme Intelligence governing the world, Truxillo launches the reader into a heavenly voyage through the orbs. But although one of the wondrous sights of this cosmic excursion is Mercury, Venus, and Mars orbiting around the Sun, his narrative seems to presuppose a geocentric universe.[45]

A thesis on philosophy sponsored in Córdoba by the Franciscan Elías del Carmen Pereyra in 1790 begins commending the Copernican system on the grounds that it saves the astronomical and the physical phenomena; the text affirms that the conflictive passages of the Bible should be interpreted "in metaphorical sense and as accommodated to the uneducated." But Fr Elías concludes that heliocentrism could not be taken as a thesis, for it cannot be demonstrated— James Bradley's heliocentric arguments based on the aberration of starlight are deemed "not even probable."[46] In his course on *Physica particularis* (undated, probably 1782–1783), the Franciscan Cayetano Rodríguez begins declaring Tycho superior to Copernicus, but in the end he proclaims the *Systhema oscillationis* better than either of the former.[47] Cayetano Rodríguez argues that the Biblical passages in conflict with the Copernican view should be read literally and

---

[43] "Quod semper in hypothesi esse permissum." Riva, *Cursus*, fols. 314v–315r.
[44] Manuel María Truxillo, *Exhortación Pastoral* (Madrid: Viuda de Ibarra, 1786), 170, 186.
[45] Ibid., 174.
[46] Francisco J. Martínez de Aldunate and José Elías del Carmen Pereyra, *Conclusiones sobre toda la filosofía* (Buenos Aires: Imprenta de los Niños Expósitos, 1790), in Enrique Martínez Paz, "Una tesis de filosofía del siglo XVIII de la Universidad de Córdoba," *Revista de la Universidad de Córdoba* 6 (1919): 277–78.
[47] The oscillatory system, the work of an obscure Spanish author, proposed that the Earth was at the centre of the universe and subjected to oscillatory movement, see Joseph Santiago de Casas, *Relox universal de péndola y en él nueva idea de la estructura del universo* (Madrid: Herederos de la Viuda de Juan García Infanzón, 1758).

adduces a number of quotations from the Fathers to support his statement.[48] It should be remembered here that the removal of the clause condemning all books teaching heliocentrism by the Congregation of the Index in 1757 was ineffectual since the 1758 edition of the Index still mentioned the five treatises originally banned on this account.[49] Teachers in Río de la Plata took the safest path.

There are hints that suggest that some of them could have held a more independent position with respect to church authority. At least, this is what transpires from a Logic course taught in 1783 in Buenos Aires by the secular priest Luis José Chorroarín (1757–1823), one of the most accomplished intellectual figures in the decades surrounding the Revolution of 1810. He studied philosophy and theology in the College of San Carlos and was ordained in Buenos Aires; in 1783 he became professor of philosophy in the school from which he graduated.[50] In 1811, Chorroarín became director of the recently founded Public Library of Buenos Aires to which he donated his own books; there was a sizable number of scientific and technical works in this lot.[51] Chorroarín devotes a section of the Logic course he taught at San Carlos to arguing about reason and the authority of the Church in matters of natural philosophy.[52] He states that, in cases such as the negation of the antipodes by Augustine, the honour and respect due to the Fathers of the Church should not be an impediment to the close examination of the matter at hand.[53] When touching on questions of natural science the doctrines of the Fathers are to be taken on a par with those of the savants; the authority of the former should not go beyond that of the philosophers whom they followed. Chorroarín taught that what the Fathers had said about the natural world had not been inspired by the Holy Ghost: regarding disputed questions of natural philosophy the opinions of the Fathers should be followed only in

---

[48] "Copernicanum systhema adversatur apertissimis divinarum Scripturarum sententiis, nec veritati consentaneum est, nec ut tale propugnari potest." Cayetano Rodríguez, *Secunda phisicae pars seu Phisica Particularis*, liber I, quaestio 7 (not foliated), FACJA, Buenos Aires.
[49] Maurice A. Finocchiaro, *Defending Copernicus and Galileo. Critical Reasoning in the Two Affairs* (Dordrecht: Springer, 2010), 179–85.
[50] Ludovico García de Loydi, *Una luz en la manzana de las luces: Chorroarín* (Buenos Aires: Municipalidad de la Ciudad de Buenos Aires, Secretaría de Cultura, 1973); Manuel J. Sanguinetti, *Chorroarín. El prócer olvidado* (Buenos Aires: Stella, 1984).
[51] See the list of books on science in Asúa, *La ciencia de Mayo*, 201–203.
[52] Luis José Chorroarín, "Curso de Lógica," in *La enseñanza de la filosofía en la época colonial*, trans. Juan Chiabra, vol. 2 (Buenos Aires: Coni, 1911), 142–45.
[53] The advice to put aside Augustine's negation of the antipodes as an example of following the natural sciences in cases of evident fact or established proof could be found in the commentary on Genesis by the sixteenth-century Spanish Jesuit Benito Pereyra, see Richard J. Blackwell, *Galileo, Bellarmine, and the Bible* (Notre Dame: University of Notre Dame Press, 1991), 20–21n24.

the absence of "powerful reasons or positive experiments" against them. In medicine, the authority of a distinguished physician should be preferred to that of the Fathers just as in questions of natural philosophy we should follow Newton or his likes. It is evident that he moved comfortably within the framework of the Augustinian approach, famously elaborated in Galileo's *Letter to Cristina*.[54]

When it came to conflicting issues surrounding natural philosophy, the scholastic teachers in Río de la Plata did not challenge the authority of religious pronouncements. But there were nuances. While Riva was straightforward in his support of Copernicanism provided it was considered as a hypothesis, the Franciscans were more cautious. All of them took seriously the (since 1757) ambiguous ban on heliocentrism. While still within the bounds of orthodoxy, Chorroarín seems to have granted natural science a larger degree of autonomy.

Historians of different persuasions have taken Fr Elías to task for introducing theological questions in his course on *physica*.[55] Theological issues do certainly crop up in his course but his position on spiritual and divine action preserves the integrity of nature.[56] To the question whether angels and demons could move corporeal bodies naturally, he answers that since these are immaterial beings they could not possibly act naturally upon material entities.[57] In the *Conclusiones ex universa philosophia*, he considers "unworthy of philosophical reason" the idea that gravity is the result of God's direct action.[58] Fr Elías also cautions against using theology as the basis for philosophical judgements. He implicitly denies that the dogma of the Eucharist could be called upon to defend the existence of absolute accidents (quality, quantity) as separable from substance and

---

[54] See Ernan McMullin, "Galileo's Theological Venture," in *The Church and Galileo*, ed. Ernan McMullin (Notre Dame: University of Notre Dame Press, 2005), 88–116.

[55] Furlong considered Fr Elías's course "an indigestible mixture of philosophical and scientific problems" and saw the discussion of theological questions as a sign of "philosophical decadence." Furlong, *Nacimiento y desarrollo*, 267. The philosopher Alejandro Korn talked about "the curious mixture of sterile theology and rudimentary science in the lessons of Fr Elías del Carmen." Alejandro Korn, "La evolución de las ciencias en la República Argentina," in idem, *Obras completas*, vol. 3 (Buenos Aires: Claridad, 1949), 604. Fr. Elías del Carmen's theological excursions have also been considered a sign of philosophical backwardness by José Carlos Chiaramonte, see idem, *La Ilustración en el Río de la* Plata, 54–55.

[56] Lértora Mendoza has transcribed the Latin titles of the questions and the determinations, see idem, *La enseñanza de la filosofía*, 190–96.

[57] Elías del Carmen Pereyra, Física, in *La enseñanza de la filosofía en la época colonial*, trans. Juan Chiabra, vol. 2 (Buenos aires: Coni, 1911), 285–96.

[58] Martínez de Aldunate and Pereyra, *Conclusiones sobre toda la filosofía*, in Martínez Paz, "Una tesis de filosofía," 251–52. Shifting to Cartesianism, Pereyra explains the movement of the planets in terms of a "very fluid and ethereal matter, in which the celestial bodies and the fixed stars are immersed." Ibid., 276.

censures those authors who "without reason and in vain...fiercely defend...[absolute accidents] as something intangible and appeal to the Venerable Sacrament of the Eucharist and the sacred Councils of the Church."[59] Many teachers might have been cautious and conservative but, as a whole, natural philosophy in Río de la Plata was not confused with theology. As discussed in the following section, the case of geology and the Sacred Scriptures proved an exception.

Biblical Geology

The treatment of the history of the Earth by clergymen in Río de la Plata could illuminate the question of how they conceived the relationship between natural philosophy and the Bible. Elías del Carmen Pereyra declares that our world was created in six "physical and real" days in the boreal spring equinox and not "in spiritual days, in order to the knowledge of the angels," which is Augustine's doctrine in *De Genesi ad litteram*.[60] This amounts to choosing the literal over the allegorical reading of Genesis. The Spanish secular priest Melchor Fernández, who taught at San Carlos, also followed the idea that the world was created in six real days, during the spring equinox in Palestine.[61] In contrasting opposition to this, Cayetano Rodríguez followed Augustine when he claimed that "God created [the world] in a single moment; the six days were spiritual, not corporeal."[62] Juan Fernández, who taught at the Convent of the Recoleta in Buenos Aires and depended much on his fellow Franciscan Cayetano Rodríguez, also adhered to Augustine's opinion that the six days described by Moses should be understood as spiritual.[63] It should be noticed that Elías del Carmen Pereyra rehearsed the legitimacy of the "metaphorical" reading of the Bible in the Copernican question while rejecting Augustine's allegorical interpretation in the case of the creation narrative. Exactly the inverse can be said of Cayetano Rodríguez.

---

59 Ibid., 274.
60 Ibid., 275.
61 Gregorio García de Tagle, Dámaso Larrañaga, and Melchor Fernández, *Theses ex universa philosophia* (Buenos Aires: Apud Typographiam Regiam Parvulorum Orphanorum, 1792), in Juan C. Zuretti, "Tesis sobre filosofía y ciencias, defendidas en 1792 en el Real Colegio de San Carlos de Buenos Aires," *Revista de la Universidad de Buenos Aires* 44 (1948): 541.
62 Rodríguez, *Secunda physice pars*, Liber I, quaestio 3 (not foliated).
63 Juan M. Fernández, *Conclusiones publico-historico-dogmatico-scholastico-phisico-theologicae ex praecipuis Sacrae Theologiae tractatibus depromptae* (Buenos Aires: Apud Typographiam Regiam Parvulorum Orphanorum, 1803), 8.

Most of these authors also tackled the question of the origin of mountains. Contradicting Thomas Burnet's *Telluris Theoria Sacra* (1681), Fr Elías affirms that there were mountains *before* and after the Deluge; they could have been formed by the Flood, the subsiding of the waters, the action of rivers, the petrifaction of lava, or the elevation of the earth due to the subterranean fire—any of these hypotheses could also explain the occurrence of fish, shells, and other marine petrified rests in the summit of mountains. This, he says, is firmly established "in the texts of the Sacred Scripture."[64] In a group of theses from a course taught at San Carlos, the secular priest José Valentín Gómez embroiders on the theme adding the names of other seventeenth-century authors occupied with theories of the Earth: John Woodward and William Whiston; he refers to Genesis 7 when he claims that the mountains were covered by the rising waters of the Flood (thus, they did not result from it). Valentín Gómez attributed the formation of fossils to the Flood, earthquakes or the action of the sea.[65]

In a series of theses defended in Córdoba, the Franciscan Anastasio Mariano Suárez repeats almost verbatim Fr Elías's argument about the origin of the mountains.[66] He explained that the origin of fossils and stones should be sought in a subterranean fire, whereas in the view of Melchor Fernández fossils were deposited by the sea on the peaks of mountains and afterwards nourished by a suitable sap.[67] Although Juan Fernández supported the Augustinian interpretation of creation in a single instant, he simultaneously claimed that angels were created on the first day while in the five subsequent days God created the rest of the universe. As to mountains, he affirms that the more prominent of them originated on the third day as a result of the return of the waters.[68]

The teaching of these authors was moulded upon a literal reading of the Bible. Most of them preferred the literal interpretation of Genesis (creation in six days) and affirmed against Burnet that mountains existed before the Deluge,

---

[64] Martínez de Aldunate and Pereyra, *Conclusiones sobre toda la filosofía*, in Martínez Paz, "Una tesis de filosofía," 281.

[65] José Valentín Gómez, *Conclusiones ex universa philosophia* (Buenos Aires: Apud Regiam Parvulorum Orphanorum Typographiam, 1802), 16–17. Valentín Gómez, born in Buenos Aires, studied in San Carlos, Córdoba (Theology) and Chuquisaca (Civil and Canon Law); he was ordained in 1799.

[66] Anastasio M. Suárez, *Asserta ex universa philosophia* (Buenos Aires: Apud Typographiam Regiam Parvulorum Orphanorum, 1792), 8.

[67] García de Tagle, Larrañaga, and Fernández, *Theses ex universa philosophia*, in Zuretti, "Tesis sobre filosofía y ciencias," 547.

[68] Fernández *Conclusiones publico-historico*, 9–10.

although they could have been formed also after it by different mechanisms.[69] Even those who followed Augustine's allegorical interpretation of Creation, such as Fr Elías and Fr Juan Fernández, made a point of affirming that mountains had been originated in the third day. The origin of fossils is variously explained through hypotheses consonant with Biblical geology. The solutions are rather formulaic and sometimes inconsistent. The teachers of natural philosophy were friars and secular priests charged with teaching the philosophical triennium as a stepping stone in their ecclesiastical careers and did the best they could; the courses did not amount to much more than the more or less articulate rephrasing of scholastic handbooks.

The secular priest Dámaso Larrañaga (1771–1848) was altogether different from this. Born in Montevideo to a family of Basque ascent, he was the foremost naturalist in that region during the period.[70] During his long life he participated actively in the political and military events of the Banda Oriental (future Uruguay), was the founder of cultural and charity institutions in Montevideo, and was eventually apostolic vicar in his native city. He was at the centre of a circle of enlightened priests on both shores of Río de la Plata interested in natural history; they kept cabinets of curiosities and large libraries, interchanged specimens, promoted vaccination, and were heavily involved in political activities (more on them in the next chapter).[71] Larrañaga studied philosophy and theology at the College of San Carlos (1789–1795) and was ordained as a priest in Rio de Janeiro on 16 December 1798. In 1813, he was sent as a representative of Montevideo and its hinterland to the assembly of the United Provinces of Río de la Plata gathered in Buenos Aires. While in that city he was named deputy director of the Public Library whose director at that time was Chorroarín. Later and with the support of the Uruguayan political leader José Artigas, Larrañaga created the Public Library in Montevideo; in 1837, he took a leading part in the creation of the Museum of Natural History in that city. Larrañaga's main concern was taxo-

---

69 See, for instance, Fortunatus a Brixia, *Philosophia sensuum mechanica*, vol. 4 (Venice: Ex Typographia Remondiniana, 1755), 5–26; Joseph A. Ferrari, *Philosophia peripatetica*, vol. 3 (Venice: apud Thomas Bettinelli, 1754), 13–16 and 156; Antonio Goudin, *Philosophia iuxta inconcusa, tutissimaque Divi Thomae Dogmata*, vol. 3 (Venice: Typis Dominici Lovisa, 1729), 11–14.
70 For Larrañaga, see Rafael Algorta Camusso, *El padre Dámaso Antonio Larrañaga* (Montevideo: Barreiro y Ramos, 1922); Alfredo R. Castellanos, "La biblioteca científica del Padre Larrañaga," *Revista Histórica* (Montevideo) 16 (1949): 589–626; Edmundo Favaro, *Dámaso Antonio Larrañaga. Su vida y su época* (Montevideo: Universidad de la República, 1950); Miguel A. Klappenbach, *Larrañaga y el Viejo Museo*, Publicación Extra del Museo Nacional de Historia Natural y Antropología, no. 53 (Montevideo: MNHN, 2004). For his writings, see Dámaso Larrañaga, *Escritos*, 5 vols. (Montevideo: Instituto Histórico y Geográfico del Uruguay, 1922–30).
71 Asúa, *Ciencia de Mayo*, 117–29.

nomical. In an 1837 letter he tells how living "in a virgin and very fertile country" he saw the need to be "like Adam [and] to impose a name to all the productions that I found, in order to be understood by the learned."[72] Larrañaga was among the first to use Linnaeus in Río de la Plata.[73] In a letter to Aimé Bonpland from 26 February 1818, shortly after the latter's arrival in the region, he told that he had had "the audacity of undertaking the vast project of describing scientifically the three kingdoms of Nature in this Country, following Linnaeus's *Systema Naturae*, edited by Gmelin."[74]

In his "Geologic Memoir on the formation of Río de Plata" (c. 1819) Larrañaga describes three distinct geological formations on the Uruguayan shore containing marine fossilized shells.[75] The first was clay and quartzose sand with unidentifiable minute shell fragments; the second, deposited toward the mouth of the Uruguay River, was composed of *Mya* (soft-shelled clams); the third, near Montevideo, was made of *Mytilus* (mussels). On the basis of these observations, he hypothesizes that the river was formed from alluvial deposits which gradually filled what had originally been a marine gulf: upon the subsiding of the water caused by a displacement toward the east of the ocean, a flood of the Uruguay and the Paraná Rivers provoked the perishing of the *testaceae* whose fossils he found in the north-eastern littoral of the River Plate. Larrañaga renounces to classify the shells for it had already been done by William Maton. He disagrees with one of the latter's descriptions and renames the shell in his honour: *Matonia antiqua*.[76]

Larrañaga's interpretive framework is that of Biblical geology, with an age of the Earth of a few thousand years and the Biblical Flood as the only cataclysmic event in geological history: "That the sea has simultaneously covered all the regions of the Earth, as Moses says, cannot be denied."[77] Larrañaga explicitly dis-

---

[72] Larrañaga to the Apostolic Vicar in Montevideo (28 October 1837), in Alfredo R. Castellanos, "Contribución al estudio de las ideas del Pbro. Dámaso Larrañaga," *Revista Histórica* (Montevideo) 17 (1951): 44.
[73] Miguel de Asúa, "Linneo entre nosotros," *Ciencia Hoy* 18, no. 104 (2008): 19–27. Haenke, the editor of the eighth edition of Linnaeus's *Genera plantarum* (1789) lived in the Andean region of Cochabamba (present-day Bolivia) and died in 1816; Azara followed Buffon and did not use Linnaean nomenclature, despite the urgings of Antonio de Pineda, one of the naturalists of the Malaspina expedition. See Asúa, *La ciencia de Mayo*, 117–42.
[74] Larrañaga to Bonpland (26 February 1818), in Larrañaga, *Escritos*, 3:260.
[75] Dámaso Larrañaga, "Memoria geológica sobre la reciente formación del Río de la Plata deducida de sus conchas fósiles," in idem, *Escritos*, 3:7–20.
[76] George W. Maton, "Description of Seven New Species of Testacea," *Transactions of the Linnean Society* 10 (1811): 325–33.
[77] Larrañaga, "Memoria geológica," 9.

tances himself from Cuvier's theory of successive revolutions. But he refers approvingly to the latter's endorsement of the ideas of Jean-André Deluc and Déodat de Dolomieu that the last revolution could not have taken place more than 6,000 years ago.[78] The naturalist priest entertains the notion of a "slow, silent, and general cause" which would be responsible for phenomena like the tides and the precession of the equinoxes. In his view, this cause should be related to the action of a fluid like light, galvanism, or the principle of gravitation, which in antediluvian times might have been stronger than in the present era. In the end, he rejects this idea, but not before calling upon a Biblical argument that could strengthen its verisimilitude: Moses, he says, tells us that in ancient times "the human being was more energetic" (probably a reference to the Nephilim in Gn 6: 1–4). This gives occasion to Larrañaga for an apologia of the Bible as a source of natural knowledge: "We should be more considered with the sacred books; meditating upon them with serenity, much light can be found that can help us to explain phenomena which so far seem incomprehensible."[79]

In the opening of a short fragment on geology, Larrañaga paraphrases an article by Richard Kirwan, "who resorts to the theory of the Flood, proposed by the very same Moses, taken in its literal sense, as the only one that fits perfectly the phenomena known at that time."[80] He concludes his long comment on Kirwan's essay underscoring that "no matter what the instrumental physical cause of the Flood was, no matter the mode of its operation, it is necessary on this occasion to invoke the divine intervention and its energy." Thus, he goes on, "it is not possible to accept Buffon's opinion which pretends to explain the universal Flood and its physical causes through natural means...to fulfil His purposes, the Almighty suspended some of the laws of nature, as we see He did on other occasions."[81]

Soon after writing his memoir on the geology of Río de la Plata, Larrañaga sent a note which mentioned his findings of a femur and a tail with plates to the French botanist Auguste de Saint-Hilaire, who had botanized in Uruguay be-

---

[78] Ibid., 13; cf. Georges Cuvier, *Discours sur les révolutions du globe*, ed. Paul Bory (Paris: Berche and Tralin, 1881), 173.
[79] Larrañaga, "Memoria geológica," 10–11.
[80] Dámaso Larrañaga, "Tierra," in idem, *Escritos*, 3:30; cf. Richard Kirwan, "On the Primitive State of the Globe and Its Subsequent Catastrophe," *Transactions of the Royal Irish Academy* 6 (1797): 233–308. For Kirwan's history of the Earth, see Charles C. Gillispie, *Genesis and Geology* (New York: Harper & Row, 1951), 49–56; Martin J. S. Rudwick, *Bursting the Limits of Time. The Reconstruction of Geohistory in the Age of Revolution* (Chicago: Chicago University Press, 2005), 334–37; David Oldroyd and Sally Newcomb, "Richard Kirwan (1733–1812)," *Earth Sciences History* 31 (2012): 287–314.
[81] Larrañaga, "Tierra," in idem, *Escritos*, 3:36.

tween late 1820 and early 1821. The priest believed that the fossil rests corresponded to a "megatherium" (which he took to be a sort of giant *tatú*) and named the species accordingly as "*Dasypus* (Cuvier's Megatherium)." In his letter, Larrañaga affirms that "these rests are found on the surface, in alluvial terrains which indicate a very recent epoch"—a claim consistent with his views on the Biblical chronology of the history of the earth.[82]

Larrañaga's geological ideas are somewhat hazy: it is not clear whether he saw the Flood as the effect of a "general cause" associated with some kind of fluid and the mode of action of this hypothetical cause is far from clear. He did assert that the Deluge was due to a special divine intervention, looked with sympathy upon Kirwan's Neptunism, rejected Cuvier's theory of multiple revolutions, and was firmly anchored on Biblical chronology. Larrañaga understood the Bible as a source of natural knowledge and as the teachers of natural philosophy in Buenos Aires, he held fast to a literal interpretation of Genesis.

A Clergyman's Natural Theology

Larrañaga's most significant work is his *Diary of Natural History*, kept for many years with daily entries since 1808. The work is a mixture of diary and commonplace book: it contains original descriptions of animal, plant, and mineral species; abundant notes taken from the literature on natural history; meteorological and astronomical observations; and commentaries on the author's trips and encounters.[83] One of the thematic threads through the Diary is the role of Providence in guiding nature and in providing humans with all that is necessary to

---

[82] [Dámaso Larrañaga], "Note sur le Megaterium de Cuvier, l'Hydromis, et une variété nouvelle de Maïs," *Bulletin des Sciences (de la Société Philomatique de Paris)* 1823, 83. The letter was reproduced in the chapter devoted to the megatherium, in the third edition of Cuvier's *Recherches*, see "Extrait d'une lettre de D. Damasio [sic] Larrañaga, curé de Montevideo, à M. Auguste de Saint-Hilaire," in Georges Cuvier, *Recherches sur les ossemens fossiles*, 3rd ed., vol. 5, pt. 1 (Paris and Amsterdam: G. Dufour and E. d'Ocagne, 1823), 191n. What Larrañaga found were the rests of a glyptodont. See also Irina Podgorny, "Fossil Dealers, the Practices of Comparative Anatomy and British Diplomacy in Latin America, 1820–1840," *British Journal for the History of Science* 46 (2013): 647–74.

[83] Dámaso Larrañaga, *Diario de Historia Natural, 1808–1814*, Buenos Aires, Instituto de Historia Argentina y Americana Emilio Ravignani, University of Buenos Aires. It has been edited as Dámaso Larrañaga, *Diario de Historia Natural: 1808–1814*, ed. Ariadna Islas (Montevideo: Biblioteca Artigas, 2015) and idem, *Diario de Historia Natural: 1813–1824*, ed. Ariadna Islas (Montevideo: Biblioteca Artigas, 2017); this second part is also transcribed in Larrañaga, *Escritos*, vol. 1 (1922), 1–122. I shall quote from the manuscript.

well-being. This topic is frequently interwoven with encomia to the diversity and richness of the local natural world. Reprehending those who complained against a prolonged drought, Larrañaga reminds them of "the blessings that God poured upon these lands, which can produce everything insofar as we are able to benefit from irrigation."[84] The Peruvians should be grateful that "the bountiful author of nature has granted them the Cinchona bark [for the intermittent fever]."[85] Those living in Río de la Plata are in no need of "exotic plants for the healing of diseases, even severe ailments such as mania."[86] The potato is seen as a gift of the New World to the Old; with it, "the beneficent hand of the Creator" has given us "abundance and an economic advantage in the midst of famine and sterility."[87] When talking about the halcyon, Larrañaga rallies against Rev William Fordyce Mavor, who had qualified as a "superstitious simplicity" St. Ambrose's claim that it was Providence and not Aeolus who bore responsibility for the days of fair weather required by the bird to propagate itself.[88] He was baffled by "a naturalist (and moreover, a Protestant Christian, as seems to be W. Mavor) failing to admire that everywhere radiates the extreme care of the Creator in the conservation of the meanest of insects or the propagation of species." He urges "the insolent Mavor" to open his Bible where he will find proof that will help him in understanding this—Larrañaga refers to Lc 12: 24 ("Considerate corvos, qui non seminant…").[89]

In Larrañaga, the idea of the divine care for nature is tied to a feeling for the natural equilibrium evocative of Linnaeus and a felt need to preserve all species.[90] Plants such as the stinging nettle, he says, are usually despised, although "they grow near our houses not in vain…looking always for the [company of] human beings, in spite of the fact that these turn away from them and seek to eradicate them." We should proceed instead more "philosophically" and in Christian like manner and with a higher notion of divine Providence, preserve

---

[84] Larrañaga, *Diario*, fol. 83r.
[85] Ibid., fol. 100v.
[86] Ibid., fol. 103r.
[87] Ibid., fol. 127r.
[88] William Mavor, *Natural History for the Use of Schools* (London: Printed for R. Phillips, 1800), 200.
[89] Larrañaga, *Diario*, fol. 59r.
[90] I am referring to Linnaeus's "economy of nature," see Carolus Linnaeus, *Miscellaneous Tracts Relating to Natural History, Husbandry, and Physick*, ed. Benjamin Stillingfleet, 3rd ed. (London: R. and J. Dodsley, 1762), 37–129; cf. Charles H. Pence and Daniel G. Swaim, "The Economy of Nature: the Structure of Evolution in Linnaeus, Darwin, and the Modern Synthesis," *European Journal for Philosophy of Science* 8 (2018): 435–54.

them.⁹¹ Also, we should not complain of Providence for the proliferation of ants. The human being should bewail "the effects of his [sic] destructive hand; it is he himself who perturbs the order of Providence...by destroying their antidotes." Larrañaga laments the elimination of anteaters and *tatús* which destroy the ants:

> We alter the equilibrium established by the author of nature among living beings and it is only just that we should experience the consequences of this disorder...Man is truly the King of Nature, but he should not be a despotic king, he should govern...under those principles dictated by reason and even by our convenience...he should look for the conservation of species ... Insofar as these are linked and each one depends upon the others, [the destruction of any of them] will generate an irreparable harm with uncountable consequences. We are still on time: let us establish a severe law against those who destroy these innocent animals [anteaters and *tatús*].⁹²

Larrañaga's idea of human stewardship on nature, his awareness of the price that human beings should pay for altering the natural equilibrium, and his proposal for the passing of legislation aimed at the preservation of endangered species sound eerily contemporaneous. His sense of the conservation of the natural balance was tied to the notion that this balance was a manifestation of a beneficent God: "We should imitate the Dutch and other peoples who punished those who killed a stork which fed on serpents. Thus, balance will be restored and we shall experience all the good that Providence has shown us in these beneficial animals."⁹³

Rehearsing the notion of the chain of being, Larrañaga defends the idea that nature does not operate *per saltum*, "everything in her is well joined and linked, and forming at the same time a harmonic whole. Everything constitutes a thread which cannot be cut; it takes its origin in God and ends in him."⁹⁴ In this connection, he affirms that no natural system of classification could be possible for what nature generates are individual creatures; those so-called "natural systems...are creations of the human defective intellect and never of wise nature or, what is equivalent, of God."⁹⁵ In his/her inability to discern the taxonomy hidden

---

91 Larrañaga, *Diario*, fols. 201v–202r.
92 Ibid., fols. 487r–487v, 488v.
93 Ibid., fol. 488v.
94 Ibid., fols. 138v–139r.
95 Ibid., fol. 143v. Larrañaga's position on the issue of classification was unstable to say the least. In a 1821 letter to Auguste de Saint-Hilaire, the former claimed he had known the French-Spanish botanist Louis Née and other "blind sectarian followers of Linnaeus," who had not paid much attention to the "natural method" of botanical classification (for which he meant Antoine L. de Jussieu's system); see Larrañaga to Saint-Hilaire (16 February 1821), in Larrañaga, *Escritos*, 3:280–82. Larrañaga made extensive notes on Cuvier's comparative anatomy

in the mind of the divinity, the naturalist has to settle for an artificial system. When describing his own observations of the metamorphosis of a caterpillar (an "admirable transformation"), Larrañaga chides those philosophers who "imagine laws and principles to which they wish to subject everything, even the divine laws" and cites (supposedly) St. Ildephonsus: "divine laws do not depend on nature, on the contrary, it is proved that natural things and natural laws emanate from divine laws."[96]

Larrañaga's excursions into natural theology are related to his confidence in the omnipresent action of divine Providence and the natural equilibrium among species. On occasion, he suggests that the complexity of a natural process, such as insect metamorphosis, is a testimony of a divinely legislated universe. More than inferring God's existence from the complexity, beauty, or harmony of the natural world, he sees these as confirming God's care for his creatures. Larrañaga's underlining of the continuous action of Providence sets him apart from those who employed the physico-theological argument to demonstrate the existence of a quasi-deist Architect of Nature. It is to these we now turn.

Non-Clerical Natural Theology

Together with the teaching of natural philosophy carried out by secular priests and members of religious orders, a non-clerical culture interested in the sciences developed in Buenos during late-colonial times. Enlightened lawyers, bureaucrats, merchants, and military thought about mathematics, chemistry, and natural history as a means conducive to agricultural improvement, technical innovations, economic advance, and social progress: they did not see the sciences as an argument undermining the foundation of religion. Even a radical thinker such as Mariano Moreno bowdlerized his translation of Rousseau's *Du Contrat social* claiming that "the author was disgraceful enough to talk nonsense on religious

---

and drafted tables of the mammals of Río de la Plata classified according to Cuvier's system; see Larrañaga, *Escritos*, 2:297–340. In his lecture on the occasion of the inauguration of the Public Library of Montevideo, he made clear his eclecticism when praising "Linnaeus, [nature's] most beloved, most faithful son, whom she has revealed all her arcane secrets; Buffon, the French Pliny, her eloquent eulogizer." Dámaso Larrañaga, "Oración inaugural," in idem, *Escritos*, 3:143; cf. Phillip R. Sloan, "The Buffon-Linnaeus Controversy," *Isis* 67 (1976): 356–75.

96 "Quia non ex natura rerum divinae leges pendent, sed ex divinis legibus naturae rerum et leges manare probantur." Larrañaga, *Diario*, fol. 376v, cf. [St. Ildephonsus], "Opusculum de partu Virginis," in *SS. PP. Toletanorum quotquot extant Opera*, ed. Francisco de Lorenzana, vol. 1 (Madrid: Joaquín Ibarra, 1782), 295. The work is currently attributed to Paschasius Radbertus of Corbie.

matters."⁹⁷ The *économistes* of pre-independent Buenos Aires proclaimed the credo of physiocracy and economic liberalism and fostered the cultivation of useful sciences and industry. Manuel Belgrano, the most articulated of them, attended the College of San Carlos and pursued further studies in Spain (1786–1793) where he became familiar with the ideas of François Quesnay, Antonio Genovesi, Ferdinando Galiani, Adam Smith, and of the Spanish enlightened reformers Gaspar Melchor de Jovellanos and the Count de Campomanes.⁹⁸ Upon his return, he was nominated secretary of the Consulate in Buenos Aires (an institution somewhere between a chamber of commerce and a commercial court) and in this capacity created the Nautical School (1799–1806), in charge of the Spanish military engineer Pedro A. Cerviño whose plan of studies envisaged an extensive foundation on mathematics and physics. In an article on "Education" published in 1810 in a periodical founded by him, Belgrano joins the chorus of the critics of scholasticism claiming that this kind of philosophy is not useful "to the king, homeland, and religion." He recommends instead the use of Condillac's *Logic* and invites to seek the truth "in the concatenation of experiences and observations." But he takes care to exempt from any possible criticism "what belongs to the dogmas, to the decisions of the Church, and to our legislation."⁹⁹ In another article, he recommends that after being taught logic, instead of proceeding to natural philosophy (*physica*) students should learn metaphysics for the latter "aims at the knowledge of the divinity on whom we depend and also of the soul, which is the noblest part of ourselves."¹⁰⁰ In Belgrano's view, natural theology takes precedence over natural philosophy. Moreover, he explicitly urges the teaching of the foundations of "our saint and sacred religion" (i.e. religious doctrine and morals), without which neither politics nor society would be possible.

Pedro Cerviño was born in Galicia (Spain) and trained in one of the military academies founded as a result of the Bourbon effort to improve the profession-

---

97 [Mariano Moreno], "El editor a los habitantes de esta América," in *Del Contrato Social...por el ciudadano de Ginebra Juan Jacobo Rousseau*, ed. Mariano Moreno (Buenos Aires: Real Imprenta de Niños Expósitos, 1810), no pagination.
98 Luis R. Gondra, *Las ideas económicas de Manuel Belgrano* (Buenos Aires: Rosso, 1923), 71; cf. Oreste Popescu, *Studies in the History of Latin American Economic Thought* (London: Routledge, 1997), 157–71; Chiaramonte, *La Ilustración en el Río de la Plata*, 72–74.
99 [Manuel Belgrano], "Educación," *Correo de Comercio* 17, 23 June 1810, 136 and 18, 30 June 1810, 137–41.
100 [Manuel Belgrano], "Metafísica," *Correo de Comercio* 22, 28 July 1810, 174–76 and 23, 4 August 1810, 177–78.

alization of the Spanish army and navy.[101] He arrived in Buenos Aires in the 1780s together with Azara, as a member of the large group of engineers and topographers in charge of the demarcation of limits with Brazil.[102] All of them were capable officers in the imperial service educated in the ideas of the Bourbon enlightenment. Cerviño's lecture opening the exams at the Nautical Academy in January 1806 is perhaps the first public pronouncement in favour of the heliocentric theory in Río de la Plata. Copernicus is mentioned in it as the intellectual hero who "stumbled upon the opinion of the learned men of his time and tried to persuade them that everything is illusory, because the Sun and stars stand still and that which moves is the mass of the globe we inhabit."[103] Cerviño expatiates on a popular version of celestial mechanics and exalts the glories of the "immortal Newton," Clairaut, and d'Alembert. The learned engineer also grants astronomy a role in natural theology, for the science of the heavenly realm "heightens the imagination and captivates and improves the spirit, filling it with magnificent ideas about nature and with a notion worthy of the wisdom, the power, and the greatness of the Creator."[104] In a later speech about an Academy of Mathematics planned by the Consulado of Buenos Aires in 1813 (which apparently never materialised) Cerviño condemns metaphysics as an useless pursuit of "monsters, chimeras, doubts, and illusions" and recalls that the only study that can raise ourselves toward the truths of Revelation is "the study of nature, of the admirable order which reigns in her revealing everywhere the wise and omnipotent hand which arranged it and which, directing us toward intelligence of the creatures, holds up those great aims for which we have been set amongst them."[105] The human being, he goes on, is the only creature which, "engrossed upon the contemplation of so many marvels, can ascend to the throne of the Creator and intone ardent hymns of gratitude and praise."[106] In an earlier speech

---

101 Enrique Udaondo, *Diccionario biográfico colonial argentino* (Buenos Aires: Huarpes, 1945), 245–46; Antonio Lafuente "Las Academias militares y la inversión en ciencia en la España ilustrada (1750–1760)," *Acta Hispanica ad Medicinae Scientarumque Historiam Illustrandam* 2 (1982): 193–209.
102 Furlong, *Matemáticos argentinos*, 100–107.
103 Pedro Cerviño, "Prolusión académica. Discurso en que se procura que para ser buen piloto es necesaria la astronomía," in Nicolás Besio Moreno, *Las fundaciones matemáticas de Belgrano* (Buenos Aires: Instituto Nacional Belgraniano, 1995), 176. This Academy of Mathematics is different from the one created in 1810 and mentioned above.
104 Cerviño, "Prolusión académica," 178.
105 Pedro Cerviño, "Discurso o memoria sobre la importancia de la Academia establecida por el Consulado de Buenos Aires," in Besio Moreno, *Las fundaciones matemáticas*, 185.
106 Cerviño, "Discurso o memoria," 186.

extolling the benefices of commercial navigation, Cerviño expounds his version of the Pauline argument about the natural knowledge of God (Rom 1) in a prose with enlightened overtones. It was the acquaintance of the West with exotic peoples made possible by seafaring, he says, which made evident the absurd mores of those who "groaned under the yoke of superstition, for they disregarded the ray of light emanating from the Supreme Being." The worship human beings owe to their Creator and the love we owe our fellow humans "are natural feelings and virtues written in a well-formed soul" so that the "regard of error could never obscure this divine light."[107] A few paragraphs later Cerviño defends the right to analyse and criticise any opinion whatever on grounds of its usefulness, with the caveat that the homage of believing something without questioning should be paid only to "the sacred truths of our religion."[108]

Cerviño's discourse confirms that the representatives of Enlightenment in Río de la Plata saw no conflict between science and religion. If it were not for his accent on the truths of Revelation, some of his pronouncements could be interpreted as implying a near-deist natural religion. Surely he was a person of enlightened ideas (he did not talk of "God," as would be natural for a Catholic, but used the characteristic expression "the Supreme Being"), but he was a member of the tertiary order of the Franciscans (he was buried in the Convent of Saint Francis in Buenos Aires) and his exaltation of modern science and disregard for metaphysics were consonant with orthodox belief.[109]

In other cases, the frontier between natural theology and natural religion was more blurred. Manuel José de Lavardén (also spelled Labardén) was the son of a colonial bureaucrat and poet. He is locally known for his "Ode to the River Paraná" and *Siripo*, the first verse drama on a non-religious theme written in Río de la Plata; he also authored a work on commerce in the region.[110] A celebrated representative of the local Enlightenment, Lavardén studied law in Chuquisaca and Spain without graduating.[111] Shortly after returning to Río de la Plata from Spain in 1778, he was appointed to an examination tribunal at the College of San Carlos. In a lecture pronounced on the occasion, Lavardén affirms

---

[107] Pedro Cerviño, "El tridente de Neptuno es el cetro del mundo," in Besio Moreno, *Las fundaciones matemáticas*, 161–62.
[108] Ibid., 165.
[109] For the social importance of tertiary orders in colonial Río de la Plata, see Peire, *El taller de los espejos*, 141–55.
[110] Manuel José de Lavardén, *Nuevo aspecto del comercio en el Río de la Plata*, ed. Enrique Wedovoy (Buenos Aires: Raigal, 1955).
[111] See Julio Caillet-Bois and Roberto F. Giusti, "Manuel José de Lavardén," in *Historia de la literatura argentina*, ed. Rafael A. Arrieta, vol. 1 (Buenos Aires: Peuser, 1958), 239–49.

that "one of the ways in which the sciences facilitate the knowledge of God is the study of nature." He departs from the premise that "when contemplating a beautiful machine nobody can doubt that it has been made by an artificer" to reach the conclusion that "the beauty of the world tells of course that there is a God." The observation of "a little insect" and the awe before "the extended firmament" confirm us in this path.[112] Lavardén's lecture has been the object of contrasting interpretations. Writing in the 1860s, the anticlerical Juan María Gutiérrez interpreted that Lavardén's use of the argument from design was a result of his frequentation of "forbidden authors."[113] Against Gutiérrez's reading, which sees in Lavardén a fellow deist, is the latter's reference to the action of "the Lord's wise Providence."[114] In his speech, Lavardén affirms that "the sciences, which were in former times imprisoned in a corner of the Orient, travel now in liberty." The Jesuit historian Furlong took this as a cue to speculate that Lavardén might have been acquainted with "Brahmanism," which he could have learned in a Spanish translation printed in Buenos Aires of the anonymous *Economy of the Human Life, translated from an Indian manuscript, written by an ancient Bramin* (1750).[115] One of Lavardén's literary compositions got him into troubles with the censor.[116] It is likely that he was a Freemason and he was certainly no friend of the Jesuits. In any case, Lavardén's heterodoxy matters less for our argument than the fact that he provided an abstract of the design argument, which seems to have been a staple of enlightened discourse in Río de la Plata.

It is always possible to argue that in their employment of natural theological arguments the enlightened economic thinkers in pre-revolutionary Río de la Plata were paying lip service to conventional values. This could be the case with Lavardén, but as concerns Cerviño and Belgrano, the burden of the proof seems to remain on those who question the possibility of the conciliatory position characteristic of Catholic Enlightenment. In the discussions around the acquisition of a physics cabinet by the University of Córdoba discussed at the beginning of this chapter, we have seen how the crown officials and heralds of

---

112 Quoted in Furlong, *La filosofía en el Río de la Plata*, 415–16.
113 See Juan M. Gutiérrez, "D. Juan Manuel de Lavardén," in idem, *Estudios biográficos y críticos* (Buenos Aires: Imprenta del Siglo, 1865), 50. Gutiérrez mixes up the biographies of Juan Manuel Lavardén and his son Manuel José.
114 Quoted in Furlong, *La filosofía en el Río de la Plata*, 415.
115 Furlong, *La filosofía en el Río de la Plata*, 416–20. The work has been attributed to Robert Dodsley and to Lord Chesterfield, but current criticism set for the former, see the discussion in Harry M. Solomon, *Robert Dodsley. Creating the New Age of Print* (Carbondale, IL: Southern Illinois University Press, 1996), 300–301n69.
116 Lavardén to Manuel Basavilbaso, 3 January 1789, in Chiaramonte, *La Ilustración en el Río de la Plata*, 208–11.

Bourbon reformism did not confront religion; they rather sought to reach a compromise. Natural philosophy was seen by them as an ally of natural theology and a way to a more "purified" religion concerned with morality and the education of the people. It should be noticed that in this case, the divide between tradition and novelty—between scholasticism and the new philosophy—did not map onto that between reactionary priests and enlightened laypersons: there were clerics who welcomed the winds of change and royal officials entrenched in conservative positions.

Conclusions

In colonial and early-independent Río de la Plata, natural philosophy and science were cultural manifestations of a society impregnated by religious institutions and meanings: the relationships between science and religion could still be seen largely as harmonious. As we shall discuss in the next chapter, the first spark of conflict took place in the third decade of the nineteenth century in Buenos Aires with the secularisation (expropriation) of church property put to scientific uses as part of an ecclesiastical reform along Gallican lines. "Harmony" between science and religion should here be taken as a general description: a closer analysis discloses occasional hidden tensions between them, like those discussed above. Compromise between not entirely consistent positions is a characteristic of the "Catholic Enlightenment," a canvas term which designates a global movement whose aim was to employ the new philosophy and new science to defend the beliefs of the Church and to reconcile Catholicism with Modernity.[117] Among the historical actors discussed in this chapter, some can be primarily seen as "enlightened," while most are better described as "Catholic;" others fall in between. The new experimental science was the terrain of conceptual transactions between tradition and novelty as was evident in the eclectic nature of the courses of natural philosophy.

For the purpose of their courses, the large majority of teachers examined here read the Bible in a literal sense (although Fr Elías, at one point, approved of Augustine's famous allegorical interpretation of the narrative of creation). The Council of Trent had not specifically taken an stance on the issue of the interpretation of the sacred text, but by the beginning of the seventeenth century the immense totality of Catholic authors followed the literal reading.[118] Larrañaga inter-

---

117 Lehner, *The Catholic* Enlightenment, 7.
118 Blackwell, *Galileo*, 12, 21.

preted his empirical geological research within the framework of the Biblical account of the history of the Earth. He distanced himself from Cuvier's revolutions and altogether rejected Buffon's naturalistic account. His approving paraphrases of Kirwan's Neptunism are consonant with his conviction about the legitimacy of the Bible as a reliable source of natural knowledge. Insofar as the public pronouncements examined here were subjected to the ideological control of crown and church, there was no conflict between the Bible and the findings of natural science. But a closer reading of an author such as the deist Lavardén opens a space of doubt.

While natural theology flourished in Britain and Protestant Continental countries, in Catholic Europe it remained a lesser though by no means insignificant tradition—a point early made by John H. Brooke.[119] The most successful work of the rich eighteenth-century French literature on the subject was the abbé Noël-Antoine Pluche's *Spectacle de la nature*, soon translated into Spanish and reedited twice in that language before 1800.[120] Although not strictly a physico-theological treatise, the work on natural philosophy by the Portuguese Oratorian priest Teodoro de Almeida (with many editions in Spanish) also catered to an audience receptive to arguments of natural theology, as did the works by Benito Feijóo.[121] All of these books could be easily found in Río de la Plata. For example, the library of Pedro Altolaguirre, a crown official in Potosí brother of the above mentioned Martín, held 16 volumes of Pluche's *Espectáculo*, 17 volumes of Feijóo's works, five volumes of Almeida's *Recreación* and 10 volumes of the Spanish translation of Buffon's *Histoire naturelle*.[122] The Public Library in Buenos Aires founded by the revolutionary government held two complete sets of Pluche's work.[123] In a lot of 365 works captured in Montevideo and auctioned in Buenos Aires in 1815, we find a complete set of Almeida's *Recreación* and practi-

---

119 Brooke, *Science and Religion*, 197–98.
120 Noël-Antoine Pluche, *Espectáculo de la naturaleza*, trans. Esteban Terreros, 16 vols. (Madrid: Gabriel Ramírez, 1753–55). Cf. Adam Drozdek, "Noël-Antoine Pluche and the Spectacularity of Nature," *Roczniki Theologie* 67, no. 4 (2020): 41–59.
121 Teodoro de Almeida, *Recreação filosófica*, 10 vols. (Lisbon: Officina de Miguel Rodrigues, Regia Officina Typographica, 1780–91); Benito J. Feijóo, *Teatro Crítico Universal*, 1st ed., 9 vols. (Madrid: Imprenta de Lorenzo F. Mojados and Imprenta de Francisco del Hierro, 1726–40) and *Cartas eruditas y curiosas*, 1st ed., 5 vols. (Madrid: Imprenta de los herederos de Francisco del Hierro and Joaquín Ibarra, 1742–60). For natural philosophy in Spain, see Federico A. López Silvestre, "Dios y la naturaleza. El desarrollo de la teología natural en España y en Galicia en el s. XVIII," *Semata. Ciencias sociais e Humanidades* 14 (2002): 417–53.
122 José Torre Revello, "La biblioteca que poseía en Potosí don Pedro de Altolaguirre (1799)," *Historia* (Buenos Aires) 1 (1956): 153–62.
123 Asúa, *Ciencia de Mayo*, 59.

cally all of Feijóo's works.[124] Literature like the "Oration on behalf of the study of natural sciences" by the enlightened Asturian Gaspar Jovellanos is very close in tone to the speeches pronounced by Cerviño in Buenos Aires.[125]

Scholastic teaching of natural philosophy in the University of Córdoba did not touch on natural theology, except for occasional general remarks such as "we shall prove God's existence deducing it from the admirable order, beauty, and variety of beings."[126] Traces of this kind of argument can be found in the late eighteenth-century Jesuit works about the natural world of Paraguay and Río de la Plata written in the Italian exile. In his multi-volume encyclopaedia, Sánchez Labrador claims that "in so magnificent Theatre of Nature…the divine hand, which can be admired in so beautiful creatures, has not wished to draw the curtain of its effects before human reason."[127] The section on insects opens with the affirmation that "in the industry and symmetry of their parts shines the wisdom of the author who formed such beautiful and small creatures."[128] After referring to Basil's *Hexaemeron*, the fifth book of Aristotle's *De partibus animalium* (the locus classicus on the worth of studying lower animals) and Friedrich Christian Lesser's *Theologie des insects* (The Hague, 1742), Sánchez Labrador explains that the craft of the artificer "shines most in his smaller works: He who stretched out the heavens and set limits to the sea is the same who gave the bee its sting, the conduct of a painful humour."[129]

The enlightened lay authors (bureaucrats, *économistes*, and military engineers) made more frequent use of a basic form of the physico-theological argument. But it is not always easy to discern whether they talk in terms of natural theology or of natural religion. This ambiguity has been pointed out by Daniel Mornet in his account of French natural theology: "Physico-theology could offer the same arguments to the Parsis or the Chinese, whose wisdom was cele-

---

124 César A. García Belsunce, *Pertenencias extrañas. Libros en Buenos Aires en 1815* (Buenos Aires: Academia Nacional de la Historia, 2013), 127, 170–71.
125 Gaspar Melchor de Jovellanos, "Oración pronunciada en el Instituto Asturiano sobre el estudio de las ciencias naturales," in idem, *Obras publicadas e inéditas*, ed. Cándido Nocedal, vol. 1 (Madrid: Rivadeneyra, 1858), 335–42.
126 Martínez de Aldunate and Pereyra, *Conclusiones sobre toda la filosofía*, in Martínez Paz, "Una tesis de filosofía," 265.
127 Sánchez Labrador, *El Paraguay natural*, Introducción, Paraquaria 16, iv.
128 Sánchez Labrador, *El Paraguay natural*, pt. 4, bk. 3, Paraquaria 19, 174.
129 Ibid., 174.

brated by the Encyclopédistes."¹³⁰ Equivocal phrasings might have been the refuge of unorthodox positions.

Among the authors considered here, Larrañaga was the only one with a firsthand acquaintance with the natural world: his diary exudes a genuine feeling for the investigation of nature. The same ring of authenticity surrounds his religious statements about Providence as expressed in the order and balance of nature—in contradistinction with the more formulaic calls to admire the Creator in his works that can be culled from the official documents of the colonial period. In that cultural milieu, Larrañaga was perhaps an exception in that he had a working command of English—while reading his *Diary*, it is difficult to avoid imagining him as a Uruguayan Gilbert White. A peculiarity of the ports of Montevideo and Buenos Aires was their incipient cosmopolitanism fostered by the settled British and French merchant communities.¹³¹ Larrañaga's familiarity with English literature on natural history might have contributed to shape his ideas on natural theology. But the core of his views seems to have derived from a sense of natural equilibrium nurtured by his assiduous reading of Linnaeus.

Some, though not all, aspects of Catholic Enlightenment in Río de la Plata were similar to those that set the intellectual tone of the reign of Charles III, with the difference that in the years surrounding the revolution of 1810 this set of values and practices began to serve the ideal of independence instead of the idea of empire. There were cultural newspapers that published articles on useful science and medicine (geography, chemistry, agriculture, medicine, natural history, inventions); the plans of the Nautical Academy and the Academy of Mathematics for the military foresaw an updated training in the exact sciences; a near craze for electricity galvanised the city: Franklin was well known and read, the books of Pieter van Musschenbroek, the abbé Jean-Antoine Nollet, and Joseph-Aignan Sigaud de la Font circulated widely, and Altolaguirre's collection of scientific instruments suggests that something of the tradition of enlightenment in the Basque country—embodied in the Royal Basque Society of Friends of the Country and the spirit of its founders, the "little lords of Azcoitía"—was

---

130 Daniel Mornet, *Les sciences de la nature en France au XVIIIe siècle* (Paris: Armand Colin, 1911), 35–36. Cf. Ann Blair and Kaspar von Greyerz, eds., *Physico-Theology. Religion and Science in Europe, 1650–1750* (Baltimore: Johns Hopkins University Press, 2020).
131 Vera Blinn Reber, *British Mercantile Houses in Buenos Aires, 1810–1880* (Cambridge, MA: Harvard University Press, 1979); David Rock, *The British in Argentina. Commerce, Settlers, and Power, 1800–2000* (Cham: Palgrave Macmillan, 2019): 1–41; Jacques Duprey, *Voyage aux origines françaises de l'Uruguay* (Montevideo: Instituto Histórico y Geográfico del Uruguay, 1952).

transferred to Río de la Plata.[132] The circle of clergymen naturalists, led by Larrañaga, was the living example of a fruitful coexistence of science and religion.

Catholic Enlightenment can be seen as the assimilation into the Catholic framework of those aspects of Enlightenment that did not clash overtly with the traditional view of nature and society. Natural philosophy and the new experimental science were the terrain of conceptual dealings between tradition and novelty. The overall attitude seems to have favoured some kind of change, although very different and perhaps contradictory trends contributed to this: Bourbon reforms, the Franciscan tradition, new currents of economic thought expressing an alternative pattern of trade interests, Spanish-American Gallicanism, and a qualified anti-scholasticism. An atmosphere of openness, experimentation, and reception of the new is present at least in educational discourse during this period. The teachers of natural philosophy were anxious to admit "the Moderns," but always within the limits of religious orthodoxy. On the side of the lay culture that emerged in Buenos Aires around the turn of the century, enlightened ideas were expressed in public always with a view at not contradicting what seems to have been a conviction of the majority of the educated public, i.e. that the *luces* and religion were not entirely at odds. Alhough in the third decade of the nineteenth century science began to be used as a symbol of secularisation, the equilibrium between religion and the new philosophy characteristic of the period described in this chapter was not entirely disrupted by the forces that limited the action of the Church in the public sphere.

---

132 Miguel de Asúa, "Belgrano y la ciencia," in *Belgrano*, ed. Marcelo U. Salerno and Roberto L. Elissalde (Buenos Aires: Academia Nacional de Ciencias de Buenos Aires, 2020), 8–19, https://www.ciencias.org.ar/contenido.asp?id=790; idem, *La ciencia de Mayo*, 116–73.

Chapter 3
Religious and Secular Spaces in Post-Independence Río de la Plata (1820–1827). From Natural Theology to Secular Science

The design of establishing a central government based in Buenos Aires during the decade of 1810 ultimately came to nothing due to the failure of the centralist Constitution of 1819 and the inroads of the caudillos of the neighbouring provinces in February 1820. Except for a renewed attempt at central organisation between 1825 and 1827, the provinces became semi-autonomous entities united by loose federal treaties. After a period of near chaos (at one point, Buenos Aires had three governors), Bernardino Rivadavia emerged as the strongest political figure in that province. He served as minister under Governor Martín Rodríguez between July 1821 and April 1824 and was afterwards elected as the first president of the country in February 1826, upon his arrival from an eight-year stay in Europa (he occupied that position until June 1827). Early after the 1810 Revolution, Rivadavia had left his mark as secretary and at times one of the members of the "First Triumvirate" (September 1811–October 1812), a strongly centralist government of the United Provinces of Río de la Plata. Rivadavia's project of implanting an "Enlightened Republic" proved a mirage. After him, the wars between Federalists (the leaders of the interior provinces, conservative and economically protectionist) and Unitarians (the elite of the city of Buenos Aires, liberal and promoters of free trade) recrudesced. The protracted conflict led to the long autocratic rule of Juan Manuel de Rosas (1829–1832, 1835–1852), who while serving the interests of the ranchers of Buenos Aires was seen as the embodiment of the traditionalist ideology of the provinces and the restoration of Catholicism.

During his long sojourn in Europe as an envoy of the government (1814–1821), Rivadavia met Jeremy Bentham in London and Antoine Destutt de Tracy in Paris, as a result of which he became an enthusiast of utilitarianism and *idéologie*.[1] It has been scarcely noticed in this connection that the particular blend of Benthanism and "ideology" was also all the rage among the *afrancesados* ("Frenchified") in Spain at the beginning of the 1820s.[2] It is is a disputed ques-

---

[1] Klaus Gallo, *Bernardino Rivadavia. El primer presidente argentino* (Buenos Aires: Edhasa, 2012), 43–67.
[2] Marcelino Menéndez Pelayo, *Historia de los heterodoxos españoles*, vol. 3 (Madrid: Librería católica de San José, 1881), 513–18.

tion whether Rivadavia's political project could be qualified as a "utilitarian experiment" in South America; what seems undeniable is that many of his reforms bore the imprint of utilitarianism.³ In his attempt at transforming the sleepy Buenos Aires into an enlightened South American Athens Rivadavia promoted the teaching of science, higher education, *belles lettres*, and aesthetic appreciation (*el buen gusto*). Both science and religion were important aspects in his program of reform. In a letter to Bentham, he described his project as "the protection of trade, *science* and the arts...and a much needed *ecclesiastical reform* which I hope to achieve" (my emphasis).⁴

Rivadavia's most perdurable foundations were the University of Buenos Aires, created in August 1821 and the Public Museum, inaugurated in 1823. There had been plans afoot for the creation of a university since 1816; it was Father Antonio Sáenz's "stubborn tenacity" which moved the successive heads of state to take the first steps in its creation.⁵ The university was a cluster of already existing teaching institutions in the style of the 1808 Napoleonic imperial "university." It encompassed all levels of instruction (elementary, middle, and higher) and was divided in six departments: first letters, preparatory studies, medicine, exact sciences, law, and divinity. Courses of science (experimental chemistry and experimental physics) were among the novelties; Rivadavia hired a number of foreign teachers of unequal value (on whom more below) to teach them. The two really important scientists who arrived in Buenos Aires at that time were Ottaviano Fabrizio Mossotti (who was hired by him but arrived after his fall from office) and Aimé Bonpland, who stayed only briefly in the city. The new scientific societies and institutions (the Society of Physico-Mathematical Sciences, the Academy of Medicine, the Department of Topography and Statistics) although ephemeral and somewhat pretentious testify to the first conscious effort to plant the seeds of science in Río de la Plata.⁶

---

3 Jonathan Harris, "Rivadavia and Benthamite 'Discipleship'," *Latin American Research Review* 33, no. 1 (1998): 129–49; Klaus Gallo, *The Struggle for an Enlightened Republic: Buenos Aires and Rivadavia* (London: Institute for the Study of the Americas, 2006), 26–45.
4 Ricardo Piccirilli, *Rivadavia y su tiempo*, 2ⁿᵈ ed., vol. 2 (Buenos Aires: Peuser, 1960), 441–42.
5 Tulio Halperin Donghi, *Historia de la Universidad de Buenos Aires* (Buenos Aires: Eudeba, 1962), 39.
6 Juan Carlos Nicolau, *Ciencia y técnica en Buenos Aires, 1800–1860* (Buenos Aires: Eudeba, 2005); Miguel de Asúa, *Una gloria silenciosa. Dos siglos de ciencia en Argentina* (Buenos Aires: Zorzal, 2010), 47–62.

Ecclesiastical Reform

The reform of the local Catholic Church in Buenos Aires was carried out along Gallican lines and backed by a sector of the clergy congenial to the government.[7] As part of his program of centralisation, Rivadavia sought to transform the Church into one of the areas controlled by the state. This was in line with the old tradition of Spanish *patronato*, tightened by Charles III's regalism. Since Rome did not recognise the independence of the new Spanish-American republics, the vacancy of the episcopal sees (Buenos Aires, Córdoba, and Salta) paved the way for policies that made the Church dependent on the state.[8] The 1813 constitutional assembly of the United Provinces passed regalist legislation (for example, it denied the papal nuncio in Spain jurisdiction over the Church in Río de la Plata; it nominated a commissary for the affairs of the regular clergy; and it established a minimum age for taking religious vows). Pius VII's brief of 30 January 1816, *Etsi longuissimo*, exhorted the Spanish-American churches to obey the Spanish king, thus dissolving any hope of restoring relations between them and the Holy See for the near future.[9]

Américo Tonda claimed that Rivadavia experienced "the influence of the Bourbon tradition, plus the liberal currents alive in the European countries he visited;" the clergy who sided with him had obvious "Jansenist" (in the Spanish sense of the term) and regalist tendencies, whereas unbelievers and anticlerical elements joined the waggon.[10] Some have claimed that the model of reform was the regalist policy of the enlightened ministers of Charles III, in particular, the ideas of the Spanish Jansenist Juan A. Llorente.[11] Góngora thought that all the Spanish American ecclesiastical reforms of the period were redolent of Josephinism.[12] Ivereigh saw Rivadavia's reform as a combination of English liberalism and Bourbon regalism and centralism. According to him, the disruption brought

---

7 There were ecclesiastical reforms in most of the provinces; the most radical was in San Juan, with the separation of Church from state, which ended in the fall of the governor. For the reform in Córdoba and the Western provinces, see Valentina Ayrolo, *El abrazo reformador. Las reformas eclesiásticas en tiempos de construcción estatal. Córdoba y Cuyo en el concierto iberoamericano (1813–1840)* (Rosario: Prohistoria, 2017).
8 Fort the case of Argentina, see Américo A. Tonda, *La iglesia argentina incomunicada con Roma (1810–1858)* (Santa Fe: Castellvi, 1965).
9 Juan C. Zuretti, *Nueva historia eclesiástica argentina* (Buenos Aires: Itinerarium, 1972), 183–87.
10 Américo Tonda, *Castro Barros* (Buenos Aires: Academia del Plata, 1961), 189.
11 Manuel Aguirre Elorriaga, *El abate de Pradt en la emancipación hispanoamericana (1800–1830)*, 2nd ed. (Buenos Aires: Huarpes, 1946), 187. Gregorio Funes wrote his *Examen crítico de los discursos sobre una constitución religiosa* (Buenos Aires: Hallet, 1825) against Llorente's work.
12 Góngora, "Estudios sobre el Galicanismo," 129.

by the application of a liberal approach—originally born in multi-denominational societies to safeguard some form of equilibrium among diverse religious groups—to a religiously uniform society such as that in Río de la Plata, would have been the action of "marginalised intellectuals" defying widely supported values.[13] David Rock also sees Rivadavia as "combining the attitudes of Bourbon despotism with support for liberal representative government."[14] Others have interpreted the reform in the context of the larger history of the local Church.[15] No one has seriously doubted Rivadavia's Catholic orthodoxy. The liberal historian Vicente Fidel López, whose father Vicente López succeeded Rivadavia as president of the country, claimed that the latter "was not a liberal and an irreconcilable enemy of colonial traditions," but "a regalist of the best school and a sincere Catholic."[16] Catholic historian Rómulo Carbia also affirmed that Rivadavia's reform was "the consequence of a deep regalism, ruthlessly pursued."[17]

Several prominent, highly qualified, and politically astute members of the secular clergy supported Rivadavia's ecclesiastical reform. All of them held Gallican and Jansenist ideas.[18] Valentín Gómez drank from the source: he stayed in Europe as diplomatic envoy between 1818 and 1820 and while in Paris and thanks to the good offices of Rivadavia, he entered in contact with the abbé de Pradt and with European liberal thought.[19] As mentioned in the previous chapter, Gómez had taught the philosophical triennium at the College of San Carlos (1799–1802), including natural philosophy. Diego E. Zavaleta, another priest close to Rivadavia, also taught philosophy at San Carlos (1795–1797); his courses on *Physica generalis* and *Physica particularis* were a rehash of current handbooks; the sections on experimental physics were the weakest of the course.[20]

---

13 Ivereigh, *Catholicism and Politics in Argentina*, 44.
14 Rock, *The British in Argentina*, 50.
15 Nancy Calvo, "'Cuando se trata de la civilización del clero': Principios y motivaciones del debate sobre la reforma eclesiástica porteña de 1822," *Boletín del Instituto de Historia Argentina y Americana*, 3rd series, 24 (2001): 73–104; Roberto Di Stefano, "Pensar la Iglesia: el Río de la Plata entre la reforma y la romanización (1820–1834)," *Anuario de Historia de la Iglesia* 19 (2010): 221–39.
16 Vicente Fidel López, *Historia de la República Argentina*, vol. 9 (Buenos Aires: La Facultad, 1911), 131. For Rivadavia's Catholic orthodoxy, see Haydée E. Frizzi de Longoni, *Rivadavia y la reforma eclesiástica* (Buenos Aires, 1947), 133–49.
17 Rómulo Carbia, *La revolución de Mayo y la Iglesia* (Buenos Aires: Huarpes, 1945), 90.
18 Tonda, *La eclesiología*; idem, *Rivadavia y Medrano. Sus actuaciones en la reforma eclesiástica* (Santa Fe: Castellvi, 1952). See also, Carbia, *La Revolución de Mayo y la Iglesia*, 89–118; Guillermo Gallardo, *La política religiosa de Rivadavia* (Buenos Aires: Theoria, 1962), 13–32. For the priests favouring Rivadavia's reform, see Calvo et al., *Los curas de la Revolución*.
19 Aguirre Elorriaga, *El abate de Pradt*, 185.
20 Lértora Mendoza, *La enseñanza de la filosofía*, 293–343.

The brief engagement of these priests in natural philosophy did not affect their ecclesiological ideas; the teaching of philosophy was just an early stop in a political and ecclesiastical career.

The ecclesiastical reform was launched in August 1821 and culminated with a law approved on 21 December 1822 by the provincial legislature—whose chamber was shaped as a semicircle following Bentham's recommendations. Measures for expropriation of ecclesiastical property took momentum during the first months of 1821. The decrees of the executive power were contested by the judicial vicar Mariano Medrano; as a result of his vocal opposition to the reform, Medrano was displaced by the ecclesiastical *cabildo*, whose members sided with the government (he was replaced by Mariano Zavaleta). The bill submitted by Rivadavia to the Legislature failed to pass; it was modified by a special commission who, though favourable to the government, would not go as far as to condone the suppression of all the religious houses in the city (on this point, the priests held their ground). The new legislation involved the suppression of tithes and of the ecclesiastical legal jurisdiction; the seminary was replaced by a College of Ecclesiastical Studies supported by the state; a Senate of the clergy, whose members were henceforth salaried by the government, substituted the ecclesiastical *cabildo* (an institution tied to the colonial era); the state also took upon itself the creation of new parishes. Rivadavia's reform aimed at the control of the Church by the state, not at the total separation of Church from state. The constitution of 1819 declared Catholicism "the state religion" and the constitution of 1826—also a product of Rivadavia's faction—affirmed that "the religion of the Argentine Nation is Roman Catholicism."[21]

The brunt of the reform was borne by the regular clergy, which was henceforth subordinated to the secular diocesan authorities;[22] religious houses with less than 16 and more than 30 members were suppressed and their lands transferred to the state; the minimum age for religious vows was established at 25 and candidates should be subjected to diocesan approval; the authority of the religious provincials ceased to be recognised. Around 90% of the religious were secularised (their new status was rather ambiguous).[23] In the end, the government expropriated buildings and lands from the Sanctuary in Luján, the Cathedral, the

---

21 *Constituciones argentinas*, 13, 47.
22 Américo A. Tonda, *El Dean Funes y la reforma de Rivadavia. Los regulares* (Santa Fe: Castellvi, 1961).
23 Cayetano Bruno, *Historia de la Iglesia en la Argentina*, vol. 8 (Buenos Aires: Don Bosco, 1972), 411–547; Fidel Iglesias, "The Platine Clergy, Independence, and the Rivadavian Reforms, 1806–27," *Journal of Religious and Theological Information* 3, no. 2 (2000): 119–51.

hospitals of the Bethlemites, the convents of the Recollect Franciscans and the Dominicans, and the assets of nun congregations and charitable organisations.[24]

Toleration toward Protestants went hand in hand with measures aimed at the establishment of a Church financially supported by the state and controlled by it. The 1820s marked the high tide of British influence in Río de la Plata. As a result of the government's policies liberalising trade British merchants flock to Buenos Aires. With the loan negotiated from the Baring Brothers (1824), Rivadavia initiated the long and conflictive history of the financial relationships between Argentina and London; there were other ventures related to the exploitation of mines and colonisation by British concerns.[25] This climate made religious tolerance a condition for good business. In 1825, the consular commissioner to Río de la Plata Woodbine Parish, appointed by George Canning, signed a Treaty of Amity, Commerce, and Navigation which allowed British subjects to build Protestant churches and to establish a British cemetery. With only one exception, provincial legislatures other than Buenos Aires did not ratify religious tolerance. Following the arrival of Rev John Armstrong, the English began to worship in Buenos Aires in an abandoned Jesuit church until the first Protestant (Anglican) church was inaugurated in 1831; a provisory Scotch Presbyterian chapel began functioning in 1829 and St. Andrew's Scotch Presbyterian Church was inaugurated in 1835.[26]

I shall argue that in this scenario of secularisation (in its original sense of seizure of church property by the secular power), science was not a moving factor. But the new scientific institutions clearly benefited from the secularising measures, inasmuch as they were installed in the confiscated convents. The transformation of religious spaces into locales devoted to scientific and technical educational ventures was a strong message of Rivadavia's ecclesiastical reform. But it should be stressed from the start that this was a by-product of the reform and should not be counted among its causes.

---

[24] Gallardo, *La política religiosa de Rivadavia*, 79–104.
[25] Henry S. Ferns, *Britain and Argentina in the Nineteenth Century* (Oxford: Oxford University Press, 1960), 67–154; Rock, *The British in Argentina*, 43–90.
[26] W. H. Hodges, ed., *History of the Anglican Church St. John the Baptist (pro-cathedral). Buenos Aires, 1831–1931* (Buenos Aires: W. F. Story, 1932); James Dodds, *Records of the Scottish Settlers in the River Plate and their Churches* (Buenos Aires: Grant and Sylvester, 1897), 136–89; Rock, *The British in Argentina*, 53–55, 61–62.

From Convent to Cemetery and Agricultural Station

The Convent of the Recollect Franciscans was built in the first decades of the eighteenth century on the edge of the city; it consisted of several buildings (including classrooms) and a church inaugurated in 1732.[27] The convent had been used as field hospital in the battles during the "British invasions" to Río de la Plata (1806–1807). Francisco de Paula Castañeda temporarily used the premises for his School of Drawing and the Franciscans taught courses of philosophy and theology in one of the buildings. The lands of the convent were adjacent to the state of Martín José de Altolaguirre, whose brother Francisco was one of the Recollect friars.[28] The convent was confiscated on 1st July 1822. Its premises were transformed into a cemetery (the "Recoleta," which exists to this day) controlled by the lay authorities and a school of agriculture. The friars were relocated to other convents.[29]

During colonial times in Buenos Aires, corpses were buried in the churchyards or, if the relatives of the deceased were rich enough to pay the high fees, in the interior of the churches. An article published in a cultural periodical two weeks after the decree of expropriation claimed that cemeteries should be located in the outskirts of the city and rallied against the inhumations in the churches which affected public health. The author alleged that the putrefaction of corpses "corrupted the air" for they emitted gases which conveyed "miasmas or putrid germs" and affirmed that "ancient ecclesiastical constitutions, pontifical letters, the annals of the church" as well as "the legislation of the most enlightened princes, public health, and our self-preservation" demanded that the people should be "free from the infections [caused by] corpses."[30] The anonymous author cites a short work by the enlightened French physician and anatomist Félix Vicq d'Azyr—a translation of Scipione Piattoli's *Saggio intorno al luogo del seppellire* (1774).[31] D'Azyr's essay was part of a body of literature which ap-

---

[27] Andrés Millé, *La Recoleta de Buenos Aires. Una visión del siglo XVIII* (Buenos Aires: Talleres Gráficos de Domingo Taladriz, 1952).
[28] Ricardo de Lafuente Machain, *El barrio de la Recoleta* (Buenos Aires: Municipalidad de Buenos Aires, Instituto Municipal de Extensión Artística, 1945), 17.
[29] Enrique Udaondo, *Reseña histórica del templo de Ntra. Sra. del Pilar (Recoleta)* (Buenos Aires: Imprenta Mirau, 1918), 83–91. The church of the convent, Our Lady of Pilar, has been well preserved and serves now as a parish church.
[30] "Hygiene pública. Cementerios," *La abeja argentina* no. 4, 15 July 1822, 150–57.
[31] Félix Vicq d'Azyr, "Essai sur les lieux et les dangers des sépultures," in idem, *Oeuvres complètes*, vol. 6 (Paris: Badouin, 1805), 257–358.

peared in France in the 1770s concerning health issues caused by the nearness of cemeteries and which drew on theories about contaminated airs or vapours.[32]

Corpses were removed from churchyards in the city and taken to the new cemetery, in which 7410 burials were performed between 1822 and 1828.[33] George Thomas Love, the editor of *The British Packet and Argentine News* (1826–1845), tells that "scenes of confusion took place; with imprecations from near relations … upon recognizing the bodies of those they never expected to have seen again in this world."[34] The measure to forbid burials in churches was unpopular, for it was seen as "…the expulsion of the departed souls from the sacred place to which they were entitled."[35] Two chaplains were assigned to the new burial ground, which was consecrated in November 1822 with the presence of many priests.[36] In the new Protestant cemetery—the construction of which had been financed by foreign residents—60 British subjects were buried from January 1821 to June 1824.[37]

The friary of the Recollects had an adjacent orchard which was also expropriated and transformed into a school of agriculture in charge of the French gardener Pierre Béranger and an assistant. By April 1825, Béranger had planted more than 250 species of trees and plants from seeds donated by Caesar A. Rodney, plenipotentiary minister of the United States. Béranger's contract was suspended for reasons unknown and the school was closed (all the students but one had quit the courses).[38] Rivadavia, then in Europe, contacted the firm Hullett Brothers to hire a new gardener. The chosen candidate was Alexander P. Sack, "a gardener of superior education, various talents, and very great experience."[39] He embarked with an assistant, Samuel (in some documents, William) Atwell and 14 crates of seedlings and seeds bought by the government (most of the plants died during the journey). They arrived in Buenos Aires in May 1826 and were given

---

[32] Philippe Ariès, *The Hour of our Death*, trans. Helen Weaver (New York: Knopf, 1981), 479–94.
[33] Udaondo, *Reseña Historica*, 93.
[34] [George T. Love], *Five Years' Residence in Buenos Ayres, during the Years 1820 to 1825…by an Englishman* (London: G. Hebert, 1825), 119.
[35] López, *Historia de la República Argentina*, 9:127.
[36] "Cementerio del Norte. Buenos Aires, 8 de julio de 1822," *Registro Oficial* [*de la Provincia de Buenos Aires*], bk. 2, no. 20, 16 July 1822, 263–64; see also Bruno, *Historia de la Iglesia*, 8:442.
[37] [Love], *Five Years' Residence*, 119–21.
[38] Pierre Béranger, "Relaciones de los trabajos practicados en el Jardín de Aclimatación desde su establecimiento hasta la fecha en que se expresa" (15 April 1825), in Julio César Álvarez Díaz de Vivar, *Rivadavia: piedra angular de la enseñanza agrícola en las Provincias Unidas del Río de la Plata* (Buenos Aires: Suelo Argentino, 1945), 146–51.
[39] "National Botanic Garden, Buenos Ayres, 10th August, 1826," *The Gardener's Magazine* 2 (1827): 91.

living quarters in the former convent. Due to the ruinous conditions and humidity of the rooms, Sack and Atwell got ill and Rivadavia had to intervene to get them new housing; he signed the decree creating the Acclimation Garden on 6 June 1826.[40] By December, Sack was not able to get a salary increase and the attempt at digging a well deep enough for irrigation was a failure (Sack complained that the workers in charge lacked "the least of common sense").[41] This operation was conducted by the British engineer James Bevans, who had been hired by Rivadavia in 1822 to design the water system of the city and a port.[42]

Given the ongoing war with Brazil, the Treasure was almost empty and the government could ill afford the needed improvements for the functioning of the garden. In June 1827, Rivadavia quit office and in little more than a year the new government closed it. At the moment of the confiscation of the convent, its orchard had around one hundred apple trees; when Sack handed back the keys, around 8500 trees grew in it.[43] Atwell was hired in the plant nursery of one of his compatriots in Buenos Aires and Sack opened his own establishment, which evolved into a prosperous business.[44] In the decree ordering the closing of the garden, the new governor, Manuel Dorrego, alleged reasons analogous to those wielded by Rivadavia at the time of expropriating the convent. Dorrego claimed that more space was needed for the cemetery, for overcrowding of corpses could cause "severe ills" to the public; he also alleged that the garden had been a failure despite the money invested in it.[45]

This was not the first time that church land in Buenos Aires was used for agricultural experimentation. The French botanist Aimé Bonpland, who had accompanied Humboldt in his celebrated South American voyage of exploration, arrived in Buenos Aires in February 1817 with seeds and around one thousand live plants plus a large number of botany books, some of which would eventually be sold to the Public Library of Buenos Aires. He had been lured to travel to Río

---

40 "Jardinero Botánico. Decreto" (6 June 1826), in *Registro Oficial de la Provincia de Buenos Aires* (Buenos Aires: Imprenta del Mercurio, 1874), 102–104.
41 Sack to Carta and Senillosa, 8 December 1826, in Álvarez Díaz de Vivar, *Rivadavia*, 201–203.
42 "Erección del Departamento de Ingenieros Hidráulicos" (7 December 1822), in *Registro Oficial de la Provincia de Buenos Aires. Año de 1822* (Buenos Aires: Imprenta del Mercurio, 1873), 157–58. One of the assistants was José María Gutiérrez. See also Udaondo, *Reseña Historica*, 100–101.
43 Álvarez Díaz de Vivar, *Rivadavia*, 258–60.
44 "Pública utilidad y particular establecimiento de plantas y árboles frutales," *La Gaceta Mercantil* no. 3267, 21 April 1834.
45 "Decreto suprimiendo el Jardín de Aclimatación," *Registro Oficial [de la Provincia de Buenos Aires]* bk. 7, no. 2, 1st April 1828, 18–19.

de la Plata by Rivadavia, whom he had met in Europe in 1815. The government sold him an orchard of seven hectares which belonged to the Bethlemites.[46] In it, Bonpland grew the plants he brought from France and perhaps also indigenous species. The authorities had forced the friars to sell the land, but Bonpland did not advance the corresponding payments; instead, he proposed mortgaging the property and paying the loan off within five years. The legal conflict that ensued lasted until the mid-1820s, when the botanist abandoned the city for good.[47] With the expropriation of all the properties of the Bethlemites the orchard passed to the government which sold it to a private person.

The Secularisation of Hospitals

It was during the Enlightenment that hospitals began to turn from a "house of mercy and refuge" to a "house of cure and medical teaching."[48] But even while "science and medicine were challenging religion as meanings of existence," hospitals did not cease to be institutions for the caring of the poor, the disabled, and the mentally deranged on the basis of Christian charity.[49] The notion of a linear trend from a decline of charity to the ascendancy of medicalisation in hospitals is historically misleading: relationships between charity and medicine have been cyclical and often complementary.[50] As in the rest of the Spanish dominions in the New World, eighteenth-century hospitals in Río de la Plata were charitable institutions established as collaborative endeavours between the crown and church organisations such as lay brotherhoods and religious orders.[51]

---

46 Pablo Penchaszadeh and Miguel de Asúa, *El deslumbramiento/L'éblouissement. Aimé Bonpland y Alexander von Humboldt en Sudamérica* (Buenos Aires: MACN-Conicet, 2010), 99.
47 Aníbal Ruiz Moreno, Vicente A. Risolía and Rómulo D'Onofrio, "El pleito de Bonpland con los bethlemitas por la propiedad de la Quinta de los sauces," *Publicaciones del Instituto de Historia de la Medicina* 17 (1955): 6–20 (mimeo); Stephen Bell, *A Life in Shadow. Aimé Bonpland in Southern South America, 1817–1858* (Stanford: Stanford University Press, 2010), 23–42.
48 Guenter B. Risse, *Mending Bodies, Saving Souls. A History of Hospitals* (New York: Oxford University Press, 1999), 676.
49 Roy Porter, *The Greatest Benefit to Mankind. A Medical History of Humanity from Antiquity to the Present* (London: Fontana, 1997), 302.
50 Jonathan Barry and Colin Jones, eds., *Medicine and Charity Before the Welfare State* (London: Routledge, 1991), 2.
51 Guenter B. Risse, "Medicine in New Spain," in *Medicine in the New World. New Spain, New France, and New England*, ed. Ronald L. Numbers (Knoxville: The University of Tennessee Press,

The Bethlemites, a religious congregation founded in Guatemala in 1660 by Pedro de Bethancourt and granted the status of religious order by Innocent XI in 1687, arrived in Buenos Aires in 1748 to take care of the city hospitals; they also established hospitals in Córdoba and Mendoza.[52] By the 1760s, they attended around 600 patients per year in Buenos Aires and even performed surgical operations (one of them invented a new technique of cutting for the stone).[53] In 1784, there were 22 Bethlemites in Buenos Aires. The coming of independence brought with it conflicts between these friars and the new authorities. A Bethlemite was among those invoved in a conspiracy against the government and was executed in July 1812.[54] Three years later, in September 1815, a Legislative Assembly transferred the hospitals to the secular clergy. This measure proved a failure and in little more than one year the Bethlemites had regained control of their houses.[55] In Rivadavia's times, they administered two hospitals for male patients in the city, situated a few blocks away from one another. One of them was a general hospital with 200 beds, in a building which had belonged to the Jesuits; the other, Santa Catalina (the Bethlemites' original foundation), was used for convalescent and incurable patients, the mentally ill, and those affected by contagious diseases.[56] Since donations were not enough to support their benevolent activities, the Bethlemites acquired properties in the city, invested capital, and exploited large estancias; at the time of the expropriation, the order owned more than one hundred black slaves, who worked in the hospitals and in the ranches.[57] The

---

1980), 12–63; Linda A. Newson, "Medical Practice in Early Colonial Spanish America: A Prospectus," *Bulletin of Latin American Research* 25, no. 3 (2006): 367–91.
52 Guillermo Furlong, *Médicos argentinos durante la dominación hispánica* (Buenos Aires: Huarpes, 1947), 235–45. The Bethlemites administered a large number of hospitals in Lima and Mexico.
53 Alberto Meyer Arana, *La caridad en Buenos Aires*, vol. 1 (Buenos Aires, 1911), 37.
54 Meyer Arana, *La caridad en Buenos Aires*, 1:51–54.
55 Bruno, *Historia de la Iglesia*, 8:44–45.
56 For hospitals in colonial and early independent Buenos Aires, see Municipalidad de la Capital, *Documentos y planos relativos al período edilicio colonial de la ciudad de Buenos Aires*, vol. 3 (Buenos Aires: Peuser, 1910), 163–481; Meyer Arana, *La caridad en Buenos Aires*, 1:29–70; Eliseo Cantón, *La Facultad de medicina y sus escuelas*, vol. 2 (Buenos Aires: Coni, 1921), 89–122; José L. Molinari, "Hospitales coloniales," in idem, *Historia de la medicina argentina. Tres conferencias* (Buenos Aires: López, 1937), 71–110; idem, "La reforma de las instituciones médicas por la Asamblea de 1813 y en la época de Rivadavia," *Historia* (Buenos Aires) 9, no. 32 (1963): 119–35.
57 Carlos A. Mayo, *Los betlemitas en Buenos Aires: convento, economía y sociedad* (Sevilla: Publicaciones de la Excma. Diputación Provincial de Sevilla, 1991).

government sold these assets (they were worth a fortune) and continued collecting benefits for several years.[58]

In March 1822, with Rivadavia as minister of government, a commission of three persons was nominated to report about the functioning of the hospitals.[59] As a result of this inquiry the government decided to close Santa Catalina and move out the patients to the remaining facility (later known as Men's Hospital), which was henceforth run by an administrator and an accountant salaried by the state; each Bethlemite brother was granted a modest pension.[60] It was in that hospital where the students of the School of Medicine of the University received practical instruction.[61] The new arrangement was part of Rivadavia's secularising and centralising policies: soon upon the displacement of the Bethlemites the government launched a plan for transferring the administration of social assistance in Buenos Aires from religious corporate bodies to the state.[62]

The Convent of Santo Domingo

Construction of the church of Santo Domingo in Buenos Aires began around 1759; the building was designed by Antonio Masella, a Savoyard architect who also directed a remodelling of the Cathedral. After more than two decades of construction, the church was consecrated in 1783 (the right tower was added in 1849). The convent proper was begun around 1790 and works ended in 1805.[63] This building complex was the scenery of armed confrontation during the 1806 and 1807 British military expeditions to Río de la Plata. Just before the law of ecclesiastical reform (1822), 60 priests, 7 choristers, and 9 lay brothers lived in the convent. As a result of its secularisation, they were obliged to either join the secular clergy with an official salary or leave the province.[64] The exception was the Irish Dominican Father Burke, who "from motives of kindness" was

---

[58] Gallardo, *La política religiosa*, 87–90.
[59] "Comisión de visita de las casas hospitalarias" (26 March 1822), *Registro Oficial de la Provincia de Buenos Aires* 2, no. 10, 28 March 1822, 124.
[60] "Los enfermos de Santa Catalina pasan a la Residencia" (1st July 1822), *Registro Oficial de la Provincia de Buenos Aires* 2, no. 19, 4 July 1822, 237–39. See also Gallardo, *Política religiosa*, 87–90.
[61] Cantón, *La Facultad de Medicina*, 2:30.
[62] "Arreglo de la medicina" (9 April 1822), *Registro Oficial [de la Provincia de Buenos Aires]* bk. 2, no. 12, 12 April 1822, 137–54.
[63] Mario J. Buschiazzo, "El templo y convento de Santo Domingo en Buenos Aires," *Anales del Instituto de Arte Americano e Investigaciones Estéticas* 4 (1951): 62–75.
[64] Bruno, *Historia de la Iglesia*, 8:483–85.

"still allowed to occupy his apartment." It seems that the rooms of the friars and the adjoining garden made the convent "a comfortable retreat."[65] As a result of Rivadavia's reforms, the convent was turned into the city's "science centre," housing the new laboratories of chemistry and physics, a natural history museum, and an astronomical observatory.

The arrival in Buenos Aires in 1823 of physics instruments, a laboratory of chemistry, specimens of natural history, and anatomical models bought by Rivadavia from Baillot, Piet & Cie. (Paris) was a significant event for science teaching in Río de la Plata.[66] The equipment was of the highest quality and soon became a material symbol of the government's will to promote science. The instruments were displayed in the former convent under the vigilant care of the pharmacist Carlo G. Ferraris, one of the liberal exiles from Piedmont hired by Rivadavia.[67] The other was Pietro Carta Molino, a doctor recently graduated from the University of Torino, who made his way to Paris, where Rivadavia hired him as a teacher for the university.[68] The Buenos Aires government paid him to study for a year the subject he was supposed to teach and had yet to learn: experimental physics. It is likely that he attended lectures and visited laboratories in Paris.[69] Carta also recommended his countryman and fellow partisan Ferraris—both had participated in the 1821 Piedmont revolt, which led to the abdication of Victor Emmanuel I.[70]

Ferraris set up a pharmacy facing the university and also worked as curator of the instruments and the museum established in the secularised convent.[71] According to the British traveller Beaumont, "the collection consisted of a mixture of skins of birds and beasts, much damaged by ill usage and insects." Beaumont

---

[65] [Love], *Five Years' Residence*, 20.

[66] Nicolau, *Ciencia y técnica en Buenos Aires*, 62–66; Máximo Barón, "La primera cátedra de física experimental en la Universidad de Buenos Aires," *Quipu* 3, no. 3 (1986): 349–400; Carta Molino to Rivadavia, 5 September 1825, in Piccirilli, *Rivadavia y su tiempo*, 2:479–83.

[67] Giacomo C. F. Calleri Damonte, "Carlo Giuseppe Ferraris: un farmacista di adozione biellese," article published 6 May 1998, https://doczz.it/doc/1276479/carlo-giuseppe-ferraris--un-farmacista-di-adozione-biellese.

[68] Salvadore Candido, "Carta Molino, Pietro," in *Dizionario Biografico degli Italiani*, vol. 20 (Roma: Istituto della Enciclopedia italiana, 1977), https://www.treccani.it/enciclopedia/pietro-carta-molino_(Dizionario-Biografico)/.

[69] Piccirilli, *Rivadavia y su tiempo*, 3:114–26.

[70] Michael Broers, "Revolution as Vendetta: Patriotism in Piedmont, 1794–1821," *The Historical Journal* 33, no. 3 (1990): 573–97.

[71] "Decreto nombrando un encargado del cuidado de los instrumentos y objetos de física, química e historia natural" (10 April 1826), *Registro Nacional. Provincias Unidas del Río de la Plata. Año 1826* (Buenos Aires: Imprenta del Mercurio, 1874), 58.

also tells that there was "a long room" in Santo Domingo furnished with glass cases on either side, plus the "philosophical instruments."[72] Ferraris took his mission seriously: he went shooting and fishing to the River Plate and he himself stuffed the animals. He also elevated yearly reports to the government on the new acquisitions and the state of the apparatus. In 1827, repairs were done to the building, three new rooms were refurbished in the upper floor and ovens provided for the laboratory of physics and chemistry.[73] According to the French naturalist Isabelle, who visited the region in the early 1830s, the laboratories were on the ground floor and the museum, which he described as little more than a "*cabinet de curiosité* [sic]", was on the first floor. He tells that the stairs were counted "among the city marvels" and praises the zeal of Ferraris (confirmed by other witnesses). By then, the museum held 1500 rock and mineral samples, 800 animals, and many insects.[74]

Carta Molino taught only one course of experimental physics; its inaugural lecture was given on 17 July 1827, about three weeks after Rivadavia fell from power.[75] A course of chemistry was also inaugurated that week and the official newspaper remarked that both of them were dictated "in the extinguished convent of Santo Domingo."[76] Carta opened his inaugural lecture with a tribute to Rivadavia followed by a plea for patronage addressed to the new incumbent. Carta's lecture on experimental physics is a doctrinal affirmation of *idéologie* with almost no specific technical reference either to physics or to experiments. He rehearses Cabanis's idea that the moral sciences are a prolongation of the natural sciences to the point that morality rests upon physiology.[77] Carta's avowal of deism was evident. There was a time, he claims, when knowledge was just "word play...in the hands of the priestly class, which used it to maintain the people in ignorance." The "eloquent Condorcet," he goes on, told how Galileo was incarcerated for having taught humanity "to better know the works of the Supreme Being and admire the simplicity of the natural laws through which he gov-

---

[72] John Barber Beaumont, *Travels in Buenos Ayres and the Adjacent Provinces of the Río de la Plata* (London: James Ridgway, 1828), 215–16.
[73] "Razón de obras hechas e inspeccionadas por el Departamento de Ingenieros arquitectos en el mes de Junio último," *Crónica Política y Literaria de Buenos Aires* no. 49, 11 July 1827.
[74] Arsène Isabelle, *Voyage à Buenos Ayres et à Porto-Alègre* (Paris: J. Morlent, 1835), 160–62.
[75] See the transcription of the first lecture in Emilio Ravignani, "Notas para la historia de las ideas en la Universidad de Buenos Aires. El doctor Carta y la enseñanza de la física experimental," *Revista de la Universidad de Buenos Aires* 13, no. 34 (1916): 70–96. For Carta Molino in Buenos Aires, see Nicolau, *Ciencia y Técnica en Buenos Aires*, 66–80. Nicolau was the first author to make clear that Carta Molino was unqualified for his position.
[76] *Crónica Política y Literaria de Buenos Aires* no. 52, 14 July 1827.
[77] Ravignani, "Notas," 85–86.

erns the world." Mothers should teach these laws to their children in order to inspire in them "sentiments of veneration and religious respect."[78] As a result of press criticism and the lack of governmental support Carta resigned from his teaching position. He had threw his lot with Rivadavia, and power was now in the hands of the latter's enemies. One of the new ministers, Manuel Moreno, was only happy to accept his resignation. Moreno, a graduate from the School of Medicine of the University of Maryland and briefly professor of chemistry in the University of Buenos Aires, was author of an anonymous virulent attack on Rivadavia's cultural policies in which he claimed that Carta taught the youth "to doubt the existence of the spirit" and even pushed them "to the threshold of materialism."[79] Moreno was not a religious person, but Catholicism was dear to his party (the Federalists). Carta went on teaching *materia medica* to pharmacy students until 1833; he died mentally deranged in the French Hospital in Buenos Aires in 1849.

The person who substituted Carta Molino in the chair of experimental physics was the liberal Piedmontese physicist and astronomer Ottaviano F. Mossotti, born in Novara and graduated as an engineer from the University of Pavia. After working for ten years in the Observatory of Brera in Milan (at that time part of the Austro-Hungarian Empire), Mossotti escaped from the Austrian police and arrived in London where he did some work for the Admiralty. He was then hired by Hullet Brothers on behalf of Rivadavia to teach physics in Buenos Aires, where he stayed from 1827 until 1835.[80] Mossotti set up his observatory in the convent of Santo Domingo. He sent meteorological observations to François Arago in Paris and was the first person to see the return of the comet Encke, on 2 June 1832; his observations were communicated to the Royal Astronomical Society.[81] The course he taught in Buenos Aires was a standard course of phys-

---

[78] Ibid., 83–88.
[79] [Manuel Moreno], *Impugnación a la respuesta dada al mensaje del gobierno* (Buenos Aires: Imprenta del Estado, 1827), 51.
[80] Juan M. Gutiérrez, *Noticias históricas sobre el orijen y desarrollo de la enseñanza pública superior en Buenos Aires* (Buenos Aires: Imprenta del Siglo, 1868), 889–912; Máximo Barón, *Octavio F. Mossotti. En el amanecer de la ciencia Argentina* (Buenos Aires: ECA, 1981).
[81] Ottaviano F. Mossotti, "Mémoire sur le climat de la ville de Buenos-Ayres," *Comptes rendus de la Académie des Sciences* 1835, pt. 1, 283; idem, "Observations of the Transit of Mercury over the Sun's Disc, in May 1832; and of the Comet of Encke in June 1832: at Buenos Ayres," *Monthly Notices of the Royal Astronomical Society* 3 (1833–36): 37–38; idem and Thomas Henderson, "Places of Encke's Comet from Observations in Buenos Ayres," *Memoirs of the Royal Astronomical Society* 1835, 244–50.

ics.[82] Mossotti's fame rests on his later work on the theory of matter and dielectrics.[83] In a memoir published in 1834, he mentions that he "conceived the idea... in a course of lectures on natural philosophy which I gave at the University of Buenos Aires."[84] According to Beaumont, the Italian professors hired by Rivadavia had been "promised handsome salaries, and enjoying a climate in which people never died;" the disappointment they experienced upon their arrival at the dismal city "would be difficult to describe." When they tried to approach the once friendly Rivadavia, they found him unapproachable and "their dreams of bliss were at an end."[85]

Rivadavia saw science as the ultimate expression of the kind of useful knowledge that would contribute to progress in the spirit of the Enlightenment. In his short-lived "Enlightened Republic," science dislodged religion as a locus of authority—physical as well as symbolic. The installation of laboratories, an observatory, and a museum in what had been so far a religious locale was perceived as a triumph of the government's ideology. Certainly, what had been secularised was the convent, the living quarters of the friars; nothing approaching profanation or deconsecration of a church had occurred. But this distinction was not always perceived as such. A nineteenth-century liberal historian observed that previous governments had not proceeded before with the secularisation of convents because they feared "hurting the idolatrous prejudices of the timorous people who lacking self-judgement saw the convent as an integral part of the temple...as if there were no difference between material things and the infinite spirit of the creator."[86] In an article written decades later, the anticlerical intellectual Juan María Gutiérrez, at that time president of the University of Buenos Aires, recalled with relish that in the 1820s "convents had been converted into schools of natural sciences...the secularisation of individuals and buildings was a rowdy affair."[87]

---

82 See the paraphrasis in Baron, *Octavio F. Mossotti*, 99–107 and the transcription of a short fragment in Gutiérrez, *Noticias históricas*, 461–65.
83 Jed Buchwald, "Mossotti, Ottaviano Fabrizio," in *Dictionary of Scientific Biography*, ed. Charles C. Gillispie (New York: Charles Scribner's Sons, 1981), 9:547–49.
84 Ottaviano F. Mossotti, "On the Forces Which Regulate the Internal Constitution of Bodies," in Richard Taylor, ed., *Scientific Memoirs*, vol. 1 (London: Richard and John E. Taylor, 1837), 450–51.
85 Beaumont, *Travels in Buenos Ayres*, 215.
86 López, *Historia de la República Argentina*, 9:105.
87 Juan M. Gutiérrez, "Las restauraciones religiosas en 1835–1841–1875," *Revista del Río de la Plata* 11 (1875): 414.

Natural History among the Clergy

Bartolomé Muñoz was an active member of the circle of enlightened naturalist priests who supported the movement for independence in Río de la Plata. Unlike qualified European botanists who worked in the region such as Haenke and Bonpland, these clerics had been born in the country or arrived in it at a tender age; they were self-educated in natural history. In the preceding chapter, we have already examined Dámaso Larrañaga's views on natural theology. His fellow José Manuel Pérez Castellano was a priest of the Banda Oriental graduated from Córdoba, whose *Observations on Agriculture* [*Observaciones sobre Agricultura*] (1813, published in 1848) resumed 40 years of rural life in his small state in Miguelete, present-day Uruguay.[88] Bartolomé Muñoz, Larrañaga's mentor in natural history who also took the patriots' side, was an avid collector of natural objects. Saturnino Segurola, who like Larrañaga was much involved in local politics, kept a modest *cabinet de curiosités* with objects of natural history, artefacts of exotic cultures, and incongruous historical objects; he was the foremost champion of vaccination in Buenos Aires, and amassed the then largest collection of historical documents in Río de la Plata, "…chiefly enriched by the spoils of the extinct company [i.e. the Society of Jesus]."[89] It seems he possessed microscopes, magic lanterns, fragments of petrified wood, animal skeletons, a collection of butterflies, and coloured pictures of animals.[90] In a letter from 2 July 1804, Larrañaga wrote to him with admiration for "all the most brilliant that can adorn a precious Museum, a cabinet of curiosities and a select library [like yours]."[91] In that same letter Larrañaga describes his own collection in disparaging terms: "My entomological collection contains a mere hundred disgusting beetles and three dozen sad and unattractive butterflies; my ornithological collection consists of some moth-eaten skins of birds; my ichthyologic collection of some dissected fish and a few wrinkled skins."[92] Larrañaga's collection grew considerably. When he took his position as president of the Commission of the Public Library and

---

[88] Furlong, *Naturalistas argentinos*, 393–94.
[89] Alexander Caldcleugh, *Travels in South America during the Years 1819–20–21*, vol. 1 (London: Murray, 1825), 175. Cf. Santiago Torres, "Detalle de los objetos recibidos por el encargado del Museo, en cumplimiento de la resolución ministerial," *Revista de la Biblioteca Nacional* 4, no. 13 (1940): 14; see also the list of documents in "Inventario de los documentos de la donación Segurola, recibidos por el Director de la Biblioteca Pública," *Revista de la Biblioteca Nacional* 4, no. 13 (1940): 15–73.
[90] Furlong, *Naturalistas argentinos*, 395.
[91] Mario Falcao Espalter, "Cartas científicas de Larrañaga. Edición y notas," *Revista del Instituto Histórico y Geográfico del Uruguay* 2, no. 1 (1921): 295.
[92] Ibid., 296.

Museum in Montevideo, in a note from 18 October 1837, Larrañaga refers to "the mineralogical and zoological collections and all my herbals with their catalogues, which I am presenting...to our National Museum [of Montevideo]."[93]

Muñoz's cabinet of natural history was the first significant donation to the Museum in Buenos Aires, founded by Rivadavia. Muñoz was born in Spain and arrived in Buenos Aires as a child in 1776; his father was cousin of a high officer in José de San Martin's liberating army. Muñoz studied at the College of San Carlos and then theology; he was ordained in 1786. Since 1808, he was chaplain of a patriot infantry regiment in the siege of Montevideo, where he met Larrañaga. Muñoz fed the Larrañaga's interest in natural history, drew pictures of animals for him, lent him books, and made his collection accessible to him, "the only one in this city and the most complete in this province as what concerns shells."[94] Larrañaga's *Diary* testimonies to the evolution of his relationship with Muñoz from discipleship to friendship: they exchanged letters, objects, and books, and encouraged each other to continue their researches on the fauna and flora of the region while armed conflicts raged around them. Muñoz edited an ecclesiastical calendar (1816–1818) and between 1820 and 1829 a yearly "Patriotic Almanac." He carried out meteorological registers, made occasional astronomical observations, drew charts of the region, and eventually presented his scientific instruments to the Public Library in Buenos Aires.[95] The observations of a Moon eclipse he performed on 9 June 1816 were patterned on those made by Buenaventura Suárez, whom he mentions as an "American astronomer" and whose *Lunario* Muñoz used to find out the date and hour of the eclipse.[96]

While Secretary of the First Triumvirate, on 27 June 1812 Rivadavia sent a note to the authorities in the provinces which informed that "a Museum of Natural History is going to be established in this Capital [Buenos Aires]" and asked them to "collect all the rare and peculiar productions of this territory worthy of being kept in that deposit."[97] The only important contribution to the Museum

---

93 Quoted in Castellanos, "Contribución al estudio," 44–45.
94 Larrañaga, *Diario*, fol. 28v (entry corresponding to 19 January 1808). Cf. Eugenio Beck (Guillermo Furlong), "Un benemérito de las ciencias en el Río de la Plata. Bartolomé Doroteo Muñoz (1831–1931)," *Revista de la Sociedad Amigos de la Arqueología* 5 (1931): 58–59; Tonda, *La iglesia argentina*, 79.
95 Beck, "Un benemérito de las ciencias," 71.
96 They used a "good Dollond achromatic telescope with a focal distance slightly more than three feet, triple objective with an aperture of three inches, a clock with a second hand, and the Moon map of Hervás y Panduro." *La Prensa Argentina* n° 41, 25 June 1816.
97 [Bernardino Rivadavia] to the Military Command in Patagones, 27 June 1812, in "Guerra," box X.6–4–1, Archivo General de la Nación (AGN), Buenos Aires. Cf. "Anuncio oficial," *La Gaceta* n° 18, 7 August 1812, 73.

was made by Muñoz, who had managed to dispatch his collection from Montevideo to Buenos Aires in September 1813.[98] His donation was publicised in the government's newspaper.[99] Larrañaga was right: the cabinet consisted mainly of shells (five hundred of them, ordered according to Linnaeus's system) and a little less than 300 plates of animals drawn by Muñoz, plus some mineral specimens.[100]

An article published thirteen years later in a Buenos Aires periodical, two weeks before Rivadavia fell from office, affirmed that Muñoz's collection "can satisfy the expectations of an amateur, but does not fulfil the exigencies of the learned."[101] Muñoz protested against the derogatory remarks and demanded moral reparation.[102] His letter was published with a conciliatory note by the editor which suggested that the priest should be aware of the impossibility for one person to cover the whole field of science. The editor of the newspaper was Pietro de Angelis, a Neapolitan historian, bibliophile, and publicist who had travelled to Buenos Aires from Paris answering Rivadavia's search for intellectual talent. De Angelis had been preceptor of the children of Joachim Murat, husband of Napoleon's sister Caroline and king of Naples between 1808 and 1815. De Angelis held liberal views and at the time of his departure from Europe was beginning to obtain recognition from intellectual and political circles in the French capital. When Rivadavia fell from power, he put his considerable intellectual talents at the service of Juan M. de Rosas, in every respect the opposite of his former patron. When Muñoz died, de Angelis seized the opportunity to again criticise the deceased. He referred to Muñoz's diffuse and superficial knowledge while underrating his natural history collection as "a small natural history museum [he] had organised for himself."[103] There might have been personal reasons for de Angelis's less than gentlemanly behaviour and also an element of anticlericalism. After his donation to the museum, Muñoz went on collecting. In July 1826, he opened to the public his museum, which contained 5,306 pieces.[104] Upon his death in May 1831, his collection was offered in sale to the university. A commis-

---

98 Beck, "Un benemérito de las ciencias," 72–73.
99 [Untitled], *Gazeta Ministerial* no. 110, 11 June 1814, 605–606.
100 "Donativo que hace a la Biblioteca del Estado de las Provincias Unidas del Río de la Plata el ciudadano Bartolomé de Muñoz," *Gazeta Ministerial* no. 110, 11 June 1814, 605–606.
101 *La Crónica Política y Literaria* no. 36, 9 June 1827.
102 [Bartolomé Muñoz], "Artículo remitido," *Crónica Política y Literaria* no. 39, 16 June 1827.
103 [Untitled], *El Lucero* no. 496, 3 June 1831.
104 "Historia natural," *La Gaceta Mercantil* no. 805, 14 July 1826. Cf. Muñoz, "Artículo remitido," where he affirms that it took him 35 years to gather "the near six thousand items scientifically arranged" he possessed at that time.

sion valued the "objects of natural history and curiosities" in $3,500 pesos. The government's minister of finance declined to buy it.[105]

The University of Buenos Aires and *idéologie*

The two philosophical currents of thought championed by Rivadavia, utilitarianism and *idéologie*, became the official doctrine of the University of Buenos Aires: the professor of civil law Pedro Somellera based the book he wrote for his course (*Principios de Derecho Civil*, 1824) on Bentham while the chair of philosophy became a stronghold of *idéologie*.[106] Here, we shall concentrate on the course of philosophy, taught first by the young Crisóstomo Lafinur and then by the unorthodox priest Juan Manuel Fernández de Agüero. Despite his scandalous reputation, the former was in essence rather conventional; his substitute taught doctrines perceived as heretical by many.

Building upon Locke's empiricism and Condillac's sensism, the French *idéologues* stressed the physiological basis of sensation as the origin of all knowledge. Destutt de Tracy, the main proponent of this philosophy, proclaimed that "thinking...is always sensing and nothing else than sensing."[107] The physician Pierre J. G. Cabanis drew upon the tradition of French materialism at the time of claiming that the brain produces thought such as the liver filters bile and the salivary glands secrete saliva.[108] The high tide of *idéologie* was the period of the French Directory. As the dominating figure of the Council of Public Instruction between 1798 and 1800, Desttut de Tracy managed to introduce courses of *idéologie* in the French *écoles centrales* hoping they would shape a new type of citizen; he aspired to lay the basis for a natural science of morality and a republican type of social science built upon his own secularist philosophy. Napoleon, who always distrusted the *idéologues*, banished them from the political scene after signing the 1801 Concordat with Pius VII.[109] By the time *idéologie* be-

---

105 Unsigned documents related to the dealings between Muñoz's executor and the University of Buenos Aires, 1831–32, in "Universidad 1828–1835," box X-6–3–1, AGN.
106 Gallo, *Struggle*, 41.
107 Antoine L. C. Destutt de Tracy, *Élémens d'idéologie. Première partie. Idéologie proprement dite*, 3rd ed. (Paris: Courcier, 1816).
108 Pierre J.-G. Cabanis, *Rapports du physique et du morale de l'homme*, 8th ed. (Paris: J. B. Baillière, 1844), 137. See Martin S. Staum, *Cabanis. Enlightenment and Medical Philosophy in the French Revolution* (Princeton: Princeton University Press, 1980).
109 Emmet Kennedy, *A Philosophe in the Age of Revolution. Destutt de Tracy and the Origins of "Ideology," Memoirs of the American Philosophical Society* no. 129 (Philadelphia: American Philosophical Society, 1978); idem, "The Secularism of Destutt de Tracy's 'Ideology'" (Tübingen,

came the new-fangled philosophy proclaimed from the secular pulpit of the University of Buenos Aires, it had long been forgotten in Paris.

The ideological program of the *idéologues* (no pun intended) suited Rivadavia's attempt to fashion an enlightened city out of a traditional colonial village. A perceptive contemporary chronicler reported that the town centre was "European in its habits, uses, ways of seeing and arguing; the suburbs were...half civilised, and Rivadavia wanted to make them instantly European just by issuing decrees."[110] Lafinur, born to a family made rich by gold mining in the province of San Luis, attended the University of Córdoba, which he left to join the patriot revolutionary forces. He enrolled in the army commanded by Gen Manuel Belgrano, at that time fighting in the northwest. In the officers' training camp, he came to know the French naturalist and adventurer Jean-Joseph Dauxion Lavaysse, who possibly introduced him to the works of the *idéologues*. Lafinur taught his course of philosophy in Buenos Aires from 1819 to 1820. Two parts of it have been conserved, logic and metaphysics. The preface provides a short panorama of the development of philosophy concluding with Newton, whose system "lords it over all the scientific academies of the world."[111] Most of the section on logic was copied verbatim from chapters 2 through 5 of Destutt de Tracy's *Logique*.[112]

The section of Lafinur's course devoted to metaphysics begins with a demonstration of God.[113] He expounds the traditional argument of God as "the first cause of all beings" and infers his existence from the "admirable order that reigns in nature." After this reference to the physico-theological argument, he brings in the idea of the degrees of being in nature, a sequence of "gradual relationships" with extends from the human being to the mineral realm. If so far Lafinur treads on traditional ground, he soon shows that his notion of God might be closer to deist natural religion than to Christianity. He mentions Voltaire's dic-

---

2001), accessed 17 September 2020, https://www.geisteswissenschaften.fuberlin.de/v/gramm aire_generale/Actes_du_colloque/Textes/Kennedy/Emmet_Kennedy.pdf; Brian W. Head, *Ideology and Social Science. Destutt de Tracy and French Liberalism* (Dordrecht: Martinus Nijhoff, 1985), 54–66.

110 Tomás de Iriarte, *Memorias. Rivadavia, Monroe y la guerra argentino-brasileña* (Buenos Aires: Sociedad Impresora Americana, 1945), 75; quoted in Gallo, *Bernardino Rivadavia*, 105.
111 Juan Crisóstomo Lafinur, *Curso filosófico dictado en Buenos Aires en 1819* (Buenos Aires: Instituto de Filosofía de la Facultad de Filosofía y Letras de la Universidad de Buenos Aires, 1938), 60. For *idéologie* in Buenos Aires, see Mariano Di Pasquale, "La présence de l'idéologie à Buenos Aires. Un élan philosophique dans le processus de professionalisation de la médecine, 1820–1840," *Cahiers des Amériques latines* 76 (2014): 111–29.
112 Juan Carlos Torchia Estrada, *La filosofía en la Argentina* (Washington D. C.: Unión Panamericana, 1961), 75–77.
113 Lafinur, *Curso filosófico*, 112–25.

tum that if God did not exist it would be necessary to invent him and reminds his students that Rousseau claimed that philosophy should show respect for two things: God and the spiritual nature of the soul. He then proceeds to the derivation of the attributes of God from his nature and to issues of traditional apologetics. Lafinur expatiates about the immortality of the soul, concluding that on purely philosophical grounds this claim is "more than probable;" to this, he goes on, we should add "the promises of Revelation" and so reach "absolute certainty" about the matter.[114] It is difficult to discern whether Lafinur's mention of Revelation was genuine or, what is more plausible, a bow to orthodoxy. In any case, the two sources of his exposition, deism and Christianity, are presented side by side.

On 20 September 1819, Lafinur presided over a public examination of his students. The audience was discomfited by what they understood were unorthodox pronouncements by the teacher and the literary event ended in turmoil. Doctor Cosme Argerich, perhaps the most important physician at that time in Buenos Aires, published a letter addressed to Lafinur with the purpose of allaying the fears of the public while also enjoining the lecturer to better explain his meaning.[115] Argerich, a member of a medical dynasty, had studied medicine in Cervera (Catalonia); upon his return to Buenos Aires, he joined the School of Medicine of the Protomedicato and was the leading figure in the creation of the Military Medical Institute (the militarised post-revolutionary school of medicine); in 1813, he joined as a surgeon Belgrano's army.[116] In his letter, Argerich argues that the philosophy of Bacon, Locke, and Condillac, which falls within the boundaries of human reason, cannot demonstrate the spiritual character of ideas and of the soul. Two principles lead to the formation of ideas: the brain (material) and the soul (immaterial). Philosophy is unable to explain the concourse of both principles and we must have recourse to religion. Argerich mentions that although he read assiduously Cabanis and Destutt de Tracy, in his teaching of physiology he always stressed the limits beyond which philosophy needs religion. He quotes a passage from François Magendie's *Précis élémentaire de physiologie* as saying "that the brain is the sole material organ of thought and all the

---

114 Ibid., 125.
115 Cosme M. Argerich, "Un comunicado del Dr. Cosme Argerich," *El Americano* no. 27, 1st October 1819, 10–15. This polemic has also been treated in Mariano Di Pasquale, "Vitalismo, *idéologie* y fisiología en Buenos Aires. La polémica entre Cosme Argerich y Crisóstomo Lafinur en *El Americano*, 1819," *Revista Ciencias de la Salud* 13 (2015): 13–28.
116 José M. Massini Ezcurra, *Los Argerich* (Buenos Aires: Instituto Amigos del Libro Argentino, 1955), 53–128.

phenomena that constitute thinking depend evidently on the soul."[117] Argerich's Spanish rendering of Magendie's text is free enough to misrepresent its original meaning. What the French physiologist meant was an expression of his methodological credo: thought should be considered as a product of the brain.[118] (Owsei Temkin has claimed that as to what concerns the relationship between body and soul, Magendie followed the materialist approach of the *idéologues*.[119]) But Argerich's two-step view of the formation of ideas is far from the materialism of Cabanis and closer in spirit, although not identical, to the philosophy of Pierre Laromiguière (whom he does not mention).[120]

Argerich's letter cast Lafinur in a favourable light but the latter felt aggrieved and in a letter to the same periodical complained of having been accused of "impiety."[121] He contends that what he had said is that if philosophy cannot demonstrate a priori the existence of an immaterial soul, then this should be demonstrated from the existence of God (which is what he had done in his course). Sensibility could not be the result of matter, "it is the result of the presence of an immaterial being in an organized body." Lafinur sets out to explain sensibility and the moral behaviour of the human being presupposing the existence of the soul, just as the Newtonians explained the mechanics of nature presupposing the attraction of bodies. Analogously to what Argerich had affirmed, he describes perception as a two-step process, with a first, primitive impression of the external world upon the organs of perception and a subsequent affection of the sensorium, which can vary because of the action of the soul.

These polemics, parochial and derivative though they were, indicate that issues of science and religion could awaken the spirits in Río de la Plata, at that time devastated by civil wars. While he stayed in Buenos Aires, Lafinur exhibited a conciliatory mood. (His open confrontation with the Franciscan Francisco de

---

117 "Que el cerebro es sólo un órgano material del pensamiento y que todos los fenómenos que constituyen el pensamiento están evidentemente bajo la dependencia del alma." Argerich, "Un comunicado," 13.
118 "Quels que soient le nombre et la diversité des phénomènes qui appartiennent à l'intelligence de l'homme...et quoiqu'ils soient évidemment sous la dépendance de l'âme, il est indispensable de les considérer comme le résultat de l'action du cerveau, et de ne les distinguer ainsi en aucune manière des autres phénomènes qui dépendent des actions d'organe." François Magendie, *Précis élémentaire de physiologie*, vol. 1 (Paris: Méquignon-Marvis, 1816), 170.
119 Owsei Temkin, "The Philosophical Background of Magendie's Physiology," *Bulletin of the History of Medicine* 20 (1946): 19.
120 Laromiguière, a late ideologue, postulated the effects of the brain of the soul but also underlined the activity of the soul and its action on the brain. Pierre Laromiguière, *Leçons de philosophie*, vol. 1 (Paris: Brunot-Labbé, 1815), 91.
121 Crisóstomo Lafinur, "Otro comunicado," *El Americano* no. 29, 8 October 1819, 7–11.

Paula Castañeda, the fiercest critic of Rivadavia's cultural program, ended in a celebrated public reconciliation.[122]) The heir to Lafinur's chair of philosophy, Fernández de Agüero, was of a different stamp.

Juan Manuel Fernández de Agüero, born in Cantabria, arrived as a child in Río de la Plata. He studied philosophy at the College of San Carlos and theology in Santiago de Chile, after which he was ordained as a priest. In 1820, he obtained the chair of philosophy, which he held until the fall of Rivadavia in 1827; since 1823 he was elected to the Buenos Aires legislature. Fernández de Agüero's course consisted of the three parts of Ideology: elementary (logic), abstractive (metaphysics), and oratory (rhetoric).[123] Like his predecessor, Fernández de Agüero copied verbatim large portions of his lectures from the works of Destutt de Tracy and Cabanis.[124] The last chapter of the course of metaphysics discusses the nature of the soul. Fernández de Agüero's claim that knowledge of unseen things should come from "the great book of nature" and not from Revelation is a defence of empiricist philosophy as well as of natural religion.[125] Despite his protestations to the contrary, he shows himself as a follower of Cabanis when he affirms that "the physical and the moral are confounded in their origin…the moral is nothing else but the physical considered under certain points of view."[126] Fernández de Agüero declares that sensibility does not require a cause more than gravity does.[127] There is an "animating principle" in the human being, he says, that is almost undistinguishable from "the first cause of the universe."[128] His conception of the "First Cause" as a power and intelligence that animates and vivifies nature has the ring of Stoic natural philosophy, which is in line with his remark that the notion of creation in the book of Genesis is just a poetic allegory of the deity's causal powers.[129] The goodness, wisdom, and power of the deity do not recognise other limits than "those of nature, in which they are inscribed."[130] These ideas are mostly taken from Cabanis, who in his *Lettres sur les causes premières* (1806–1807, published 1824) had affirmed

---

122 Juan W. Gez, *El Dr. Juan Crisóstomo Lafinur* (Buenos Aires: Cabaut, 1907), 41–44.
123 Juan M. Fernández de Agüero, *Principios de Ideología*, ed. Jorge R. Zamudio Silva, 3 vols. (Buenos Aires: Universidad de Buenos Aires, Facultad de Filosofía y Letras, Instituto de Filosofía, 1940).
124 Torchia Estrada, *La filosofía en la Argentina*, 88–95.
125 Fernández de Agüero, *Principios de Ideología*, 2:108–9.
126 Ibid., 2:29–30, 106.
127 Ibid., 2:108.
128 Ibid., 2:109.
129 Ibid., 2:108.
130 Ibid., 2:109.

that the fundamental forces of the world were the vital emanations of a cosmic reason.[131]

Fernández de Agüero saw Jesus Christ as a philosopher fighting against the theology of his time.[132] In his view, the dogma of Christ's divinity was detrimental to the religion and moral teachings of Jesus.[133] Fernández advocated a "moral and interior" religion, based upon the knowledge of ourselves and our own hearts.[134] Furthermore, the basis of morals are physiological, not theological: Jesus's injunction to love one another was the manifestation of "a physiological principle of sympathy."[135] The rector of the university, Father Antonio Sáenz, declared these ideas as "impious doctrines contrary to the state's religion."[136] Even Rivadavia reprimanded Fernández de Agüero in private for his unorthodoxy. At the height of the controversy, in July 1824, Sáenz locked the door of the classroom to impede the unruly professor from teaching. But it sems that Fernández de Agüero had the backing of the faculty and in the end the government confirmed him in his functions.[137] At the height of the conflict, Sáenz told the government that many students' fathers had asked him to set up an alternative course of philosophy in order to protect their children from "dogmatic atheism and materialism."[138] In one of his works, Esteban Echeverría, Argentina's foremost Romantic poet, recalls his youth when "In the capital city of Buenos Aires / the professors taught us to doubt / of God, of virtue, of heroism / of the good, of justice, and of ourselves."[139] He adds a note indicating that he is referring to "Condillac's and de Tracy's sensualist philosophy and Bentham's principles of legislation...doctrines which imply materialism and atheism."[140]

---

131 Sturm, *Cabanis*, 7.
132 Fernández de Agüero, *Principios de Ideología*, 2:150.
133 Ibid., 2:154.
134 Ibid., 2:156.
135 Ibid., 2:139, 146.
136 Antonio Sáenz, letter to governor Las Heras, 30 July 1824, cited in Fernández de Agüero, *Principios de Ideología*, 3:179–80.
137 See the documents related to this polemic in Fernández de Agüero, *Principios de Ideología*, 3:177–88; Gutiérrez, *Noticias históricas*, 121–26; Halperin Donghi, *Historia de la Universidad de Buenos Aires*, 37–38.
138 Letter from Antonio Sáenz to Gregorio Las Heras, 7 September 1834, in Fernández de Agüero, *Principios de Ideología*, 3:186–87.
139 "Allá en la capital de Buenos Aires / a dudar me enseñaron los doctores / de Dios, de la virtud, del heroísmo, / del bien, de la justicia y de mí mismo." Esteban Echeverría, "Avellaneda," in idem, *Obras completas*, ed. Juan M. Gutiérrez, vol. 1 (Buenos Aires: Imprenta y Librería de Mayo, 1870), 340.
140 Ibid., 1:437. In this poem, Echeverría's own religious views move ambiguously between Christianity and natural religion.

A disadvantageous peace treaty to end the war between the United Provinces with Brazil over the territory of Uruguay precipitated the fall of Rivadavia. The change of ideological orientation brought by the coming to power of Federalist governor Manuel Dorrego marked the end of Fernández de Agüero's course, seen as "contrary to religion and morals."[141] We have seen that Manuel Moreno implied that in that course "youth [was] introduced to deism."[142] Moreno's adversaries challenged him to name but just one student who had turned into an atheist or materialist as a result of hearing Fernández de Agüero's courses and did not lose the opportunity to accuse the former of "awakening religious fanaticism with treacherous hypocrisy."[143] In this squabble between the Federalist Moreno and the mouthpieces of Rivadavia's centralist party, religion was used to serve political purposes. The fall of Rivadavia signalled a serious defeat for the Unitarians. During the following two and a half decades the Federalists would have the upper hand. The United Provinces of Río de la Plata were organised under the hegemony of Buenos Aires, ruled by the strong hand of Rosas (1829–1832 and 1835–1852).

Rosas was a cattle baron who owned huge extensions of land—as governor, he represented represented the landed interests of the cattle raisers and exporters of the province of Buenos Aires and at times opposed them.[144] On the basis of an astonishing horsemanship, a visceral instinct for power, and a neat calligraphy, he raised to power leading militias of gauchos and Indians and eventually run the province with the efficiency and the cold ruthlessness of an *estanciero*. In his road to power "he became a regional chieftain, sough and rewarded elite allies, seized the state, established a personal dictatorship, and survived through violence."[145] Rosas and the provincial caudillos presented themselves as champions of tradition and Catholic religion against the liberal Unitarians. The Church under Rosas was highly politicised—his portrait was placed on the altars and most of the priests of Buenos Aires became vocal supporters of the Federal party: bishop Medrano wore the official red badge and instructed the lower clergy to preach obedience to the government. It has been argued that the Federal-

---

141 See the transcription of documents in Fernández de Agüero, *Principios de Ideología*, 3:205 (Letter of the rector Valentín Gómez to Minister Manuel Moreno, 10 November 1827).
142 [Manuel Moreno], *Impugnación a la respuesta*, 51.
143 [Juan Cruz Varela], "Universidad," *El Granizo. Diario político, literario y comercial* no. 9, 8 November 1827.
144 Raúl O. Fradkin and Jorge Gelman, *Juan Manuel de Rosas. La construcción de un liderazgo político* (Buenos Aires: Edhasa, 2017).
145 John Lynch, *Argentine Caudillo. Juan Manuel de Rosas* (Wilmington, DE: Scholarly Resources Books, 2001), 163.

ists' defence of religion was not only a political strategy but a genuine result of "the provocative and unpopular character of the liberal reform...The political wisdom of Rosas lay in his adjustment both of the rough religiosity of the common man and his sense of nationhood."[146] Rosas's Gallicanism was different in style from that of Rivadavia, but the idea of having a state-controlled Church was the same as well as his sense of using religion to discipline society. Rosas gave part of the convent of Santo Domingo back to the Dominicans (the Museum was not moved), but in exchange for official protection he required that the friars should proceed "faithfully and decidedly in favour of the cause of Federalism."[147] Gutiérrez blamed Rosas for having "thrown out to the streets the laboratories of physics and chemistry, the Museum of natural sciences, and Mossotti's telescopes in order to free space for the cloisters and the cells of the friars."[148] This was an exaggeration arising from the author's anticlericalism, but it is true that Rosas never showed any interest in the sciences.

By the last decade of the nineteenth century, Catholic groups close to Pius IX's anti-liberalism saw favourably on Rosas and his alleged defence of religion. In 1899, a Jesuit wrote that "the families in favour of him [Rosas] were and still are Catholic. The Unitarians were all liberals and their leaders were also Freemasons."[149] Not all Catholics followed this path. José Manuel Estrada (more on him in the next chapter) who supported the tradition of political liberalism of the Unitarians, depicted Rosas as "a tyrant as to character and nature, a tyrant since childhood, a tyrant at heart, and from his mother's womb."[150]

Conclusions

In Rivadavia's regime, a double process took place: science was disconnected from religion and took on secular meanings. At least, this is what the ruling elite spreading the new ideas aspired to accomplish. The government's promotion of scientific institutions should be seen as an integral part of a larger project for an enlightened city. If science constituted the epistemological model of En-

---

[146] Ivereigh, *Catholicism and Politics in Argentina*, 45–46.
[147] Quoted in Zuretti, *Nueva historia eclesiástica*, 252.
[148] Gutiérrez, "Las restauraciones religiosas," 415.
[149] Pablo Hernández to Rafael Pérez, 11 January 1899, quoted in Cayetano Bruno, *Historia argentina* (Buenos Aires: Don Bosco, 1977), 500.
[150] José Manuel Estrada, *Lecciones sobre la historia de la República Argentina*, vol. 2 (Buenos Aires: Librería del Colegio, 1896), 445.

lightenment, it is only logical that Rivadavia would foster scientific culture to underpin his political project.

To say that Rivadavia's religious policy was "doctrinaire, provocative to his enemies, and irrelevant to his main program [of economic reform]" may be only slightly overstated.[151] But although his more radical associates would have welcomed the banishment of the religious orders, the government's ecclesiastical reform was not intrinsically anti-religious. It was an attempt at establishing an enlightened Catholic church which served two purposes: on the one hand, the new legislation subjected the church to the state along Gallican lines; on the other hand, it empowered the parish priests (seen as educators of the rural population) while undermining the power of the regular clergy, associated to baroque forms of religiosity. As a Protestant traveller remarked with satisfaction, the first thing which occurred to Rivadavia on his coming to power "was to clip the wings and everyway curtail the influence of the cloistered clergy... [while at the same time] everything was done to raise the secular clergy into the eyes of the people."[152] Convents, the lairs of lazy and ignorant friars, should be turned into lighthouses of scientific learning. The strong message of the transformation of "useless" ecclesiastical space into a citadel of science was not lost on those apostles of secular culture such as Juan M. Gutiérrez. Destutt de Tracy congratulated Rivadavia for his ecclesiastical reform, which he saw as part of the bloody war between "the supporters of superstition" and "those who use their power... [to further] the progress of *lumières*, freedom, and civilisation."[153]

Rivadavia's scientific "Renaissance" was made possible by the political situation in post-Napoleonic Europe. French and Spanish officers trained in Napoleon's army and Italian liberals escaping from the regimes of the Holy Alliance arrived in Río de la Plata seeking a living. Together with acceptable but not outstanding qualifications, they brought with them the ideologies of Enlightenment and liberalism in which they had been nurtured. The meagre results of Rivadavia's scientific policy could be attributed to the government's chronic lack of funds, the lack of interest of the local students who preferred to follow lucrative careers, and the difficulty for hiring first-class teachers. It is telling that Mossotti, the only authentic scientist among the teachers hired by the government, avoided ideological pronouncements during his tenure in the University of Buenos Aires. Carta, perhaps the least qualified among them, was the most outspoken advocate of secular values.

---

[151] Lynch, *The Spanish American Revolutions*, 76.
[152] Caldcleugh, *Travels in South America*, 1:193.
[153] Destutt de Tracy to Rivadavia, 2 December 1822, cited in Piccirilli, *Rivadavia y su tiempo*, 2:324.

The reception of French *idéologie* in Río de la Plata is indicative of the scope and the limits of the intellectual renewal stimulated by Rivadavia and his followers. *Idéologie* was a late product of Enlightenment, verging on materialism. This was the first time that a secular philosophy was taught in Buenos Aires and the arguments over it show that the challenge did not pass unnoticed to the more conservative sectors of society. It should be kept in mind that the courses taught in the 1820s were derivative versions of a current of thought which had by that time faded in Europe, swept away by the warm breeze of Romanticism. More to the point, although *idéologie* took on the trappings of science, it was not science. In his debate with Lafinur, Argerich was able to score a point when he cited (and misconstrued) a scientific authority such as Magendie.

The net results of Rivadavia's science in Río de la Plata were rather poor but their symbolic resonances were large. From that moment on, science would be vaguely associated with secularism. In a sense, this was the defeat of the project of Catholic enlightenment, which had shaped learning and science in colonial times and during the first decade of independent life of the country. Local history of science would be written as starting with Rivadavia. This kind of interpretation transformed the link between science and secularism advanced by the actors into a historical given. Juan M. Gutiérrez's history of higher teaching in Buenos Aires (1868) is a collection of documents organised according to academic fields with each section preceded by a short introduction. Gutiérrez was a talented intellectual, political figure, and promoter of the sciences whose anticlericalism increased as he grew older—Menéndez Pelayo called him "the most complete man of letters in those parts of the New World."[154] In his work, Gutiérrez laid the basis for an influential narrative about the development of learning in the country: "Physico-mathematical sciences, which reveal the created world and the greatness of God...made their appearance with the foundation of the University [of Buenos Aires]."[155] All that went before were "the dark depths of our sorrowful past."[156] (Gutiérrez omits the Jesuit tradition of astronomical observation and the circle of priests devoted to natural history in Río de la Plata.) In his view, the enlightened clerics that acted during the Independence such as Maciel, Chorroarían, and Sáenz, were unable "to follow the progressive movement of ideas for they were attached to their [religious] state and to the scholastic discipline of sacred theology." They "were alien to the observational sciences, to cal-

---

[154] Marcelino Menéndez Pelayo, ed., *Antología de poetas hispano-americanos*, vol. 4 (Madrid: Sucesores de Rivadeneyra, 1895), clxxx.
[155] Gutiérrez, *Noticias históricas*, vi.
[156] Ibid., xii.

culus, and incapable of using instruments of physics or geodesics."[157] In this view, the convent of Santo Domingo was an emblem of the new age: "Reform and voluntary secularisation had left those cloisters bereft of their former inhabitants and by a transformation in harmony with the changing times, theologial studies gave way to the study of the marvels of nature, *which declare the glory of God*, according to the expression of the Scriptures" (emphasis in the original).[158] Science, according to Gutiérrez, could only flourish in the climate of an enlightened natural religion. The next chapter will argue how in the 1880s, with the help of Darwinism and the conflict thesis, scientific discourse in Argentina would lose the last connection to deism and turn into a versatile instrument at the service of a secularist political agenda.

---

157 Ibid., 845.
158 Ibid., 909–10.

Chapter 4
The "Conflict Thesis," Darwin, and Secularising Politics in Late Nineteenth-Century Argentina

In his entry in the *Journal of Researches* dated 23 November 1833, Darwin noted that "mockery of all religion" was one of the vices that stained "the higher and more educated classes" in Buenos Aires.[1] Eighty years later, the Scottish traveller John Foster Fraser passed an even harsher judgement on the religious climate of the southern metropolis: "I have called Buenos Aires a pagan city. So it is."[2] In the last three decades of the nineteenth century, Argentina's capital city went through a process of secularisation which resulted in a decrease in the social influence of the Catholic Church. The inchoate confrontation between science and religion during Rivadavia's regime was in full fledge in the 1880s. In this chapter, I shall focus on two aspects of this situation which can be seen as two sides of the same coin: the local reception of the "conflict thesis" and of Darwin's evolutionary theory. Both contributed to building a rhetoric that drew upon the authority of science to legitimize the government's secularist agenda.

John W. Draper's *History of the Conflict between Religion and Science* arrived in the country in the midst of political battles over the shape of the future relationships between the state and the Catholic Church. In the decade between 1875 and 1885 variants of the conflict thesis were expounded, discussed, and used as rhetorical weapons in the battles over the issue of religious teaching in elementary schools. After going through some historical background, the first part of this chapter analyses the discussions over the conflict thesis between liberal secularists and Catholics in newspaper articles, public speeches, parliamentary debates, and other forms of public discourse during that period. Against the backdrop of a weak institutional church, a vigorous growth of nascent scientific institutions, and a cultural atmosphere permeated by positivism, the opposing parties argued about the conflict thesis while each reclaimed for itself the legitimacy of science. The episode permits a close look at how the intellectual leaders who conceived the project of a secularised state utilised science-based philosophies for purposes of political argument and ideological legitimation.

This secularising trend was coterminous with the arrival and assimilation of evolutionary theory, discussed in the last sections of the chapter. Research on

---

[1] Charles Darwin, *Journal of Researches* (London: Henry Colburn, 1840), 183.
[2] John F. Fraser, *The Amazing Argentine. A Land of Enterprise* (New York: Funk & Wagnalls, 1914), 29.

the reception of Darwinism in Argentina is not lacking.³ Two collections of sources in translation attest to the global interest in the fortunes of the theory in the southern cone of the Americas.⁴ Notwithstanding these important contributions, the key issue of the role played by religious and non-religious convictions in this process has remained largely unexplored. But issues of belief and unbelief bore decisively on the way in which evolutionism was received in Argentina and are far from being a sideshow. Just as the conflict thesis, evolutionism was part of the rhetorical apparatus which sought to justify the implementation of secularist educational policies.

Historical Background

On 3 February 1852, Rosas was defeated in the battle of Caseros by Justo José de Urquiza (a Federalist caudillo and powerful cattle-raiser of the province of Entre Ríos), allied with troops of Uruguay and Brasil. The constitutional assembly convened as a result voted a constitution and Urquiza was elected president of the country. Since the province of Buenos Aires was not able to impose its hegemony, it became an independent state. A few years later, in October 1859, the army of the interior provinces commanded by Urquiza defeated the troops of Buenos Aires led by Bartolomé Mitre and Buenos Aires unwillingly joined the Confederation. The ensuing unstable equilibrium came to a head when Mitre and Urquiza met again in the battlefield in September 1861. This time Mitre was the victor and he was elected president of the whole reunited nation (1862–1868) with the province of Buenos Aires now in a forceful position. He was succeeded in office by Sarmiento (1868–1874) and Nicolás Avellaneda (1874–1880).⁵

Sarmiento was the premier representative of what is usually called the "Generation of 1837," to which also belonged Juan Bautista Alberdi, Vicente F. López,

---

3 Marcelo Montserrat, "The Evolutionist Mentality in Argentina: an Ideology of Progress," in *The Reception of Darwinism in the Iberian World*, ed. Thomas Glick, Miguel Ángel Puig-Samper and Rosaura Ruiz (Dordrecht: Springer, 2001), 1–27; Adriana Novoa and Alex Levine, *From Man to Ape. Darwinism in Argentina, 1870–1920* (Chicago: Chicago University Press, 2010).
4 Alex Levine and Adrian Novoa, eds., *¡Darwinistas! The Construction of Evolutionary Thought in Nineteenth Century Argentina* (Leiden: Brill, 2012); Leila Gómez, ed. and Nicholas F. Calloway, trans., *Darwinism in Argentina: Major Texts (1845–1909)* (Lewisburg: Bucknell University Press, 2012).
5 James R. Scobie, *La lucha por la consolidación de la nacionalidad argentina, 1852–1862* (Buenos Aires: Hachette, 1964); Eduardo Míguez, *Bartolomé Mitre. Entre la nación y la historia* (Buenos Aires: Edhasa, 2018).

Juan M. Gutiérrez, Esteban Echeverría, Bartolomé Mitre, and Félix Frías.[6] These intellectuals fought with their pens against Rosas while exiled in Montevideo or Chile. Usually identified as "Romantics," most of them reacted against the credo of Enlightenment in which they had been brought up—others, like Gutiérrez, never entirely abandoned it. Having been educated in the philosophy of Condillac and Bentham, they schooled themselves in the French Romantics, Saint-Simon, the liberal Catholics Lammenais and Montalembert, the eclecticism of Victor Cousin and Théodore Jouffroy, and the philosophy of history of Herder and Vico.[7] In 1843, Sarmiento recalled his intellectual path: "Voltaire, Rousseau, Montesquieu, Diderot, Buffon and that constellation of great men seem ignorant compared to Cousin, Villemain, Hugo, Dumas, Jouffroy, Saint-Hilaire, and so many others."[8] Romanticism gave these thinkers the possibility of amalgamating the establishing of a liberal order with the building of a national identity.[9]

The members of the Generation of 1837 held different religious positions. Frías was a Catholic; Gutiérrez and López were prominent members of Masonic lodges; others were deists and supporters of conventional Christianity as a force of social order. Echeverría's *Handbook of Moral Teaching for Elementary Schools* (1846), written in Montevideo, begins with a section on God, "the creator and conservator of the universe whom we owe love and worship." He can be seen everywhere, "in heavens, on Earth, in the ants, in the human being, in all His works;" God's laws are "the laws of the order [of the universe]."[10] The poet Echeverría, a Romantic deist, saw no conflict between science and religion for they concerned different human spheres. In his *Socialist Dogma*—the manifesto of the Generation of 1837—he affirms that religion is "the moral ground on which society rests, the divine balm for the heart, the source of our future hopes," whereas science "teaches man [sic] to know himself, to penetrate the mysteries of nature, to raise his thoughts to the Creator and thus find the ways of individual and social improvement and perfection."[11] Bartolomé Mitre, a key figure in

---

[6] William H. Katra, *The Argentine Generation of 1837: Echeverría, Alberdi, Sarmiento, Mitre* (London: Associated University Press, 1996).

[7] Juan B. Alberdi, "Mi vida privada," in idem, *Escritos póstumos*, ed. Francisco Cruz, vol. 15 (Buenos Aires: Imprenta J. B. Alberdi, 1900), 294–95.

[8] Domingo F. Sarmiento, "Memoria sobre ortografía americana," in idem, *Obras*, ed. Augusto Belin Sarmiento, vol. 4 (Buenos Aires: Félix Lajouane, 1886), 41.

[9] Míguez, *Bartolomé Mitre*, 70.

[10] Esteban Echeverría, *Manual de enseñanza moral para las escuelas primarias del Estado Oriental*, in idem, *Obras completas*, ed. Juan M. Gutiérrez, vol. 4 (Buenos Aires: Imprenta y Librería de Mayo, 1873), 346.

[11] Esteban Echeverría, *Dogma socialista de la Asociación de Mayo* (Montevideo: Imprenta del Nacional, 1846), 56–57.

the consolidation of Argentina as a modern nation, kept his youthful Catholic faith while joining the ranks of Italian liberal emigrés in Montevideo.[12] Welding the gospel of progress to the warfare view of science and religion, Alberdi claimed that the new South-American republics needed "engineers, geologists, and naturalists more than lawyers and theologians. They will improve with roads and artesian wells…not…with sermons and legends."[13] The Catholic Frías held a contrasting position, in part a product of his detestation of the 1848 French revolution and his belief that economic progress should follow "a style of social and political coexistence based on religion."[14] In an 1851 letter to Alberdi written from Paris and later published under the title of "Religion and science," he affirmed that "virtue is better than science and the latter without virtue is fearsome." Criticising the belief that religion should be left to "the masses, children, and women" Frías contended that the scepticism of the ruling elite would eventually turn into "the gross materialism that is invading France;" the only sound "basis of a society aspiring to democracy," he claimed, was "the religious education of the people."[15] Frías's Romantic liberal Catholicism was a minority view. Sarmiento, Gutiérrez, and those who contributed to lay the grounds of scientific culture in Argentina set themselves against religious education, which they saw as a relic of Spanish colonial times which should be superseded by scientific progress. Their followers, the Generation of 1880, pushed even further this opposition in a positivist key.

The decades of the 1870 and 1880s were those of the definite political organisation of the country. By 1880, the year when General Julio A. Roca took the presidency, Argentina had acquired much of its final political organisation: the question of the southern frontier had been settled and vast territories opened for colonisation, Buenos Aires became the federal capital of the country, and the provincial militias and their caudillos would soon be inactivated for good. The emerging conservative order that was to rule the country for more than three decades was aptly summarized in the lemma proclaimed by Roca in his inaugural

---

12 Míguez, *Bartolomé Mitre*, 76–85.
13 Juan Baustista Alberdi, *Organización de la Confederación Argentina*, vol. 1 (Besanzon: J. Jacquin, 1858), 33.
14 Tulio Halperin Donghi, *Una nación para el desierto argentino* (Buenos Aires: Prometeo, 2005), 55. See also Horacio Sánchez de Loria, *Félix Frías. Acción y pensamiento político* (Buenos Aires: Quorum, 2004).
15 Félix Frías, "La religión y la ciencia," in idem, *Escritos y discursos*, ed. Pedro Goyena, vol. 1 (Buenos Aires: Imprenta y Librería de Mayo, 1884), 30–32.

address: "Peace and administration."[16] During his term of office and that of his successor, Congress voted a series of measures that reduced the areas of society under the tutelage of the Catholic Church, such as the law that introduced the Registry Office thus abolishing the parochial register of births and deaths (1884) and that of civil marriage (1888). The peak of the tensions between state and Church was reached in the debates conducive to the reform that regulated elementary education (1883–1884), which from that moment on excluded religious teaching for all practical purposes.[17] This secularising wave reached a climax in the expulsion of the Apostolic Delegate Msgr Luigi Matera and the rupture of diplomatic relationships between the country and the Holy See in January 1885. Ten years before that, on 28 February 1875, a mob had burned to ashes the Jesuit College of El Salvador in Buenos Aires.[18] The whole period was coloured by a heated battle between the "liberal" and "Catholic" sectors of the social and intellectual elite that ruled the country and shaped public opinion.[19]

In Argentina, as in the rest of Ibero-America, the process of secularisation run along the lines of the French model of *laïcité*.[20] By 1890—a year marked by political revolution and financial breakdown—it had begun to subside; from then on, the conservative government would seek some kind of accommodation with the Catholic hierarchy in view of the need to face the challenge of the rising socialist movement.[21] It should be noticed that "secularisation" here is taken as "the appropriation…by secularist institutions of functions that traditionally had been in the hands of ecclesiastical institutions."[22] This is Casanova's third sense of secularisation, equivalent to Taylor's first mode, namely the "re-

---

16 [Julio A. Roca], "Discurso del nuevo presidente," *La Prensa*, 13 October 1880. For the political scenario, see Natalio R. Botana, *El orden conservador. La política argentina entre 1880 y 1916* (Buenos Aires: Hyspamérica, 1986); Ezequiel Gallo, "Society and Politics, 1880–1916," in *Argentina since Independence*, ed. Leslie Bethell (Cambridge: Cambridge University Press, 1993), 79–111.
17 For an overview of the conflict, see Ivereigh, *Catholicism and Politics in Argentina*, 49–63; Michael A. Burdick, *For God and the Fatherland. Religion and Politics in Argentina* (New York: The State University of New York Press, 1995), 21–25; and Lee B. Kress, "Argentine Liberalism and the Church under Julio Roca, 1880–1886," *The Americas* 30 (1974): 319–40.
18 Bruno, *Historia de la Iglesia*, 11:105–19.
19 Néstor T. Auza, *Católicos y liberales en la generación del ochenta* (Buenos Aires: Educa, 2007).
20 See, for example, Baubérot, *Histoire de la laïcité en France*. For a sociological view of laicisation in Argentina, see Fortunato Mallimaci, *El mito de la Argentina laica. Catolicismo, política y estado* (Buenos Aires: Capital Intelectual, 2016).
21 Ivereigh, *Catholicism and Politics in Argentina*, 62.
22 Casanova, *Public Religions in the Modern World*, 13.

treating [of religion] in public space."²³ Church historians have argued that the confrontations regarding the delimitation of temporal and religious spheres of competence between 1870 and 1890 in Argentina could be seen as a result of the tensions arising from the consolidation of the national state on the one hand and the "Romanisation" of the Catholic Church as a result of a universal trend on the other.²⁴ Others have claimed that the privilege of *patronato*—inherited from the former imperial metropolis and not resolved until the Concordat of the 1960s—might have been at the base of the protracted and oscillating conflict between an "Ultramontane" church and a "Gallican" state.²⁵ Much of this literature coincides in replacing an inherited view of two neatly distinct, opposing camps (Catholic Church and state) with a more nuanced view of a complex process of political and symbolic transactions—as one author put it, "the relationship between fin-de -siècle Catholicism and modernisation in the 1880s was made of grey tones."²⁶

Three factors figure prominently in the background of our story. Firstly, the massive waves of immigration, which brought with them large groups of anticlerical southern Europeans and simultaneously a large number of priests, nuns, and members of Catholic congregations from all over Europe. The arrival of Protestant immigrants posed new challenges to a traditionally Catholic society, to which liberals and Catholics provided different answers. The former promoted this kind of immigration and in Parliamentary debates raised this issue as one of the main arguments in favour of non-confessional education and religious pluralism.²⁷ Sarmiento, who shaped much of the debate on secularisation, famously arranged the arrival in Argentina of more than fifty New England Protestant teachers.²⁸ The presence of these educationists in a secondary school for

---

23 Taylor, *A Secular Age*, 15; see also ibid., 1–2.
24 Roberto Di Stefano and Loris Zanatta, *Historia de la Iglesia en la Argentina* (Buenos Aires: Grijalbo, 2000), 332–36.
25 Ignacio Martínez and Diego Mauro, "Ctéato y Éurito. Iglesia, religión y poder político en la Argentina en el siglo XX," in *Fronteras disputadas: religión, secularización y anticlericalismo en la Argentina (siglos XIX y XX)*, ed. Roberto Di Stefano and José Zanca (Buenos Aires: Imago Mundi, 2016), 1–34. Cf. also the more extensive treatment of this approach in Ignacio Martínez, *Una nación para la Iglesia argentina* (Buenos Aires: Academia Nacional de la Historia, 2013).
26 Miranda Lida, *Historia del catolicismo en la Argentina entre el siglo XIX y el XX* (Buenos Aires: Siglo XXI, 2015), 34.
27 Di Stefano and Zanatta, *Historia de la Iglesia*, 346.
28 Georgette M. Dorn, "Sarmiento, the United States, and Public Education," in *Sarmiento and His Argentina*, ed. Joseph T. Criscenti (Boulder: Lynne Rienner, 1993), 77–89; Julyan G. Peard, *An American Teacher in Argentina. Mary Gorman's Nineteenth-Century Odyssey from New Mexico to*

girls in the city of Córdoba triggered in 1884 a political conflict between the vicar of the chapter Gerónimo Clara and Roca's minister of education, Eduardo Wilde.[29] The second aspect that should be emphasized is the important role played by the lodges of Freemasons in the battles for secularisation, which has been extensively considered in recent histories of anticlericalism in Argentina.[30] Most of the prominent liberals of the period belonged to them. In the third place, the protagonists of the Catholic side were laypeople, not priests or members of the hierarchy. They set up religious associations, founded periodicals, and launched a Catholic political party which did not survive the conflict.[31] At one point of the Parliamentary deliberations about the issue of the laicisation of education in elementary schools, in August 1883, a demonstration of aristocratic ladies was able to temporarily turn the tables of the proceedings.[32] The Catholic leaders belonged to the same social elite as their occasional contenders, from which they were separated only on account of the question of secularisation; all of them saw themselves as "liberals," in the sense of the defence of a republican democratic state with civil liberties in opposition to what they understood had been the tyranny of Juan M. de Rosas (1835–1852).[33] It was against this backdrop that Draper's book was received in Buenos Aires and that the relationship between religion and science became a staple issue in the political debate over secularisation. Draper's formulation of the conflict thesis was not the only one in scene: it coexisted and at times merged with a more diffuse notion originated in the French Enlightenment and developed by Comte's positivism, which also saw science and religion as incompatible (see the Introduction).

---

the Pampas (Lewisburg: Bucknell University Press, 2016). For Protestant immigration in this period, see Monti, *Presencia del protestantismo en el Río de la Plata*, 107–92.

29 Auza, *Católicos y liberales*, 193–96.

30 Ema Cibotti, "La masonería frente a las leyes laicas," *Todo es Historia* 34, no. 405 (2001): 77–82; Roberto Di Stefano, *Ovejas negras. Historia de los anticlericales argentinos* (Buenos Aires: Sudamericana, 2010), 197–212; Emilio J. Corbière, *La masonería. Política y sociedades secretas* (Buenos Aires: Sudamericana, 1998), 252–54; Yolanda E. de Paz Trueba, "Masonería y sociabilidad en el centro y sur de la campaña bonaerense. Fines del siglo XIX, principios del XX," *Prohistoria* no. 16 (July–December 2011): 1–10.

31 Di Stefano and Zanatta, *Historia de la Iglesia*, 352.

32 Auza, *Católicos y liberales*, 146–47.

33 It is customary to think of this elite in terms of a "generation," see for example David W. Foster, *The Argentine Generation of 1880. Ideology and Cultural Texts* (Columbia and London: University of Missouri Press, 1990). Recent approaches favour a less homogeneous view, see Paula Bruno, *Pioneros culturales de la Argentina* (Buenos Aires: Siglo XXI, 2011), 16–17; this book provides updated bibliographies of Eduardo Wilde and Eduardo L. Holmberg.

The 1878 declaration of principles that accompanied the creation of the aggressively secularist Liberal Club in Buenos Aires affirmed that "all the advances and progress of the natural sciences will serve us as a text in order to highlight the shameful errors used to alter the nature of our beliefs and disown the emancipation of modern reason."[34] Science and progress were the bywords of the times and "everybody felt entitled to lighten his [sic] cigar in the torch of civilisation."[35] The establishment of new scientific institutions between 1870 and 1900 fed the imaginary of progress and modernity heralded by the positivist-minded secularising elite. "Missionaries of science" from Central Europe, the Francophone world, and the United States were either Protestants or agnostics. Although lay Catholics in the public sphere—mostly lawyers—took a critical distance toward this scientific hustle (which they perceived as hostile and never entirely understood), they still manifested enthusiasm for science. The ecclesiastical hierarchy, fighting for the survival of a church on the defensive trying to come to grips with the ongoing Romanisation, had more pressing issues with which to deal.[36]

Science in Argentina was a key element in the material and symbolic blueprint of nation building. It would not be an exaggeration to say that Sarmiento almost single-handledy introduced modern science in Argentina, as part of the credo of progress and public education, which was the leitmotif of his political and public life.[37] In 1874, with occasion of the inauguration of a railroad in the province of Entre Ríos, he summarised his civilising program: "What we need is railways, telegraphs, education, and immigration."[38] Three years before, in the inauguration of the National Observatory, he had claimed that Argentines "should renounce the rank of nation and the right to be called a civilised people unless they take their post in the files of progress and the movement of the natural sciences."[39] By the decade of 1870, modern scientific institutions clustered around two centres: the interior city of Córdoba, an enclave of tradition and Ca-

---

34 "Prospecto del 'Club Liberal'," *El Librepensador*, 1st August 1878.
35 Carlos Guido y Spano, *Autobiografía* (Buenos Aires: Kapeluz, 1969), 40.
36 Ivereigh, *Catholicism and Politics in Argentina*, 54–55.
37 Marcelo Montserrat, *Ciencia, historia y sociedad en la Argentina del siglo XIX* (Buenos Aires: CEAL, 1993).
38 Domingo F. Sarmiento, "Inauguración del ferro-carril de Concordia (Entre Ríos)" (29 March 1874), in idem, *Obras*, ed. Augusto Belin Sarmiento, vol. 21 (Buenos Aires: Imprenta Mariano Moreno, 1899), 361.
39 Domingo F. Sarmiento, "Discurso de inauguración del Observatorio Astronómico" (24 October 1871), in idem, *Obras*, ed. Augusto Belin Sarmiento, vol. 21 (Buenos Aires: Imprenta Mariano Moreno, 1899), 326.

tholicism, and cosmopolitan Buenos Aires.⁴⁰ Córdoba was the home of the National Astronomical Observatory, headed by the American astronomer Benjamin A. Gould, and of the Academy of Sciences, a group of first-rate German scientists, who also taught at the local university. Both institutions had been created by Sarmiento, who saw in them outposts of northern civilisation in an otherwise barbaric Hispanic society.⁴¹

In Buenos Aires, the Public Museum was directed by the German naturalist Hermann Burmeister, also hired by Samiento. The Argentine Scientific Society, created in 1872 by the geographer and diplomat Estanislao Zeballos, was in the hands of engineers engaged in the transformation of Buenos Aires from colonial village to modern metropolis and of naturalists who led the exploration and geographical survey of the country, besides unearthing its paleontological treasures.⁴² Both groups joined in singing the praises of positivism and proclaimed the credo of biological evolution. In the decade of 1860, Juan M. Gutiérrez created the Department of Exact Sciences of the University of Buenos Aires (a school of engineering) and hired well-qualified Italian professors for teaching the courses.⁴³ In the 1880s, an ambitious Natural History Museum and a second astronomical observatory (for all practical purposes, a French foundation) were erected in the nearby city of La Plata, in which a national university was soon organised, with an aim at cultivating experimental science.⁴⁴ Later, in 1906, an

---

40 This paragraph is based on Asúa, *Una gloria silenciosa*; bibliographical references have been kept to a minimum.
41 Marcelo Montserrat, "Sarmiento, propulsor de la ciencia," *Ciencia e Investigación* 42, no. 5 (1988): 277–83; Lewis Pyenson, "La ciencia en Córdoba en el siglo XIX," in *Universidad Nacional de Córdoba. Cuatrocientos años de historia*, ed. Daniel Saur and Alicia Servetto, vol. 1 (Córdoba: Universidad Nacional de Córdoba, 2013), 251–81; John E. Hodge, "Benjamin Apthorp Gould and the Founding of the Argentine National Observatory," *The Americas* 28, no. 2 (1971): 152–75; Luis Tognetti, *La Academia Nacional de Ciencias. Etapa Fundacional: siglo XIX* (Córdoba: Academia Nacional de Ciencias, 2000); idem, *La Academia Nacional de Ciencias en el siglo XIX. Los naturalistas, publicaciones y exploraciones* (Córdoba: Academia Nacional de Ciencias, 2004).
42 Juan C. Nicolau, "Historia de la Sociedad Científica Argentina en el siglo XIX (1872–1900)," *Anales de la Sociedad Científica Argentina* 231, no. 1 (2002): 5–72.
43 Horacio H. Camacho, *Las ciencias naturales en la Universidad de Buenos Aires* (Buenos Aires: Eudeba, 1971), 35–61.
44 For the Buenos Aires and La Plata Museum, see Susan Sheets-Pyenson, *Cathedrals of Science. The Development of Colonial Natural History Museums during the Late Nineteenth Century* (Kingston and Montreal: McGill-Queen's University Press, 1988). For the La Plata Observatory, see Miguel de Asúa, "Historia de la Astronomía en la Argentina," in *Historia de la Astronomía Argentina*, ed. Gustavo E. Romero, Sergio A. Cellone and Sofía Cora (La Plata: Asociación Argentina de Astronomía, 2009), 1–19; Marina Rieznik, *Los cielos del sur. Los observatorios astronómicos de Córdoba y de La Plata, 1870–1920* (Rosario: Prohistoria, 2011).

Institute of Physics was established there, which staffed by German physicists enjoyed a brief spell as an international centre for theoretical physics.[45] From early on, medical doctors played a prominent role on the scientific scenario of Buenos Aires.[46] The Argentine Medical Circle, founded by rebellious students in 1872, never quite abandoned the imprint of its radical origins and just as the SCA, it remained a bulwark of secularist scientism. As to what concerns the upper echelons of the profession, Buenos Aires was a suburb of medical Paris. Most of the scientific institutions created in the 1890s were related to Public Health, an issue of primary importance as a result of urban demographic growth.

In the following sections, I explore the ways in which the conflict thesis was used in public discourse (the press, popular lectures, parliamentary debates, and official documents) in Buenos Aires during the period 1875–1885. We shall see how it was utilised to underpin a positivist worldview, in turn a legtimazing instrument of secularist politics. Catholic polemicists contested its validity and sought to claim the mantle of science for their cause. The cultural effervescence built around Draper and the conflict thesis is explained in terms of its functional adequacy with the arguments that justified a secularising political program. This debate run parallel to the reception of Darwinism in Argentina. Here, I shall consider a sample of the liberal and Catholic reactions to the advent of the theory of evolution within the context of the secularising impulses that transformed the country's urban society in the last two decades of the nineteenth century.

Debate over Draper in the Liberal Press

On 30 September 1877, the Buenos Aires newspaper *El Mosquito* [The Gnat] published an advertisement of the local "European Bookshop Jacobsen" offering for sale the recent Spanish translation of Draper's *Conflict between Religion and Science*.[47] The book was a success among the liberal readership of the satirical weekly, which took every opportunity to display a heavy handed anticlericalism. The arrival of Draper's book in Buenos Aires in the mid-1870s was marked by a series of reviews in the liberal newspaper *La Tribuna*. Their examination will

---

45 Lewis Pyenson, *Cultural Imperialism and Exact Sciences. German Expansion Overseas, 1900–1930* (New York: Peter Lang, 1985), 139–246.
46 Cantón, *La Facultad de Medicina*, vol. 3.
47 *El Mosquito*, 30 September 1877. Juan Guillermo Draper, *Historia de los conflictos entre la religión y la ciencia*, trans. Augusto T. Arcimís, with a prologue by Nicolás Salmerón (Madrid: Imprenta de Aribau, 1876).

allow us to discern three different kinds of reactions to the book. In March 1876, the Catholic daily *La América del Sud* published a two-part article that argued the inexistence of prehistoric human beings. It was reproduced without acknowledgement from a Spanish religious journal, which in turn had translated it from the original in *La Civiltà Cattolica*, the Rome-based Jesuit magazine with a strongly anti-Darwinian editorial policy.[48] The gist of the article is a mathematical argument tending to show that if human beings had come to existence in the remote past (300,000 years before Adam), the present number of individuals would exceed the spatial limits of the Earth by several orders of magnitude. *La Tribuna* treated this negation of the prehistoric past as one more example of religious "obscure night birds" smashing themselves against the shining beacon of light that was science.[49] Draper, said the author of the note, had recently given an excellent account of the fateful "antagonism between science and religion;" he confirmed this statement commenting upon the cases of Galileo and Bruno in the darkest possible tints. The answer of the Catholic newspaper consisted in a historiographical defence of the validity of the Genesis account on the basis of the testimonies of ancient civilisations. It was long (five parts) as it was naively pedantic—a favourite source of the author was Cesare Cantù's *Storia universale* (Turin, 1840–1847), on which the author drew to offer an anti-Draperian version of the Copernican story favourable to Catholic interests.[50] The liberal contender of *La Tribuna* saw fit to come back wielding again the book by "M[onsieur] Drapper."[51] Although not declared as such, his second article is little more than a translation from the review of the French edition of Draper's book that had appeared in the Parisian *Revue scientifique*.[52]

---

[48] "Algunas observaciones a los que defienden la teoría de los hombres prehistóricos," *La América del Sud*, 15 and 17 March 1876. Cf. "Ciencia prehistórica. Algunas observaciones a los que defienden la existencia de los hombres prehistóricos," *La Cruz* (Sevilla) 2 (1875): 616–26; "Una proposta ai fautori degli uomini preistorici," *La Civiltà Cattolica* 24 (1873): 265–68. For the anti-Darwinian campaign in *La Civiltà Cattolica*, see Mariano Artigas, Thomas F. Glick and Rafael A. Martínez, *Negotiating Darwin: The Vatican Confronts Evolution, 1877–1902* (Baltimore: The Johns Hopkins University Press, 2006).
[49] "La protesta contra la ciencia," *La Tribuna*, 17 March 1876.
[50] "Los sabios del día," *La América del Sud*, 18, 19, 21 and 22, 23 and 24 March 1876.
[51] "A propósito de una polémica," *La Tribuna*, 23 March 1876.
[52] John W. Draper, *Les conflits de la science et de la religion* (Paris: G. Baillière, 1875). The review in question appeared in *Revue scientifique de la France et de l'étranger*, 2nd series, 4, no. 50 (5 June 1875): 1170–72.

Six months after this first skirmish, the politician and veteran soldier Benjamín Villafañe, published a review of Draper's book in *La Tribuna*.[53] Draper is condemned as a pantheist whose materialist psychology of perception and memory conduces to denying the existence of an immortal spiritual principle in the human being. His book is qualified as incomplete philosophy and partial history, a "pamphlet" with a sectarian approach which ignores all the benefits that Christianity contributed to the progress of humanity.

Around a week later, *La Tribuna* began to publish a series of articles on Draper signed under the pseudonym "Pyreneus."[54] Its author was Alejo [Alexis] Peyret, a French republican, socialist, and Freemason who in 1852 immigrated to Argentina, where he developed a career as administrator of an agricultural colony in the province of Entre Ríos and engaged his intellectual talents in the defence of the secularist cause.[55] In his first note, the author obligingly follows the *History of the Conflict* and rehearses all of Christianity's crimes against knowledge. But he makes clear that Draper only gave literary expression to what was already known; he is not an interpreter, he just "speaks impartially." Unlike the British-American chemist-historian, the reviewer is as aggressive against Protestantism as against Catholicism. His article is less a commentary on Draper than an excuse to expound his own story, carved out of a deep anti-religious feeling. Christian religion was not only "the remora of humanity" and "anti-scientific," it is "anti-juridical," "anti-social," "anti-progressive," and "anti-economical." The last part of the review is a long argument tending to demonstrate that Christianity was no more than a development of the community of the Essenes, which in turn came out of Buddhism.[56]

---

53 [Benjamín Villafañe], "Los conflictos entre la ciencia y la religión. Juicio crítico de la última obra de Draper," *La Tribuna*, 19 and 20 September 1876. For Villafañe, see Ángel J. Carranza, "Benjamín Villafañe (nacido en Tucumán el 30 de marzo de 1819)," *Revista Nacional* 12 (1890): 259–285.
54 Pyreneus (Alejo Peyret), "A propósito de Draper," *La Tribuna*, 28 September and 6, 11, and 25 October 1876.
55 Carranza, "Benjamín Villafañe," 283. See Beatriz Bosch, "Alejo Peyret, administrador de la colonia San José," in *III Congreso de Historia Argentina y Regional, Santa Fe-Paraná, 1975* (Buenos Aires: Academia Nacional de la Historia, 1977), 53–72; Alberto Sarramone, *Alexis Peyret* (Buenos Aires: Biblos, 2016).
56 He draws upon Victor Duruy's *Histoire sainte d'après la Bible* (Paris: Hachette, 1845) and quotes second-hand extracts from Josephus and Philo of Alexandria. The passages on Oriental religions are taken from Émile-Louis Burnouf, *La science des religions* (Paris: Maisonneuve, 1876).

In his riposte, Villafañe strove to strike a middle path between Catholic devotion and anti-religious odium.[57] What he particularly dislikes in Draper's book is the disregard of the religious phenomenon as such and the reduction of the human being to an irrational animal, which "hurts the moral sense" and "proclaims the kingdom of unbridled passion." Villafañe's strictures against Draper's work are philosophical rather than historical. He reads the book through the spectacles of the French opposition between Auguste Comte's positivism and Victor Cousin's "spiritualist philosophy," here represented by the Hegelian philosopher Étienne Vacherot, the heir to the latter's chair at Sorbonne whose *La métaphysique et la science* (Paris, 1863) postulated a reconciliation between metaphysics and the sciences.[58]

La América del Sud considered Pyreneus' first article "scandalous and blasphemous."[59] But the answer of the Catholic journalist was not so much a rebuttal as an attack on the frivolous editorial policy of *La Tribuna*. Except for an aside comment, there is no specific mention of Draper. The real and intolerable issue for the author is the denial of the divinity of Christ which, he says, even non-Christian thinkers of Antiquity such as Porphyry and Celsus had at least "indirectly" recognised. A few months later, on 13 June 1877, Peyret published still another article on Draper, under his usual pseudonym.[60] More than ever intoxicated with notions of progress, he parades the historical episodes allegedly demonstrating the "repellent monstrosity" of Christianity. The essential point of the article is the author's satisfaction with Draper's dichotomic reading of history: science vs. religion, reason vs. faith, logic vs. imagination, analysis vs. sentiment, backwardness vs. progress. In view of all this, his claim that religion is a human fact and as such should be respected carries no conviction.

This time, the Catholics were able to launch a more solid counteroffensive. *La América del Sud* published a Spanish translation in 24 parts of a long critical review of Draper's book which had appeared that same year in *La Civiltà Cattolica*.[61] The original article had been written by Father Giovanni Maria Cor-

---

[57] Benjamín Villafañe, "Todavía Draper," *La Tribuna*, 27 October 1876.
[58] For the rehearsal in Buenos Aires of the polemic Comte vs. Cousin, see Pablo Ubierna, *Las humanidades. Notas para una historia institucional* (Gonnet: UNIPE, 2016), 134; cf. also W. M. Simon, "The 'Two Cultures' in Nineteenth-Century France: Victor Cousin and Auguste Comte," *Journal of the History of Ideas* 26 (1965): 45–58. For Vacherot, see Morgan Gaulin, "Refonder la philosophie en 1860: Ernest Renan critique de Vacherot," *Nineteenth-Century French Studies* 38 (2009): 52–66.
[59] "A propósito de Draper," *La América del Sud*, 1st October 1876.
[60] Pyreneus, "Los conflictos de la ciencia y de la religión," *La Libertad*, 13 June 1877.
[61] "Examen crítico de los conflictos entre la religión y la ciencia," *La América del Sud*, 1st August through 10 October 1877. I found the reference to the publication of this series in Diego Cas-

noldi SJ, a noted neo-scholastic philosopher.⁶² The Catholic newspaper eventually offered for sale the whole series in book format.⁶³ It is significant that in that same year the Spanish journal *La Ciencia Cristiana* published a different translation of Cornoldi's review, which was also edited as a book.⁶⁴ Cornoldi's text, as expected, is strongly apologetic and deals mostly with those aspects of Draper's book that touch the history of the Church and its doctrine (for instance, much space is given to the defence of Augustine). The Jesuit was a philosopher, not a historian and his preference for systematic over historical criticism is only too evident. With all its shortcomings, from a methodological point of view this was the most scholarly of the commentaries on Draper that could be read in Buenos Aires in those years and throws into relief by way of contrast the amateurish character of the others.

The three attitudes to Draper's book as expressed in the polemics carried out in the Buenos Aires newspapers between 1877 and 1878 can be summarised as the conservative Catholic position (*La América del Sud*), the radical and uncompromising secularist view of things (Peyret), and Villafañe's spiritualist perspective with no particular confessional commitment. In what follows, we shall see how the responses to Draper grew in variety and nuance of argument in a more explicitly political context.

---

telfranco, "La América del Sud (1876–1880) y las perspectivas católicas sobre el vínculo entre la ciencia y el catolicismo en la Buenos Aires de fines del siglo XIX," *Boletín del Instituto de Historia Argentina y Americana Dr. Emilio Ravignani* 47 (2017): 63–100.

62 The first part of the series was [Giovanni M. Cornoldi], "Esame critico della storia del conflitto fra la religione e la scienza di Guglielmo Draper," *La Civiltà Cattolica* 28 (1877): 142–56.

63 *La América del Sud*, 1ˢᵗ November 1877. I have not been able to find this pamphlet, which was sold in the offices of the newspaper.

64 The first part of the series was "Examen crítico de la historia de los conflictos entre la religión y la ciencia de Guillermo Draper," *La Ciencia Cristiana* 2 (1877): 512–29. The book was edited in Spain as Juan Cornoldi, *Examen crítico de los conflictos entre la religión y la ciencia de Guillermo Draper* (Granada: Librería Católica de San José, 1878). For the reception of Draper in Spain, see Alejandro Mayordomo, "La recepción en España de la obra de J. W. Draper. Una perspectiva del conflicto entre religión y ciencia," *Historia de la educación. Revista Interuniversitaria* 4 (1985): 139–58; Jaume Navarro, "The Conflicting Circulation of the Conflict Thesis," *Zygon* 54, no. 4 (2019): 1107–24.

The Conflict Thesis in the Parliamentary Debates of the 1870s

The notion of science as a symbol of progress dispelling religious obscurantism emerged in the course of the political debates conducive to the reform of the constitution of the province of Buenos Aires (1870–1873). It was a shock when Eugenio Cambaceres, one of the members of the convention convened to that end, proposed the separation of Church from state (in the end, the proposal was rejected, even by the liberal members of the assembly).[65] In his speech of 18 July 1871, which gained much applause from the more radicalised sectors of opinion, he claimed that reform was demanded by the new generation, "rising high the banner of modern science."[66] It did not matter much, he went on, if the popular masses of the country still adhered to "fanaticism," for "the great truths proclaimed by science and embodied in immortal geniuses such as Columbus, Galileo, and Fulton...have always had to fight their way against the barriers opposed by ignorance and obscurantism."[67] While decrying atheism, Cambaceres conceived religion as a "sentiment" useful to infuse a minimum of moral sense among the populace. Another member of the constitutional convention, Luis Vicente Varela, also invoked science in his abrasive allocution in favour of the separation of state and Church. If the state would seek to restrict the freedom of conscience of the citizens, he says, their soul "would rise up and compel the individual to exclaim [in unison] with Galileo: *E pur si muove*, thus translating into this sublime phrase the independence of body and spirit."[68] The image might not be entirely consistent but it does reveal how the notion of a conflict between science and religion was enmeshed with arguments about the secularisation of the state.

Seven years later, a law proposal allowing students from Catholic schools to take their final exams at state institutions was debated in the national Chamber of Deputies. Catholics invoked the principle of freedom; liberals such as the noted historian Vicente Fidel López retorted that freedom of teaching was associated with freedom of cults which, as Draper and others had shown, was tied to "the emancipation of individual reason," to which a Catholic could never con-

---

65 [Provincia de Buenos Aires], *Debates de la Convención Constituyente de Buenos Aires, 1870–1873*, vol. 1 (Buenos Aires: La Tribuna, 1877), 481–91.
66 Ibid., 1:489.
67 Ibid., 1:490.
68 Ibid., 1:567.

sent.[69] López further claimed that the hidden goal of the legislative proposal was to allow the Jesuits to present to state examinations students who had not taken philology. Jesuits, argued López, saw with disfavour this discipline on the grounds that philology exposed the translation of the Catholic Bible as corrupt and undermined belief "in the uniform or single creation of the human being [monogenism] now discarded by science" (just like in the preceding century they had boycotted "the teaching of cosmography and the natural sciences," condemned by the Church in the "saintly person of Galileo").[70] In the course of his anti-Jesuit diatribe, López allowed himself a second passing mention of Draper's "famous" book.[71] The challenge of the liberal representative was taken up by the Catholic Félix Frías, who had lived several years in France and was on personal terms with Montalembert. Against the conflict thesis, he would argue that "the most famous savants of our time are Catholics" (or nearly so, for his list includes Claude Bernard, who was at least not "an enemy of revealed religion").[72]

The 1879 Parliamentary Debate over Science and Religion

The anthropologist and geographer Francisco P. Moreno is perhaps the most famous nineteenth-century explorer of Patagonia. He was a pioneer in the creation of national parks and the all-powerful director of the Museum of La Plata. *Travel to Southern Patagonia* (1879), the account of one of his expeditions, includes many passages which presuppose the evolutionary origin of human beings, deny historical value to the narrative of Genesis, and disregard core Christian beliefs. The debate in 1879 in the Senate of the Nation of a bill that proposed the purchase by the state of 500 copies of Moreno's book to be distributed among public libraries shows once again how questions of science and religion were hotly discussed in the political arena shortly before the secularist thrust of the 1880s. The background issue was whether the state should financially support

---

69 Congreso Nacional, Cámara de Diputados, *Diario de sesiones... Año 1878*, vol. 1 (Buenos Aires: Imprenta y Librería de Mayo, 1879), 579.
70 Ibid., 1:583–84. For the Vatican position on evolution and monogenism (the idea that the whole human race descends from a single couple), see Pawel Kapusta, "Darwin from *Humani generis* to the Present," in *Darwin and Catholicism. The Past and Present Dynamics of a Cultural Encounter*, ed. Louis Caruana (London: T&T Clark, 2009), 29–32.
71 Cámara de Diputados, *Diario de sesiones... 1878*, 1:586.
72 Ibid., 1:616–17.

a scientific work some of whose statements were deemed offensive to religious persons.

The book in question is a report of Moreno's 1876–77 expedition to the Santa Cruz River and the Southern Andes, with much about the aboriginal peoples and their origin. The author presents himself as an anthropological pioneer, in the steps of Paul Broca. The tale of Adam and Eve as told in the "poem of Paradise," he says, has now been superseded by science, which has revealed the antiquity of "fossil man" from the remains hidden in the "great cataclysms of creation."[73] Moreno defends polygenism and censures the "ultramontane routine," which reduced some human races to a non-human status. He considers belief in the afterlife as a "mythological idea" of the "savages" and affirms that only science can confirm the belief "that all is ended with our disappearance from the terrestrial stage."[74] In a passage which encapsulates his cosmological views, Moreno reflects that

> the Monera paved the way for the human being. The same laws that rule the celestial bodies and the living beings in them rule the human spirit—everything answers to the sublime law of harmony. The same genesis, the same evolution that rules matter rules intelligence. Without the gradual development of the brain, the gradual development of thought cannot be explained, neither can the influence of the latter over the former be denied.[75]

Three positions could be distinguished in the course of the parliamentary debate over the official endorsement of Moreno's work: intransigent Catholics, liberals, and conciliatory Catholics. One of the Catholic senators accused the book of having "anti-Christian, materialist, and even pantheist ideas" and of asserting that the Bible had been disproved by science.[76] He was rebuked with the argument that more than two and a half centuries ago, Galileo, "condemned for heresy for affirming that it was the Earth which moved [around the sun]," had claimed that the Sacred Scriptures "were not intended to teach astronomy."[77] Gerónimo Cortés Funes, a Catholic senator from Córdoba (he was the nephew of Dean Gregorio Funes), made a long exposition in which he deployed many of the issues involving the relationships between science and Christianity at that time, in particular those related to the impact exerted by archaeology and anthropology on

---

[73] Francisco P. Moreno, *Viaje a la Patagonia Austral*, vol. 1 (Buenos Aires: Imprenta de La Nación, 1879), 3–4.
[74] Ibid., 85, 4 and 188.
[75] Ibid., 213.
[76] Congreso Nacional, Cámara de Senadores, *Diario de sesiones... Período de 1879* (Buenos Aires: Imprenta de la Honorable Cámara de Diputados, 1931), 475–76.
[77] Ibid., 477.

the Biblical account of human origins. He began enumerating what he found amiss in the book: Moreno considered that matter was eternal, supported polygenism, claimed that human beings had appeared on the Earth long before Adam and Eve, and denied the immortality of the human soul.[78] Cortés Funes contested each of these points. He did not entirely deny some kind of cosmic evolution, but he claimed that "man" [sic] cannot arise from matter, for there is "an unbridgeable chasm between inorganic and organised matter" and intelligence is the exclusive property of the spirit; spontaneous generation was a "mere hypothesis" lacking any kind of experimental support, as the "famous materialist" Rudolf Virchow had shown. Cortés Funes held that science had refuted polygenism and affirmed that scientific chronology confirmed Biblical chronology (he admitted that the Biblical days of creation should be understood as "long periods of undetermined duration").[79] He saw that a naturalist account of human origins clashed with the literal reading of the Bible and also conflicted with the Christian conception of human beings, as free, intelligent creatures with an immortal soul.

A liberal senator, Aristóbulo del Valle, shifted the grounds of the discussion. Declaring himself a "spiritualist," he argued that "the Congress was neither the place to arbitrate on scientific controversies" nor was it "the Congregation of the Index." Moreno's work was a scientific work "entirely alien to religion" and those who believed that the Bible held scientific as well as religious truths should await the verdict of science on the value of the book under discussion.[80] Manuel D. Pizarro showed himself a tolerant Catholic. He did not find that Moreno's book threatened his faith. Evolutionary ideas were not knew and had been shown false by the Bible and by science itself. Moreno's book was not addressed to "the masses" and "enlightened believers" would not be scandalised by what they could find in it. Pizarro criticised the conflict view when he affirmed that Catholic religion "had never avoided the discussion of the great scientific truths;" on the contrary, "it had provoked controversy and had emerged triumphant from the test."[81] The conservative Catholic senator from Córdoba Luis Vélez took a very defensive position. He claimed that "science's dry and arid formulae" had been thrown "against those dogmas which have ruled and will continue to rule the world...The present attack against the Church...is carried out in the pompous name of science." Vélez mentioned the case of Samuel Morton as

---

78 Ibid., 477–78.
79 Ibid., 479, 482.
80 Ibid., 485.
81 Ibid., 490.

an example of the connection between polygenism and racism and affirmed that anti-religious scientific theories would be "swept away by science itself." He added that the negation of the soul and the affirmation of the eternity of matter (ideas propagated by German materialists such as Büchner and Moleschott) plunged society "into the deepest abyss."[82]

This was not the first neither would it be the last time when questions of science and religion were discussed in Congress in Argentina. The larger issue looming behind this debate was the religious identity of the country: was it a secular state? Should a secular state judge upon the supposedly anti-religious contents of a scientific book? In the discussion about human origins there were frequent references to the conflict view, which shows once more how evolutionism, Draper's ideas, and the project of a lay state were inextricably linked in Argentina around 1880. In the end and as a sign of things to come, the bill was approved.

Estrada

Juan Manuel Estrada, who had studied law in Buenos Aires and was the great-grandson of Jacques [Santiago] de Liniers, the hero in the defence of the city of Buenos Aires during the British Invasions of the River Plate (1806–1807), was one of the most articulated voices in the confessional camp in the 1880s. His opposition to evolution—the biological meaning of which he barely understood—was founded in his perception that Darwin's theory went against the notion of the human being as created in God's image. In 1871, with opportunity of the reform of the constitution of the province of Buenos Aires, Estrada summarised his view of the relationships between the state and the Church in Montalembert's famous dictum: "A free church in a free state."[83] Estrada's first public encounter with Darwin took place in the early 1860s.

On 27 January 1862, the young liberal Italian Gustavo Minnelli inaugurated a "Course on Universal History" at the University of Buenos Aires, under the patronage of the recently nominated rector of the institution, Juan María Gutiérrez. Minelli was an adventurer with little formal education who had followed Garibaldi in his 1840 March over Rome and after much travelling through Greece and

---

[82] Ibid., 491–93.
[83] José M. Estrada, "La iglesia y el estado," *Revista Argentina* 10 (1871): 232. For Estrada, see Francisco S. Tessi, *Vida y obra de José Manuel Estrada* (Buenos Aires: Peuser, 1928) and Horacio M. Sánchez de Loria, *El pensamiento político de José Manuel Estrada. Del liberalismo católico al utramontanismo* (Buenos Aires: Torre de Hércules, 2021).

Italy sailed to South America. He presented himself as a doctor and worked as such in the frontier in Patagonia, among the natives and the gauchos.[84] Minelli's course did not extend beyond four of five meetings and no record of it has remained other than the inaugural speech.[85] Gutiérrez stopped attending it after the first lecture, but the charismatic Italian provoked a small scandal—he was even excommunicated by the archbishop.[86] Minelli's inaugural lecture is a speculative account of human origins patterned after the Enlightenment model of history. After explaining that religion began with polytheism, progressed through pantheism, and in the end evolved into rationalism, Minelli intimates a polygenetic theory of human origins: although all humans belong to "the same lineage, they are sons and daughters of different families and types."[87] He hopes that with the help of the natural sciences "the ridiculous fairy tales" of primitive ages (i.e. the Biblical narrative) will be replaced by "the historical facts of the origins of our lineage."[88]

Estrada answered with a series of articles criticising Minelli's evolutionary explanation of human history.[89] For this he drew upon apologetic authors such as Nicholas Cardinal Wiseman and Auguste Nicolas.[90] Wiseman's third edition of *On the Connection between Science and Revealed Religion*—a series of lectures given at the English College in Rome and later reworked for the general public—was available in Spanish and French translations.[91] Auguste Nicolas's popular *Études philosophiques sur le christianisme* devotes a section to Moses

---

[84] Mario Cavriani, "Minelli, Antonio," in *Dizionario Biografico degli Italiani*, vol. 74 (Roma: Istituto della Enciclopedia italiana, 2010), http://www.treccani.it/enciclopedia/antonio-minelli_(Dizionario-Biografico)/.
[85] Gustavo Minelli, *Curso de Historia Universal. Discurso preliminar* (Buenos Aires: Imprenta La Tribuna, 1862).
[86] Néstor T. Auza, "Racionalismo y tradicionalismo en el Río de la Plata. Gustavo Minelli-José Manuel Estrada," *Teología* 37, no. 73 (1999): 99–121.
[87] Minelli, *Curso de Historia Universal*, 22, 24–25.
[88] Ibid., 4.
[89] Originally published in *La Tribuna*. See José Manuel Estrada, *El génesis de nuestra raza. Refutación de una lección del Dr. D. Gustavo Minelli sobre la misma materia* (Buenos Aires: Imprenta de la Bolsa, 1862).
[90] The liberal press criticised the speech for being a patchwork of other authors' writings. Pedro Goyena, "José Manuel Estrada," in idem, *Crítica literaria* (Buenos Aires: La Cultura Argentina, 1917), 139. For the reception of the polemic in the Buenos Aires newspapers, see Auza, "Racionalismo y tradicionalismo."
[91] Nicholas Patrick S. Wiseman, *On the Connection between Science and Revealed Religion*, 2 vols. (London: J. Booker, 1836); idem, *Discursos sobre las relaciones que existen entre la ciencia y la religión revelada*, 2 vols. (Madrid: Palacios, 1844).

and the sciences.⁹² Following Wiseman's lead, Estrada claimed that science was on the side of religion since its "foremost representatives" supported with their authority "the great truths of Christian theology."⁹³ His claim in favour of the single origin of the human race is based upon the results of ethnography, anthropology, and history (a claim which turns on its head Minelli's position). It is science which supports the Biblical threefold division of the races (the generations of Noah), which is consistent with the idea of a common ancestor for all humans. At the time of defending the Biblical account, Estrada appeals to the Augustinian view that the six days of Creation "were not [calendar] days but epochs of indefinite duration," which should not contradict "Mosaic chronology."⁹⁴ The section devoted to the Flood is that in which Estrada seems to follow Wiseman more closely, who in turn depends upon Rev William Buckland's interpretation of the findings of the Kirkdale Cave in Yorkshire as evidence of the Flood: "the truth of the Universal Flood, says Estrada, has been finally demonstrated by modern science…so much so that not believing in the Flood is just a geological heresy."⁹⁵ Estrada's unsurmountable objection to the evolutionary account of the human being is its incompatibility with the dignity of God's children. He contrasts the "grandiose picture" of God's creation with the "dirty and discoloured smudge" of those who claim that the human being descends "from a split-tailed porpoise or from a monkey with a cold which lengthens its nose;" in any case, "a vile, obscure, and contemptible origin."⁹⁶ A person of his times, Estrada sees the dignity of the human being not only in the possession of an immortal soul but also in his/her capacity to discover and to invent: electricity, the steam engine, and the telegraph are testimonies of human excellence, which could not be explained if the human race descended from "the megatherium discovered by the marquis of Loreto" in Buenos Aires in 1766.⁹⁷ Estrada's conclusion is an apt synthesis of his line of argument: "I have used science to convince the scholar and philosophy to persuade the rational man."⁹⁸

---

92 Auguste Nicolas, *Études philosophiques sur le christianisme*, 4 vols. (Paris: A. Vaton, 1842–1845); idem, *Estudios filosóficos sobre el cristianismo*, 3 vols. (Madrid: Rivadeneyra, 1845) and other editions. See Corinne Marion, "Auguste Nicolas. Un laïc apologiste au XIXᵉ siècle," *Communio* no. 231 (2014): 121–32.
93 Estrada, *El génesis de nuestra raza*, 68.
94 Ibid., 17.
95 Ibid., 67. Cf. William Buckland, *Reliquiae Diluvianae* (London: John Murray, 1823).
96 Estrada, *El génesis de nuestra raza*, 36.
97 Ibid., 49.
98 Ibid., 68.

Estrada's later speeches addressed the pressing political issues implied in Darwin's evolutionism. In one of them (1878), he trod the narrow path between a defence of freedom and a condemnation of liberalism arguing that this is inevitably revolutionary and violent and as a result "sterile for freedom."[99] Darwinism is presented as a legitimazing tool of social violence either conservative or revolutionary. Estrada brings up Bismarck's *Kulturkampf* as the kind of political doctrine that "turns a Nation into an army, blind and powerful like a machine, and submissive like a flock of sheep." In his view, this political order is a consequence of "the hateful doctrine of those who attribute to the human being the genealogy of the monkey and the law of the panther."[100] Evolution theory is not only connected with the *Kulkturkampf* of the 1870s. It is also related to the social revolutions and upheavals of that decade in the Old World in which "societies venture into paganism and destroy themselves in the struggle for existence that Darwin attributed to animal species."[101] In a speech given to the Catholic Club on 21 August 1880, Estrada addressed the issue of the human being as an image of God in connection with the relationships between science and religion.[102] Science is "excellent," he says, for it makes possible the "human dominion over nature," but "it is also a law for us to aspire to immortal goals beyond [those perceived by] the senses."[103] Estrada hails science as an instrument of human advancement so far as it does not enter into conflict with the tenets of religion. He even grants the "rare talent for observation of men like Darwin, no matter how repugnant the villainous consequences derived [from his doctrines]".[104] Evolution was wrong not only because it was incompatible with a Christian notion of the human creature; it was science overstepping itself and straying beyond the appointed limits of observation; science indulging in fancies such as "Haeckel's indecent superstitions," the consequence of "the gullible and ridiculous pride of those who discovered in a grain of phosphate of lime the germ of the Genesiac anchovy."[105]

There are no denunciations of science in Estrada's speeches and writings, in which a whiff of the technological optimism celebrated by his liberal contempo-

---

99 José M. Estrada, "La libertad y el liberalismo," in idem, *Obras completas*, ed. Alberto Estrada, vol. 12 (Buenos Aires: Compañía Sudamericana de Billetes de Banco, 1905), 170.
100 Ibid., 163.
101 Ibid., 164.
102 José M. Estrada, "El naturalismo y la educación," in idem, *Obras completas*, ed. Alberto Estrada, vol. 12 (Buenos Aires: Compañía Sudamericana de Billetes de Banco, 1905), 207–29.
103 Estrada, "El naturalismo," 208–09.
104 Ibid., 212.
105 Ibid., 212.

raries is perceptible. "Darwinism" and "evolutionism" were for him ideological labels akin to "positivism," "naturalism," and "materialism." All these were refracted aspects of the secular worldview he set out to fight, in defence of what he saw as human dignity and a free Christian society. His notion of evolution was a stereotyped formula: Darwin taught that humans descended from base animals whereas Haeckel went even further and affirmed that they had originated from a lump of matter. In all, evolutionism authorised an agonistic view of society. As a lawyer, Estrada possessed neither the disposition nor the learning to engage in scientific debates. What was at stake for him were the grounds of the social and political order.

Darwin and the Conflict Thesis on Stage

There were six theatres in Buenos Aires in the 1880s.[106] In one of them, *El Nacional*, inaugurated in 1882, the Círculo Médico Argentino (Argentine Medical Circle, from now on CMA) staged popular lectures that catered to the educated and progressive sectors of a city of 345,000 inhabitants.[107] In two of these events the discussion of the relationships between science and religion, the conflict thesis, and Darwin played a leading role.

The politician and journalist Onésimo Leguizamón was one of the main figures among the ranks of those who fought for secularism and against religious teaching in the schools. He had already manifested his thought on the matter in the official reports he wrote as ministry of Justice, Cult, and Public Instruction (1874–1877). In a 1875 memory presented to Congress, at the time of reporting the yearly results of the Astronomical Observatory in Córdoba directed by Gould, Leguizamón saw fit to remember that modern astronomy had contradicted the ancient legends, the sacred books, and the medieval academies: Confucius, Joshua, and Ptolemy "had been refuted in the past centuries by Philolaus, Copernicus, and Galileo."[108] A few years later, he fleshed out this skeletal argument with an account of the historical relations between science and religion.

---

[106] Michael G. and Edward T. Mulhall, *Handbook of the River Plate*, 5th ed. (Buenos Aires: Mulhall; London: Trübner, 1885), 246.

[107] Ibid., 238. The CMA was an influential organization of young graduates and medical students with liberal leanings opposed to the current authorities of the Buenos Aires School of Medicine.

[108] Onésimo Leguizamón, *Memoria presentada al Congreso Nacional de 1875 por el Ministro de Justicia, Culto e Instruccion Publica* (Buenos Aires: Imprenta Americana, 1875), cxxxvi.

The occasion was a lecture given on 29 April 1885 in the above mentioned theatre on distributing the prizes in a contest organized by the CMA.[109] Leguizamón begins contrasting the pursuits of religion and science. The former has to do with "imagination and sentiment," while the latter is the realm of facts and experimentation.[110] Leguizamón was a Freemason and a deist who revered "the only author and legislator of the Universe."[111] What he offers is the Enlightenment view of history of science: it was born in the Greek world, somehow survived through the Middle Ages thanks to "patient monks" and the Arabs, and flourished again in the Renaissance. The focus of his story is technological: the invention of the printing press, of paper, and of the compass, which launched humanity in the age of discovery. The Reformation is seen sympathetically: it was a cry for freedom and Luther's free exam eventually gave new life to the positive sciences. The orator betrays the French provenance of the source he is drawing upon when he affirms that Newton's greatest achievement was that he "allowed Laplace to explain the regular movement of all planets and comets around the Sun."[112] The usual references to the technical marvels of the nineteenth century are there, but what arouses the speaker's enthusiasm is that humanity now lives "in a sea of light" and that "Revelation and miracles have ceased to have meaning for science!".[113] Leguizamón was an influential cultural commentator and his conceptualization of the relationships between science and religion was representative of the large sector of political opinion he embodied. There are no Hypatias or Galileos in this almost gentle account of progress. From this perspective, science did not fought a long battle against religion: it gradually and naturally superseded it.

Two years before Leguizamón's lecture, on 19 May 1882, the CMA had organised in the same theatre in Buenos Aires an homage to Darwin, exactly one month after his death.[114] The main speaker was Sarmiento, former president of the country and at that time a staunch opponent of Catholicism and religious teaching.[115] Besides his intellectual calibre and his age, three traits set Sarmiento apart from those who fought for the liberal cause. Firstly, he belonged to a for-

---

109 Onésimo Leguizamón, "Discurso del doctor Onésimo Leguizamón," *Anales del Círculo Médico Argentino* 8 (1885): 219–40.
110 Ibid., 220.
111 Ibid., 223.
112 Ibid., 233.
113 Ibid., 232, 234.
114 "Homenaje a Darwin," *Anales del Círculo Médico Argentino* 5 (1882): 546–47.
115 See for example, Tulio Halperin Donghi, Iván Jaksic, Gwen Kirkpatrick and Francine Masiello, eds., *Sarmiento: Author of a Nation* (Berkeley: University of California Press, 1994).

mer, Romantic generation and although a propagandist of progress, he was never fond of positivism. In the second place, he had an immense admiration for the United States paralleled by a disregard for French culture.[116] Finally, it has already been remarked that he introduced modern science in Argentina. The other lecturer was the young naturalist and author Eduardo L. Holmberg, one of the main protagonists of the reception of evolution theory in Argentina. The first part of Sarmiento's conference is a celebration of evolution; its second part is related to what he calls "the evolution of thought" and amounts to a chronology of Western "civilisation" (one of his driving political ideas), which bears a resemblance to chapters six through eight of Condorcet's *Esquisse de'un tableau des progrès de l'esprit humain* (1795).[117] We should now look at each of these parts.

Far from being a follower of Darwin, Sarmiento was an avowed disciple of Spencer. That evening, he professed that he believed in evolution in "general terms, as a method for the spirit," because he needed "to rest upon a principle that is both harmonious and beautiful, in order to dispel doubt, which is the torment of the soul."[118] There was not a single reference to the *Origin* in the lecture (although he did mention the *Voyage of a Naturalist* and included a long fragment from one of Huxley's *Six Conferences for Workers*).[119] In Sarmiento's speech, two anti-evolutionist authors, Louis Agassiz and Hermann Burmeister, received as much praise as Darwin himself. Sarmiento had came to know the former in Boston, during his stay there as Argentina's minister plenipotentiary. As mentioned above, he had appointed Burmeister director of the Public Museum of Buenos Aires and also as the organizer of the country's Academy of Sciences in Córdoba. Sarmiento's ambiguities are also manifest when it comes to geology: he interprets the paleontological exhibition at the local Museum as "a genuine cemetery of past creations."[120] Side by side with his catastrophism (also evident in his major work *Conflicts and Harmonies of Races*, published in that year) he affirmed without blushing that the geologists (including Burmeister) "have become convinced that the phenomena that produce the present structure of the

---

116 See, for example, Elda C. Patton, *Sarmiento in the United States* (Evansville, IN: University of Evansville Press, 1976).
117 Domingo F. Sarmiento, *Darwin en una conferencia seguido de El Congreso de Tucumán* (Buenos Aires: El Nacional, 1882); idem, "Darwin: Lecture Given in the Teatro Nacional, Following the Death of Charles Darwin," in *Darwinism in Argentina. Major Texts*, 113–38. The pagination in the notes refers to this English translation.
118 Sarmiento, "Darwin: Lecture," 125.
119 Ibid., 117–19.
120 Ibid., 119.

Earth are neither the result of present convulsions, nor some terrestrial energy that determines repeated catastrophes."[121] The fact that Sarmiento could get away with these contradictions and still be celebrated as a harbinger of Darwinian ideas can be accounted for only partly in terms of his impressive oratorical skills. We have also to take into account the willingness of his audience to see him (and Darwin) as symbols of a secularising ideology of science and progress. Admittedly, Sarmiento had an idea—no matter how vague—of the mechanism of natural selection.[122] But his main concern was "to relate to Darwin's theory in order to explain the social influence exerted over our times by such shifts in thought," to spread the power of liberation inherent in the doctrine of evolution "here, in our country, in the pampas, and Patagonia."[123]

Sarmiento's grand story of the growth of secular learning in the history of humanity developed in the second part of his speech does not differ much from Leguizamón's account, except in its more open animadversion against Catholicism and its explicit political focus. A crucial milestone in the ascending trajectory from the monkey to Edison Sarmiento propounds comes after the invention of printing: "*Common universal education for all so they can read what is written. The priests cease to teach in the cathedral schools.* INSTRUCTION IS SECULARISED" (italics and small caps in the original).[124] The emphasis should not be a surprise: Sarmiento used his Darwin lecture to campaign on the issue of the reform of elementary education. But his disquisition is also a manifestation of his dream of integrating the narrow historical circumstances of Argentina into the world scene. As he saw it, large chapters of the book of nature were being written in his country by the world-class scientists he himself had hired during his term in office: Gould and Burmeister. Together they constituted a dipole field organised around the two main scientific institutions of the country: the National Observatory in Córdoba and the Public Museum in Buenos Aires. But they were also complementary in a deeper sense. Sarmiento saw Gould's catalogues of the southern sky and Burmeister's multi-volume (and unfinished) *Description de la République Argentine* as a Herculean attempt at surveying both heavens and earth: the former was the continuator of "Hipparchus' inventory," while the latter was "Pliny's successor."[125] In a contemporary article, Sarmiento had made clear his though on this point: "Newton, Burmeister, Darwin will be the new the-

---

121 Ibid., 136.
122 Ibid., 133.
123 Ibid., 125, 122.
124 Ibid., 129, with slight modifications of the translation.
125 Ibid., 137; see Miguel de Asúa, "El apoyo oficial a la *Description Physique de la République Argentine* de H. Burmeister," *Quipu* 6, no. 3 (1989): 339–53.

ologians, those who reason (*logos*) on the work of God (*Theos*)."¹²⁶ In his imaginative conception Argentina was becoming the home of those engaged in a new scientific and secular description of the whole of creation: Burmeister was writing "the *true*" (my emphasis) *History of Creation*—he is referring to Burmeister's *Geschichte der Schöpfung* (Leipzig: Otto Wigand, 1843) a natural history of the world in the Humboldtian tradition—and Gould had "restored the movement of the stars, as Copernicus did for the Earth."¹²⁷ Shortly before that he had affirmed that thanks to Copernicus "the hand of God and the firmament are no longer needed to hold every sun and planet...*The sciences and those who teach them are no longer clergymen*" (emphasis in the original).¹²⁸

Sarmiento superimposed his own secular values on the material he commented upon. Burmeister could have felt that this interpretation of his work was near the mark. In the last edition (1872) of his *History of Creation*, the nebula which gave origin to the universe evolved out of "forces inherent to matter;" after successive cataclysms, the human being "appeared on Earth" as "the crown of creation and its master" on account of his spiritual life, self-reflection, and moral freedom.¹²⁹ These ambiguous statements are compatible with Burmeister's radical unbelief.¹³⁰ The case is different with Gould. His 1869 address as retiring president before the American Association for the Advancement of Science (an opportunity in which "the 'conflict between science and religion' filled the air") is a long reflection on this question.¹³¹ Distancing himself in the same measure from "the bigot and the atheist," he sought to reconcile the "ministry of religion" with the "ministry of science."¹³² The result of his philosophical excursion is "a middle term" between the phenomena of matter and those of spirit,

---

126 Domingo F. Sarmiento, "El Congreso de Tucumán," in idem, *Darwin en una conferencia*, 72.
127 Sarmiento, "Darwin: Lecture," 137; Burmeister had been appointed director of the Public Museum in 1862. The catalogues of the southern skies published during Gould's period at the front of the observatory were *Uranometría Argentina* (Buenos Aires, 1877–1879), *Catálogo de Zonas Estelares* (London, 1884), and *Catálogo General Argentino* (London, 1886).
128 Sarmiento, "Darwin," 129.
129 Hermann Burmeister, *Geschichte der Schöpfung*, 8th ed. (Leipzig: Otto Wigand, 1872), 646–47.
130 Herbert Ewe, "Hermann Burmeister," in idem, *Bedeutende Persönlichkeiten Vorpommerns* (Weimar: Hermann Böhlaus Nachfolger, 2001), 33–40.
131 George C. Comstock, "Biographical Memoir Benjamin Apthorp Gould, 1824–1896," *Memoirs of the National Academy of Sciences* 17 (1924): 163; Benjamin A. Gould, "Address of the President," *Proceedings of the American Association for the Advancement of Science...Eighteenth Meeting...1869* (Cambridge, MA: Joseph Lovering, 1870), 1–37.
132 Gould, "Address of the President," 23, 11.

which he situates in the class of energies associated with "life and the will."[133] Whatever the strand of liberal Protestantism he endorsed, it is clear that Gould's views were far away from the attempts of secularising public figures such as Sarmiento and Leguizamón to recruit the labours of the National Observatory (Córdoba) for their own propagandistic purposes.

Holmberg was 28 years old at the time of delivering his speech on Darwin. Named after his grandfather Eduard Ladislaus Kaunitz Baron von Holmberg (a military engineer who had arrived in the wake of the Napoleonic wars and had fought in the wars of Independence from Spain), Eduardo Ladislao would become one of the most important naturalists in nineteenth-century Argentina. Although he had obtained his doctor's degree only two years before his lecture, he enjoyed a considerable scientific reputation built upon his research expeditions to the northern and southern regions of the country. His fame had been strengthened by the publication in 1875 of a rumbustious satirical *roman à clef* which celebrated Darwinian evolutionism against the entrenched opposition of Burmeister.[134] As many in his generation, Holmberg was distinguished in more than one field: he pioneered in the writing of science fiction in Argentina and is considered the creator of the genre of "scientific fantasy" in Latin America.[135]

Holmberg's 1882 speech is an exposition of Darwinism articulated with a proclamation of the author's materialist convictions—in this he was far apart from Sarmiento.[136] The whole is bracketed by claims about the relationship between science and religion. His position is sharpened with the help of a rhetoric built upon the use of pairs of corresponding opposites (science vs. religion, reason vs. feeling, brain vs. heart, cold vs. heat, elite vs. rabble). The speech opens with a strong statement about science and religion, "two things perfectly isolated from each other," which "[never] could march along together."[137] Coming full circle, the lecture ends where it began, with a delimitation of the boundaries between religion and science—a distinction which implies a hierarchical ordering of things with science at the top. The scheme depends upon the traditional figurative ranking of bodily organs. The "brain" is the seat of reason, the instrument of science that the sophisticated members of society use as a tool in their inter-

---

[133] Ibid., 27–28.
[134] Eduardo L. Holmberg, *Dos partidos en lucha*, ed. Sandra Gasparini (Buenos Aires: Corregidor, 2005).
[135] Sandra Gasparini, *Espectros de la ciencia. Fantasías científicas en la Argentina del siglo XIX* (Buenos Aires: Santiago Arcos, 2012), 95–148.
[136] Eduardo L. Holmberg, *Carlos Roberto Darwin* (Buenos Aires: El Nacional, 1882). This edition includes the text and copious notes by Holmberg.
[137] Holmberg, *Carlos Roberto Darwin*, 7.

pretation of reality. The "heart" harbours feelings and emotions, the matter out of which religion is made; this is the world of *hoi polloi*, who should be educated out of their mystification into a colder but more real image of the world. Holmberg's son and biographer interprets his father's utterances as pointing toward a peaceful coexistence between science and religion. To the astonishment of "a numerous and qualified public who had read Draper's *Conflicts between Religion and Science*," he says, Holmberg would have denied that such conflicts existed.[138] If that was the case (which I do not think), the denial was bought at the prize of disqualifying religion even as a valid contender.

Holmberg is particularly hard on the Judaeo-Christian tradition: the [Christian] Church is "the eternal suicide victim of history."[139] In one of the notes added to the published version of the lecture, he affirms that either Moses was an impostor or he did not exist at all.[140] In his claim that the author of the Pentateuch was a plagiarist of Hindu traditions, he depends on Louis Jacolliot's *La Bible dans l'Inde. Vie de Iezeus Christna* (1869); he is indebted to the same source when he affirms that the Biblical account of creation is based on an intentional misinterpretation of the Hebrew *barah*.[141] As opposed to Sarmiento, education is not Holmberg's main target; he is after bigger game. The natural sciences are the only "impregnable bastion" of liberalism. Only when all the elementary schools of Argentina could have a microscope and when all the amphitheatres could rise "the saintly altar of the corpse," then the "God of Newton, Kepler, and Bernardin de Saint-Pierre could feel [him/herself] God in the temple of the best feelings."[142] In contradistinction to the concept of *creatio ex nihilo*, Holmberg proclaims "the eternity of matter," proudly reclaiming as his own "Büchner's abject materialism."[143] He is not one who admitting a spiritual dimension could feel comfortable in a deist or agnostic position. All through his life he was something between a materialist and a pantheist and as a result his discourse is more confrontational than his senior's.[144]

---

138 Luis Holmberg, *Holmberg, el último enciclopedista* (Buenos Aires, 1952), 117.
139 Holmberg, *Carlos Roberto Darwin*, 36.
140 Ibid., 83–84.
141 Ibid., 12, 24. Cf. Louis Jacolliot, *La Bible dans l'Inde. Vie de Iezeus Christna* (Paris: Librairie Internationale; Brussels: Lacroix and Verboeckhoven, 1869).
142 Eduardo L. Holmberg, "Viajes a las Sierras del Tandil y de la Tinta (1880)," in idem, *Excursiones bonaerenses*, ed. Juan C. Chebez and Bárbara Gasparri (Buenos Aires: Albatros, 2009), 55–56. The original was published in 1884.
143 Holmberg, *Carlos Roberto Darwin*, 21.
144 In a speech given in 1905, when he was 53 years old, he claimed: "If I believe with all the force of a sincere religiosity in the eternity of matter, of force, of energy, should that not be respected?". Quoted in Holmberg, *Holmberg, el ultimo enciclopedista*, 118.

Darwinism is treated in the third section of Holmberg's lecture, almost as an afterthought. Most of the examples which the speaker uses to illustrate Darwin's theory of evolution address the social, political, and economic anxieties of his audience. The notion of variation by domestication is illustrated with the example of the infanticides of the Spartans and the alleged "degeneration" of Buenos Aires male population on account of the stronger individuals having been killed in the prolonged intestine wars.[145] His first example of natural selection is the charm of the Buenos Aires belles. Massive immigration meant a greater variety of human types. Since the dowry system was quickly falling into disuse, males had opportunity to choose their partners according to their beauty, which resulted in females of "the beautiful, intelligent, and graceful type."[146] Another illustration he gave of the principle of natural selection was of an exotic species of thistle which at that time was invading the estancias and which he deemed a dangerous menace to the country's economy.[147] Holmberg's third example is the extermination of the "Indians," the native population of the pampas, carried out by "the civilised ones, the Christian ones, the ones armed with Remingtons." The frontier war gave him opportunity to address the question of morality: it does not matter if "the Indian's cause" was just; he "defends his land, which we have stolen from him; he wounds us, he kills us, he robs us." The conflict is subjected to Malthus's law, "as we too fight for our lives, with good ideas, with good weapons...we do nothing more than put into practice our upper hand."[148] Holmberg sets Malthus in explicit contraposition with divine Providence. Parents tend to take more care of sickly children than of those that are strong and robust; as a result, the former live longer, although miserable lives and frequently reach old age, while it is not infrequent for the latter to die young. This caring for the weak is "the inversion of the natural order."[149] Immigration, racial conflict, sexual behaviour, economic production, all these red-hot issues were sweepingly reviewed by Holmberg in terms of Darwin's principle of natural selection and the notion of "struggle for life," which was also invoked as

---

145 Holmberg, *Carlos Roberto Darwin*, 56–57.
146 Ibid., 58.
147 Ibid., 58–62.
148 Ibid., 65–66. Holmberg had been the editor of the section devoted to arachnids in the scientific proceedings of Roca's expedition. See Eduardo L. Holmberg, "Arácnidos," in *Informe oficial de la Comisión científica agregada al Estado Mayor General de la Expedición al Río Negro*, vol. 1 (Buenos Aires: Ostwald y Martínez, 1881), 117–68. See Lewis Pyenson, "Athena's Retinue: Nineteenth-Century Scientists Embedded in the Army," *British Journal for the History of Science* 45, no. 3 (2012): 377–400.
149 Holmberg, *Carlos Roberto Darwin*, 64.

the foundation of a conception of morality innocent of Hume's naturalistic fallacy: in Holmberg's moral order "what is" takes precedence over "what ought to be."

At no point did Holmberg provide a proper explanation of natural selection —his two quotations from the *Origin* in the notes come from the first ten pages of the New York 1871 edition. He avowed that he had only had three weeks to prepare the lecture and, though he considered himself a "Darwinist," confessed that his particular field was not evolution, but the collection, description, and classification of the flora and fauna of Argentina. As an introduction to evolutionary thought he recommends Haeckel's *History of Creation*, which he deemed suited "to our [Argentine] literary tastes."[150] (As discussed below, he was right.) Holmberg was a capable field naturalist with a talent for systematics. A closer reading of the lecture with its sprawling footnotes shows that more than celebrating Darwin's scientific achievements, he was bent on spreading his ideological credo while feeding his reputation as a literary wit.

Leguizamón, Sarmiento, and Holmberg were key protagonists of the Buenos Aires cultural, educational, and scientific world in the early 1880s. Their views were celebrated manifestations of liberal opinion and did much to give form to discourse on the relationships between science and religion and to popularise Darwin's ideas. Each of them is tied to a particular attitude toward religion (deism in the first two cases, pantheism or materialism in the third). One thing these speeches have in common: the occasions, setting, audience, motives and circumstances in which they were pronounced were manifestations of social forces committed to a secularising political program.

Catholic Answers

The only Catholic answer to the Darwinian *soirée* was the lecture delivered two months after the event, on 15 July 1882, by the 22-year-old medical student Pedro S. Alcacer. Born in Santa Fe, he had studied in the Jesuit College of that city; later in Buenos Aires he joined the Literary Academy of the Plata, a cultural club of the Jesuit College of El Salvador. The argument of Alcacer's anti-evolutionist lecture was built upon the premise that evolutionism is not a scientific doctrine, for it goes against observational and experimental science. For the most, Alcacer's arguments were taken from French anti-Darwinian authors (Jean-Pierre Flourens and Jean-Louis Quatrefages) through the intermediary of French Catholic apolo-

---

[150] Holmberg, *Carlos Roberto Darwin*, 69–70.

gists such as Constantin James and the Jesuit physicist François-Marie Moigno.[151] He quotes from the French translation of Darwin's *Origin* by Clémence Roger (it is likely that he quotes second hand).[152]

Alcacer deemed the hypothesis of a proto-organism situated at the base of the tree of life gratuitous, because this creature had not been found.[153] He also raised the by then well-known objections against transformism, such as the non-viability of transitional forms and the gaps of the fossil record. In the end, Alcacer granted the existence of "an order and a progressive development in life" called forth both by religious faith and by paleontological science—he even asserts that "modern transformism is right."[154] This apparently contradictory claims were based on a distinction between morphological changes in the species (which he called "variations" and deemed acceptable) and physiological changes (which he called "transmutations" and rejected).[155] This differentiation did not seem to dovetail with his previous denial of the possibility of apes evolving into humans on the grounds that "the feet of the former are prehensile while those of the latter are fit for perambulation" (clearly, a morphological change).[156]

A whole section of Alcacer's lecture is devoted to defend vitalism against materialism. He dismisses the positivistic definition of life in Littré's and Robin's edition of Nysten's medical dictionary on the grounds that it is impossible to define life without resorting to metaphysics.[157] His own definition is that life is "the spirit with which God animates organised matter."[158] Alcacer merges explicitly science and religion: "Life for Revelation and for the natural sciences is only just one word, *creation*, God's spirit" (italics in the original).[159] The vital force is not a property of matter, it is a gift of the Creator (I shall come back to this in the next section).

---

[151] For this literature, see Harry W. Paul, *The Edge of Contingency. French Catholic Reaction to Scientific Change from Darwin to Duhem* (Gainesville: University Presses of Florida, 1979), 64, 135–36.
[152] Charles Darwin, *De l'origine des espèces*, trans. Clémence-Auguste Roger (Paris: Masson, 1862).
[153] Pedro S. Alcacer, *La vida y el transformismo moderno* (Buenos Aires: Imprenta de la Penitenciaría, 1882).
[154] Ibid., 64, 67.
[155] Ibid., 69.
[156] Ibid., 49.
[157] "C'est l'ordre d'activité de la matière à l'état d'organisation." [Pierre-H. Nysten], *Dictionnaire de médecine*, ed. Émile Littré and Charles Robin, 13th ed. (Paris: Baillière, 1873), 1672.
[158] Alcacer, *La vida y el transformismo*, 23.
[159] Ibid., 29.

In his talk, the young medical student censured Sarmiento's speech for presenting Agassiz as "an ardent defender of Darwin."[160] Hitting the nail of the head, he also quoted a strongly anti-Darwinian passage from Burmeister, this time playing on Sarmiento's incongruous praise of his anti-evolutionist friend.[161] A month and a half after Alcacer's lecture, Sarmiento responded to these barbs. Darwin, he said, "could have made mistakes, but it is not proper that a student should stand up to correct him…following the indications of Father Moigno… who talked about faith, not about science."[162] Very much like Sarmiento, Holmberg rebuked Alcacer for relying on religious authors instead of citing Haeckel; he adds that in the young Catholic's lecture "only a few French authors are mentioned, no German at all, and among the English only Erasmus Darwin."[163] Holmberg's criticism lost much of its sting once it became evident that the French and German authors he mentioned in his own lecture were taken from those discussed in Haeckel's *History of Creation*. Moreover, his allegation that Alcacer's quotation of Burmeister's *Description physique* was contrived was unfounded. The way in which Holmberg dispatched his obscure Catholic colleague with a few condescending taunts is in tune with his own lackadaisical treatment of Darwinism.

Inocencio Torino, a recently graduated physician contemporary of Alcacer, also reviewed the latter's lecture.[164] He called attention to Alcacer's fuzzy distinction between morphological and functional changes in the species and reprehended his colleague's dismissive treatment of Darwin's figure and personality. These and other points were well argued; it was the penultimate sentence which revealed the motive of the reviewer's uneasiness: "Science is not religion."[165] To be sure, Alcacer's speech was laden with metaphysical ballast and was strongly derivative, both flaws his critics were quick to point out. But they stood on no surer ground; even Holmberg, that young bright thing with a number of accomplishments to his credit, trod on thin ice when he talked about evolution. In the

---

160 Ibid., 46.

161 Ibid., 45. Certainly, Burmeister denied the possibility of species transmutation for it did not satisfy the requirements of a strictly empiricist view of science. See Hermann Burmeister, *Description physique de la République Argentine. Tome troisième. Animaux vertébrés. Partie 1. Mammifères vivants et éteints* (Buenos Aires: Coni; Paris: Savy, 1879), 13–14.

162 Domingo F. Sarmiento, "La inteligencia en la vida argentina," in idem, *Obras*, ed. Augusto Belin Sarmiento, vol. 46 (Buenos Aires: Imprenta Mariano Moreno, 1900), 183–85.

163 He devotes a long note of the published version of his own lecture on Darwin to upbraiding the Catholic medical student, see Holmberg, *Carlos Roberto Darwin*, 104–107.

164 Inocencio Torino, "La vida y el transformismo moderno," *Anales del Círculo Médico Argentino* 6 (1882): 34–36.

165 Ibid., 36.

year preceding Darwin's death, Torino had delivered a lecture in which he sketched out a theory connecting Pasteur's enzymatic and microbiological discoveries with the evolution (appearance and vanishing) of diseases.[166] Its precarious conceptualisation reveals how the young physicians who stood for evolution and secular science were nearly as flimsy in their arguments as their antievolutionist counterparts.

Excursus on "Madness and Crime"

A sequel to Alcacer's discussion shows how the debate on evolution was intimately linked with other issues of science and religion which touched the secularist sensibilities of the medical circles of Buenos Aires. In 1883, two doctoral dissertations with the same name ("Madness and Crime") were presented to the School of Medicine of the University of Buenos Aires, by students who had been medical interns at the Penitentiary, a recently created model institution.[167] One of them was by Alcacer, the other by Alejandro Korn, an Argentine of Prussian descent who would in the event become the director of a large mental hospital in the city of La Plata, a centre-left politician linked with the 1918 University Reform movement, and one of the first original philosophers of the country (more on him in the next chapter).[168] In his dissertation, Alcacer—who introduces himself as a Catholic—rehearses his previous definition of life: "Life is not a mere manifestation [of physico-chemical forces], on the contrary, it is something

---

[166] Inocencio Torino, "Las teorías evolutivas y la ciencia médica," *Nueva Revista de Buenos Aires* 2 (1881): 241–57.

[167] *Memoria de la Penitenciaría Nacional correspondiente al año 1894 acompañada de un retrospecto* (Buenos Aires: Penitenciaría Nacional, 1895). Convicted women were not consigned to secular state prisons but put under the supervision of the religious congregation of The Good Shepherd, see Lila Caimari, "Whose Criminals Are These? Church, State, and *Patronatos* and the Rehabilitation of Female Convicts (Buenos Aires, 1890–1940)," *The Americas* 54, no. 2 (1997): 185–208. For criminology in the period, see also Julia Rodríguez, *Civilizing Argentina: Science, Medicine, and the Modern State* (Chapel Hill: The University of North Carolina, 2006) (it should be read with care, for it has some historical factual mistakes).

[168] Pedro S. Alcacer, *Locura y crimen* (Buenos Aires: Imprenta de la Penitenciaría, 1883); Alejandro Korn, *Locura y crimen* (Buenos Aires: Imprenta de la Penitenciaría, 1883). For Korn, see Luis Farré, *Cincuenta años de filosofía en Argentina* (Buenos Aires: Peuser, 1958), 101–33; W. J. Kilgore, "Latin American Philosophy or the Place of Alejandro Korn," *Journal of Inter-American Studies* 2, no. 1 (1960): 77–82.

absolute which the metaphysician recognises as such."[169] This vitalistic definition is related to a concept of mind irreducible to matter: "The dominion of pure physiology, says Alcacer, should be kept apart from the dominion of physiological psychology."[170] It is "the unity of our self, of our most intimate sense" which argues against a purely physiological interpretation of mind.[171] Since he discusses the thorny question of moral and legal responsibility in persons convicted of violent crimes, he sees it necessary to affirm that "morality does not depend upon our [physiological] organisation."[172]

The Catholic daily *La Unión* congratulated the brand new doctor "for carrying his religious convictions to the terrain of the sciences."[173] The next day, the same newspaper devoted its front-page editorial to review his dissertation. The well argued piece—obviously written by a lawyer, probably Pedro Goyena—claims that Alcacer's work "does not break the unity of the sciences ... [for] ... medicine cannot be a science or becomes a truncated science if its conclusions contradict morality and law." The journalist's point is that a materialist conception of mental illness leads to the denial of human responsibility and puts into question the very basis of society: "Our most sacred and legitimate rights would disappear because of a cut of the scalpel, a microscopic exam, or a chemical analysis."[174] Very different was the review by the editor of an annual bibliography, who complained that "there was poetry in excess" in Alcacer's "romantic" work whose only valuable section was the set of psychiatric case studies at the end of it.[175] The same reviewer had dismissed Alcacer's lecture on Darwin as "a speech full of flowers and metaphors" with arguments which were a rehash of "Father Llanas's books" (Eduardo Llanas Jubero was a Spanish Piarist apolo-

---

169 Alcacer, *Locura y crimen*, 24. This conception is buttressed by quotations extracted from Claude Bernard, *La science expérimentale* (Paris: Baillière, 1878). Bernard's views on these questions were not those ascribed to him by Alcacer; see, for example, Sebastian Normandin, "Claude Bernard and *An Introduction to the Study of Experimental Medicine*: 'Physical Vitalism,' Dialectic, and Epistemology," *Journal of the History of Medicine and Allied Sciences* 62, no. 4 (2007): 495–528.
170 Alcacer, *Locura y crimen*, 25.
171 Ibid., 38.
172 Ibid., 47.
173 "La tesis del Dr. Alcacer," *La Unión*, 13 June 1883.
174 "El Dr. Alcacer," *La Unión*, 14 June 1883.
175 Alberto Navarro Viola, review of *Locura y crimen*, by Pedro S. Alcacer, *Anuario Bibliográfico de la República Argentina* 5 (1884): 284.

gist).[176] The critique contrasted Alcacer's dissertation with Korn's, which was a "properly scientific [work], based on solid ground and with precise and sensible conclusions."[177]

In his work, Korn had claimed that "psychological phenomena, if judged *by a scientific criterion*, cannot be considered but the result of a physiological process subjected ... to the operations of the body organs and the influence of pathogenic causes" (my emphasis). Korn declares that it is not his intention "to prejudge over the essence of the intellectual faculties" and he takes distance from Jean-Bernard Luys's nosography of mental illnesses based upon histopathology for "the importance of anatomical pathology should not be exaggerated once we are in front of a living human being."[178] But he then goes on to affirm that "thought is a physiological fatality" so that "crime in all cases is the unavoidable outcome of pre-existent conditions and what is left to the forensic psychiatrist is to declare if those conditions depend either on physiological or on pathological causes."[179] Moreover, he affirms that the only difference between the reflex arc and the reflexive act is the greater complexity of the former so that "moral sensibility depends upon the integrity of the nervous system."[180]

What matters for our argument is how the question of the relationships between science and religion subtends this discussion. Korn represented the positivist default position among doctors. During Alcacer's defence of dissertation, two of the members of the examination board felt compelled to make an avowal of their materialistic beliefs. In turn, Catholic lawyers warned about the legal implications of a deterministic view of human action. Alcacer addressed scientific questions starting from a religious point of view, which went against the grain of the times. In his lecture on evolution, he concluded that all that is left to us is "the dogma written in Moses' *Genesis*;"[181] and he declared that his dissertation was "the authentic expression" of his "scientific and religious beliefs."[182] Catholics and secularists were at that time highly polarised: the presentation of

---

176 Alberto Navarro Viola, review of *La vida y el transformismo moderno*, by Pedro S. Alcacer, *Anuario Bibliográfico de la República Argentina* 4 (1883): 215–16. Eduardo Llanas Jubero, *Los seis días de la Creación* (Barcelona: Fidel Giró, 1889).
177 Alberto Navarro Viola, review of *Locura y crimen*, by Alejandro Korn, *Anuario Bibliográfico de la República Argentina* 5 (1884): 285.
178 Korn, *Locura y crimen*, 9–10, 22–23. Cf. Jean-Bernard Luys, *Traité clinique et pratique des maladies mentales* (Paris: Delahaye and Lecrosnier, 1881). Luys was heavily criticised by Alcacer.
179 Korn, *Locura y crimen*, 25.
180 Ibid., 25–26. Korn would later drastically change his position, see Alejandro Korn, *La libertad creadora* (Buenos Aires: Losada, 1944).
181 Alcacer, *La vida y el transformismo*, 73.
182 Alcacer, *Locura y crimen*, 14.

two dissertations on the same subject and with the same title but supporting opposite viewpoints on sensitive issues of science and religion was one more expression of this state of affairs.

The Great Debate

The biennium 1883–1884 marked the peak of the confrontation between liberals and Catholics in Buenos Aires. The debate over the law of educational reform remains as a milestone in the process of secularisation that reshaped Argentina's social and political life. Around those years, the liberal majority of the government pushed forward a project—eventually sanctioned as a law—that made elementary school obligatory, free, and co-educational. The proposal also eliminated religious education (the teaching of Catholic catechism) from the school room (pupils could still receive religious instruction from persons designated by their churches outside the school timetable, which never realised).[183]

Two examples from the public discourse of that period will show how notions associated with Darwin and evolution were used in the political debate over secularisation. In the first months of 1883, two controversies involving secularism had Sarmiento as the protagonist and in both of them evolutionary rhetoric played a part. The refusal of a group of Irish nuns running a school in the province of Buenos Aires to submit the examination programs to the local Board of Education prompted Sarmiento to write a series of articles against religious education, against the Catholics, and against the Irish, "the most backward people coming from Europe."[184] The scathing answer by the Catholic leader Pedro Goyena prophesied (in Lamarckian fashion) that Sarmiento would eventually become "blind and deaf," if he had "not eyes for light and ears for truth (Darwin)."[185] Goyena's tirade was compared by the French-Argentine contemporary literary critic Paul Groussac to the "scalping of an Iroquois warrior":[186] an apt quip, although it should be mentioned that when it came to the conceptual dissection of the theories of evolution, Goyena's knife was rather blunt.

---

[183] In practical terms, the law applied only to the city of Buenos Aires and the very scarcely populated "national" territories (Patagonia and the dense forests of the north east); the provinces retained their autonomy in questions educational.

[184] Domingo F. Sarmiento, "Lágrimas de cocodrilo," in idem, *Obras*, ed. Augusto Belin Sarmiento, vol. 48 (Buenos Aires: Imprenta Mariano Moreno, 1900), 184.

[185] Pedro Goyena, "Contra mentira, verdad," *La Unión*, 2 March 1883.

[186] Paul Groussac, *Los que pasaban* (Buenos Aires: Taurus, 2001), 107.

Sarmiento's confrontation with Manuel D. Pizarro is further evidence of the careless use by Catholics of the evolutionist vocabulary, employed mostly for purposes of abuse. The already mentioned Pizarro was a Catholic lawyer, journalist, and Roca's former minister of Cult, Justice, and Public Instruction who quit office in 1882, at the height of the secularist controversy. Darwin, evolution, and apes crop up frequently in his 1883 articles in the Catholic newspaper *La Unión*, collected as *The Simian School and the National Constitution*.[187] Pizarro identified the separation between Church and state defended by Sarmiento with "the positivist or materialist school" and with those who "together with Darwin, reject the *divine genesis* of man and wish to make of him an improved beast descending from the monkey and only slightly superior to it in the zoological scale" (emphasis in the original).[188] He compared a secularised commonwealth to "a society of apes, who neither need nor are able to conceive any government but the whip and the bread of whoever would have them dancing in the streets."[189] In his speech in the Catholic Congress, convened in 1884 with the aim of prompting Catholics to participate in public life, Pizarro again pitted the Christian society against "Darwin's evolutionary ethics [and] Bentham's utilitarian ethics."[190] Transposing the evolutionist theme into a Christian key, he commended Catholic participation in politics as the "true *struggle for life*," a contest of love, charity, and fraternity in which the privileged orders of society would help their inferiors, instead of letting them perish. This, he said, is "the law of evolution and *mystical selection*," which points to life eternal (emphasis in the original).[191] In his own contribution to the Congress, Alcacer accused the Argentine liberals of tracing the origin of the human race to "a ridiculous quadrumanous creature."[192] Estrada decried Spencer and Darwin as prophets of liberalism, while moaning the century's "universal secularisation of everything."[193] In his toast at the banquet which concluded the event, he further

---

[187] Manuel D. Pizarro, *La escuela simiana y la Constitución Nacional. Relaciones de la Iglesia y del Estado* (Buenos Aires: Imprenta de La Unión, 1883). For Pizarro, see Gaspar Ferrer, "Manuel Demetrio Pizarro," in *Controversias políticas del ochenta* (Buenos Aires: Club de Lectores, 1964), 131–152; Luis R. Frías, "Manuel D. Pizarro," in *La Argentina del ochenta al Centenario*, ed. Gustavo Ferreri and Ezequiel Gallo (Buenos Aires: Sudamericana, 1980), 199–214.
[188] Manuel D. Pizarro, "El gobierno puramente civil'," in idem, *La escuela simiana*, 39.
[189] Manuel D. Pizarro, "Teología trascendental," in idem, *La escuela simiana*, 62–63.
[190] Manuel D. Pizarro, "La religión y la política," in idem, *Miscelánea*, vol. 1 (Córdoba: La Minerva, 1897), 238.
[191] ibid., 255.
[192] *Diario de sesiones de la Primera Asamblea de los Católicos Argentinos* (Buenos Aires: Igon Hermanos, 1885), 272.
[193] Ibid., 450, 458.

linked evolutionary theory with the government's secularist policy, proclaiming that a "naturalist and mad" philosophy degrades human beings into bestiality, out of which result "crowds of fools, cunning as foxes, lewd as monkeys, garrulous as parrots, and if they hold public office, cruel and rapacious as beasts of prey!".[194] The audience, who easily decoded the denigrating allusions to individual liberal politicians, burst into enthusiastic applause.

Discussion of the theory of evolution in Buenos Aires during the nineteenth century had a strong political valence. In the course of the long debate that took place in the course of July 1883 in the National Chamber of Deputies, liberals raised once and again questions of the relationship of science and religion in long-winded arguments that presupposed sundry variants of deism, naturalism, and a positivistic view of human progress. A survey of this polemic will show the rhetorical function served by the conflict thesis in these political debates.

To begin with, Emilio Civit, a young party-liner deputy from the province of Mendoza, affirmed that to show that the Catholic Church was an enemy of every single liberty, it would only be necessary "to remember the life of Newton, of Galileo, of Kepler, the name of Torquemada, and Saint Bartholomew's night."[195] This was not an isolated statement. Little more than a year later, at the time of voting for a proposal to supress the government's financial support of the country's Catholic seminaries, Civit argued that the education provided in those establishments was not "scientific, since it has been demonstrated elsewhere and also in this Chamber of Deputies, that science goes contrary to many of the dogmas and beliefs of the Church."[196] Civit was being provocative (he immediately added that the teaching in the seminaries was "anti-religious and anti-Christian"), but his affirmation that the conflict thesis had been "demonstrated" by a political deliberative assembly in Buenos Aires (of all places) is revealing of the mood of the times.

We should return to the 1883 controversy. Pedro Goyena rose up to Civit's challenge. Goyena was a gentlemanly and fearful adversary with a sharp mind. In his answer to Civit, he took upon himself the (admittedly quite difficult) task of arguing that the *Syllabus* of 1864 did not condemn progress and civilisation as such, only their "modern" forms. It has been a "senseless pride," he says, which made science take the wrong turn: all the marvels of the modern world— the steamboat, air balloons, the telegraph, germ theory, and the discoveries of

---

[194] Ibid., 476–77.
[195] Congreso Nacional, Cámara de Diputados, *Diario de sesiones...Año 1883*, vol. 1 (Buenos Aires: Imprenta La Universidad, 1884), 515.
[196] This took place in October 1884. Congreso Nacional, Cámara de Diputados, *Diario de sesiones...Año 1884. Tomo Segundo* (Buenos Aires: Stiller and Laass, 1885), 226.

astronomy—were not enough to make better persons. A taste of disenchantment surfaces in the prose of the late Romantic that was Goyena: "The sky does not harbour a promise of hope anymore; the bright stars ceased being a symbol of faith."[197] He sets himself squarely against the conflict theory: "There has not been a single scientific conquest of the human spirit...which has been ignored or rejected by the Church. Nobody could affirm...that the Church has been an enemy of the light and the high flights of intelligence." Like other Catholics before and after him, Goyena strove to show that the famous protagonists in the history of science were Catholics or at least "obedient to the laws of Christianity": Newton, Copernicus, Euler, Volta, Faraday, Ampère, Arago, and the Jesuit Secchi.[198]

Eduardo Wilde, then national minister of Public Instruction and the undisputed champion of the secular cause, pronounced a long speech before the Chamber of Deputies that soon became famous.[199] Wilde was a doctor, a politician, a satirical journalist, and a bright writer with an uneasy mixture of cruel wit and sentimentality.[200] One of his pronouncements is an accomplished version of the conflict thesis: "Between science and religion there is an open and obvious contradiction—there has to be and this contradiction cannot but exist." The reason for this is that "the dogmatic affirmations of religion" and "the conquests of the human intellect" cannot coexist together.[201] The conflict is inevitable for the reason that the image of the world due to science is in perpetual change, whereas the notions of Christian religion were based upon modes of thought prevalent at the time of its origin. It is no surprise that science should be in contradiction with religion, because it is in contradiction with itself. This is due to progress: "The law of progress has to be necessarily obeyed: progress is everywhere."[202] In his speech, Wilde goes on a long detour through physics, chemistry, natural history, and medicine in order to show that progress in the sciences means that the new findings contradict the old.[203]

---

197 *Diario de sesiones...Diputados...1883*, 1:527.
198 Ibid., 1:528.
199 It was reprinted as a pamphlet: Eduardo Wilde, *La cuestión religiosa en el Congreso Argentino* (Buenos Aires: La Tribuna, 1883). In what follows, I shall refer to the parliamentary acts of the debates.
200 For a comparison of Holmberg and Wilde as physician writers, see Bonnie Frederick, "A State of Conviction, a State of Feeling: Scientific and Literary Discourses in the Works of Three Argentine Writers, 1879–1908," *Latin American Literary Review* 19 (1991): 48–61.
201 *Diario de sesiones...Diputados...1883*, 1:563–64.
202 Ibid., 1:563–64.
203 Ibid., 1:565.

Despite its trenchant and focused initial formulation, Wilde's spelling out of the relationship between science and religion is unstable. Together with his idea of an unsurmountable conflict, he simultaneously affirms that though different, they can be held together in the mind of an individual if kept separate.[204] The reason for this is that science concerns "reason cultivated by study" whereas religion refers to "the feelings."[205] This second interpretation of the relationship between science and religion is very much that of Holmberg's.

The Catholic lawyer, politician, and publicist Tristán Achával Rodríguez retorted to Wilde's argument along two lines. He saw no difficulty in disposing of his opponent's conception of scientific progress as a contradiction between successive scientific doctrines—these are "very different" things, he said. And even under the hypothesis that there was a disagreement between fundamental scientific principles as a result of progress, this did not imply that they should necessarily be in conflict with the doctrines of the Church—pressing his point, the Catholic deputy lamented that Wilde had not been able to mention any particular case of opposition between the Holy Writ and modern scientific knowledge.[206] Achával calls attention to the traditional objection that according to Genesis God created light in the first day whereas the Sun was created only two days later. The explanation he provides for his query is that if we accept the wave theory of light, then we conclude that what God did in the first day was to create "the vibrations of the ether," the cosmic matter which, according to [Alexander von] Humboldt, is that which was created in the very first place.[207] In like manner as Sarmiento made the scientific achievements of Benjamin Gould (who was not averse to natural theology) part of a political secularising discourse, Achával transformed Humboldt's monism into an argument for Christianity.[208]

The relationship between science and religion was a much discussed issue in the debate over the educational reform. It was perceived by the liberals as a powerful rhetorical weapon to be used against their adversaries, who resorted to counter-strategies which negated or at least diluted the force of the conflict

---

204 Ibid., 1:565.
205 Ibid., 1:565–66.
206 Ibid., 1:613. See Ludovico L. Lóizaga, *Tristán Achaval Rodríguez. Su vida y su obra* (Buenos Aires: L. J. Rosso, 1930).
207 *Diario de sesiones...Diputados...1883*, 1:614. Cf. Alexander von Humboldt, *Kosmos*, vol. 3 (Stuttgart and Tübingen: Gotta'scher Verlag, 1850), 42–44.
208 Nicolaas Rupke, "Alexander von Humboldt and Monism," in *Monism. Science, Philosophy, Religion, and the History of a Worldview*, ed. Todd Weir (New York: Palgrave Macmillan, 2012), 79–80.

thesis. It should not be lost sight of that this debate was ultimately resolved in terms of political power. In the following section we shall see how the contest over the conflict thesis petered out over the next few years once the victorious secularising crusade lost much of its momentum.

Sequels

The liberal conquest in Parliament was celebrated with a demonstration with the support of masonic lodges and the participation of a troupe of Italian immigrants evocative of Garibaldi.[209] The crowd turned up at Sarmiento's house. An engineering student, who boldly declared himself to be an atheist, came up with an overdramatic version of the conflict theme: "Science and religion are antagonistic things, things pitted against each other... Science has dictated a death sentence and Catholicism writhes in anguished agony." Picking up on the interchange between Wilde and Achával Rodríguez, he proceeded to invalidate the latter's exegetical excursion on the wave theory of light (which he likely considered his preserve).[210] Samuel Gache, who would eventually become an important hygienist and president of the CMA, culminated his address to Sarmiento, the "venerable old man," inviting him to "emulate Galileo's *Epur si muove*...and claim: The world moves on despite what the Sacred Scriptures say!".[211] Untouched by these manifestations of respect, Sarmiento thought wise to set himself apart from the radicalised students, whom he advised to "be respectful of the religious beliefs inherited from our education and historical tradition."[212]

Once the defeat of the "Ultramontane" sector of opinion was guaranteed, the liberal newspaper *La Nación* published a celebratory article which aspired to curb the last rests of Catholic resistance with "the words of science." The note amounts to little more than an almost literal paraphrase of some pages from Herbert Spencer's *Education: intellectual, moral and physical* (1861).[213] The author is

---

209 Auza, *Católicos y liberales*, 150–52.
210 *Discursos pronunciados en la manifestación liberal de la juventud universitaria de Buenos Aires* (Buenos Aires: El Nacional, 1883), 9–13.
211 Ibid., 29.
212 Domingo F. Sarmiento, "Alocución aceptando la visita de los jóvenes estudiantes de la Universidad y Colegios," in idem, *Obras*, ed. Augusto Belin Sarmiento, vol. 22 (Buenos Aires: Imprenta Mariano Moreno, 1899), 196–200.
213 It had been translated to Spanish as Herbert Spencer, *De la educación intelectual, moral y física* (Seville: Administración de la Biblioteca Científico-Literaria; Madrid: Librería de Victoriano Suárez, 1879). Cf. Adriana Novoa, "The Rise and Fall of Spencer's Evolutionary Ideas in Ar-

far from endorsing the conflict thesis; on the contrary, he seeks to argue that religion and science are not at odds if the former is understood as pure, "essential religion" instead of "devotion" or "superstition." This was nicely condensed in two quotations, one from Spencer ("not science, but the neglect of science is irreligious"), the other from Huxley ("True science and true religion are twin-sisters, and the separation of either from the other is sure to prove the death of both").[214] Spencer's arguments about the "religious aspect" of science are carefully rehearsed and the whole is capped with a quote from Tyndall commending a humble approach in the investigation of nature.[215] All this is geared to support the final conclusion of the editorialist that the educational reform approved in the Chamber of Deputies is in consonance with "the last conclusions of paedagogical science."

At that point, "Draper" had turned into something of a bypass word in Buenos Aires, used without particular concern for its precise meaning. For instance, in his answer to the editorial note referred above, the Catholic Goyena agreed that "religion and science do not exclude each other, on the contrary, they complement themselves," in the sense that Francis Bacon had granted this thesis: solving the secrets of nature shows us the necessity and the foundations of faith.[216] But he decries the thesis of the "positivists...the postulates of Comte, Draper, and all that school," according to which science will substitute religion. Goyena blurs the not insignificant differences between positivists and materialists in their common rejection to metaphysics. His occasional contender Sarmiento did no better in this respect. Commenting upon the recent demonstration

---

gentina, 1870–1910," in *Global Spencerism. The Communication and Appropriation of a British Evolutionist*, ed. Bernard Lightman (Leiden: Brill, 2015), 173–91.

214 Herbert Spencer, *Education: Intellectual, Moral and Physical* (New York: Appleton, 1896), 89–92. Huxley's phrase was also taken from this book. It had been pronounced in one the Lectures to Working Men at the School of Mines, see "Science and Religion. Museum of Geology," *The Builder* 17 (1859): 35–36.

215 It was taken from Spencer's *Education*. Cf. the original in [John Tyndall] "What Knowledge is of Most Worth. Lectures on Education Delivered at the Royal Institution of Great Britain. London, 1855," *The Westminster Review* 141 (July 1859): 1–23.

216 [Pedro Goyena], "La ley nefanda y los apóstatas," *La Unión*, 20 July 1883. Goyena quotes in Spanish "A little philosophy inclineth man's mind to atheism; but depth in philosophy bringeth men's minds about to religion." Francis Bacon, "Of Atheism," in *Essays*, ed. M. A. Scott (New York: Charles Scribner's Sons, 1908), 71.

of liberal students, he criticises Goyena for boasting of "not reading any book of the contemporary savants: Darwin, Renan, Huxley, Spencer, and Draper."[217]

Less than two months after these events, in an influential essay signed on 12 September 1883 criticising the secularist bent of the educational reform, the senator and former president of the country Nicolás Avellaneda argued that if the state should control teaching in order to warrant that "only science will be learnt," then science becomes "a contradiction if not an imposture."[218] He portrays the situation as a polar conflict. On the one hand, "Christ in the school" as "the law of duty and the independence of consciousness," on the other hand the overpowering of the person by the state through "science," understood as the "sad sect" which calls itself positivism, "because it counts, weighs, and smashes the atom, without ever finding God at the end of its alchemical manipulations."[219] Far from being an enemy of science, during his presidential tenure Avellaneda had continued Sarmiento's efforts at institutionalising science in Argentina. Commenting on a book of the explorer and naturalist Luis J. Fontana, he affirmed that "the atmosphere resulting from the scientific spirit raises the national mind and strengthens the convictions and prosperity of the people… There is no consistent civilisation without scientific spirit."[220] But as a Catholic, he took his distance from positivism.

In his book reviews, the journalist Santiago Estrada (brother of José Manuel) employs the anti-evolutionist arguments circulating at that time among Catholics. The narrative poem *Prometeo* (1878) by Olegario V. Andrade tells the story of the titan confronting the "outdated" Jupiter as a symbol of humanity, the "Promethean race." The poem refers to "the real Genesis of the world/which the Biblical legend falsifies" and pictures "the old Galileo" defying Rome's "grim fanaticism."[221] Estrada criticises the implicit evolutionism in Andrade's poem on scientific grounds: he claims that all "the learned naturalists, independently of their religious beliefs, contradict that system" (he enumerates 22

---

217 Domingo F. Sarmiento, "Las manifestaciones," in idem, *Obras*, ed. Augusto Belin Sarmiento, vol. 48 (Buenos Aires: Imprenta Mariano Moreno, 1900), 324; originally published in *El Nacional*, 25 July 1883.
218 Nicolás Avellaneda, "La escuela sin religión," in idem, *Escritos y Discursos*, vol. 3 (Buenos Aires: Compañía Sudamericana de Billetes de Banco, 1910), 225.
219 Ibid., 230–231.
220 Nicolás Avellaneda, "*El Gran Chaco*. Libro de don Luis J. Fontana," in idem, *Escritos y Discursos*, vol. 1 (Buenos Aires: Compañía Sudamericana de Billetes de Banco, 1910), 125.
221 Olegario V. Andrade, "Prometeo," in idem, *Obras Poéticas* (Buenos Aires and La Plata: Peuser, 1887), 105 and 107.

of them, including Burmeister).²²² In the 1860s, the liberal historian Vicente Fidel López published a series of articles (translated into French by the Egyptologist Gaston Maspero) in which he argued that the Incas came from the Pelasgians on the ground of alleged similarities of the Quechuan language to Greek.²²³ López's project was a local attempt at using comparative linguistics to invalidate the Biblical account of ancient history; his arguments and thesis were criticised on linguistic grounds by the French-Argentine linguist Juan Mariano Larsen, who had been a Jesuit between 1839 and 1843.²²⁴ We have seen that in the Parliamentary discussion on school reform, López had criticised the Catholic disregard for teaching ancient tongues. In an introduction to a book of the psychiatrist José María Ramos Mejía (on whom more in the next chapter) he repeated his accusation that the Church had been an enemy of the science of language. His prologue is an apology of materialism and evolution: the real "Architect of the universe" is the "natural, eternal, and immovable law." Priestly religious cult, López goes on, will eventually "cease to be worship and turn into an idea, into science, into awe before the universal order."²²⁵ In his review of Ramos Mejía's book, Santiago Estrada recapitulates all the efforts done by the Church in favour of linguistics and the study of ancient tongues (his sources were the apologetic writers familiar to Argentine Catholics in the 1880s, such as Wiseman and Nicolas). Estrada resorts to scientific arguments to disqualify evolution: if transformism were true, all apes should have been transformed into human beings and human beings into angels, but "where is the bureau for refining monkeys in an epoch when even the stars are subjected to laboratory analysis?".²²⁶ Journalistic aftershocks of the educational reform continued calling upon Draper and the conflict thesis, although with diminishing intensity and increasing imprecision.

---

222 Santiago Estrada, "Olegario V. Andrade," in idem, *Miscelánea*, vol. 1 (Barcelona: Herder, 1889), 142.
223 Vicente F. López, *Les races aryennes du Pérou. Leur langue, leur religión, leur histoire*, trans. Gaston Maspero (Paris: Franck, 1871). Cf. Juan A. Ennis, "Las novedosas ciencias del lenguaje y la política de sus usos: Vicente Fidel López en la *Revista de Buenos Aires* (1863–1869)," *Boletín de la Sociedad Española de Historiografía Lingüística* 12 (2018): 53–74.
224 Juan Mariano Larsen, "Filología americana: la lengua quichua y el doctor López," *Revista de Buenos Aires* no. 84 (1869): 409–31; "Larsen, Juan Mariano," in *Gran Enciclopedia Argentina*, ed. Diego Abad de Santillán (Buenos Aires: Ediar, 1956–1963), 4:345–46.
225 Vicente F. López, "Introducción," in José María Ramos Mejía, *La neurosis de los hombres célebres en la República Argentina* (Buenos Aires: La Cultura Argentina, 1915), 80–81.
226 Santiago Estrada, "La neurosis de los hombres célebres en la historia argentina," in idem, *Miscelánea*, vol. 1 (Barcelona: Henrich, 1889), 310–11.

Conclusions

Except for Holmberg, none of the characters in our story could be called a scientist. The cultural atmosphere of late nineteenth-century Buenos Aires was permeated with Comtean or Spencerian positivism but this was a different thing from the burgeoning scientific system which, no matter how sympathetic to these philosophies its leaders could have been, minded its own business. The relevance of our case study rests in that Argentina is a scale model, a condensed image refracted by the local circumstances of much that was going on at that time in Romance-speaking Europe, where the pattern of secularisation was different from that of the anglophone world. What we have here is a mutualistic interplay between a long-trend, politically-driven attempt at curtailing the power of the Church in society and science-based philosophies used for purposes of ideological legitimation and rhetorical expediency.

In his research on the reception of Draper and of Andrew D. White's *History of the Warfare of Science with Theology in Christendom* in anglophone Protestant societies, Ungureanu distinguishes two kinds of response. On the one side, liberal Protestants recognised in these works their own inclination to the separation of theology from religion, on the other side orthodox Protestants and some secularist unbelievers ironically converged on a rejection of the books, although for different motives.[227] This scenario is different from the case under consideration, in which Draper's *Conflict* was uniformly used either as a tool of legitimation of secularising positions by liberals and radicals, or abused as an enemy of religion by advanced and conservative Catholics alike. As has been shown by Jaume Navarro, the reception of Draper in Spain took place in the context of the opposition between liberal Krausists and conservative Catholics, in the years that marked the passage from monarchy to Republic and back to monarchy; much of its import concerned freedom of teaching, the rising of secular educational of higher learning, and the so-called "polemic over Spanish science."[228] The extent to which the reception of Draper in Argentina was independent from that in Spain (a country toward which at that time the elites of Buenos Aires showed an airy disregard) is shown by the above-mentioned fact

---

227 James C. Ungureanu, "Science and Religion in the Anglo-American Periodical Press, 1860–1900: A Failed Reconciliation," *Church History* 88, no. 11 (2019): 120–49.
228 See Navarro, "The Conflicting Circulation;" Mayordomo, "La recepción en España;" and Ernesto and Enrique García Camarero, eds., *La polémica de la ciencia española* (Madrid: Alianza, 1970).

that Cornoldi's review of Draper's books was translated anew in the southern capital.[229]

Many variations upon the theme of conflict between science and religion played up to the liberal audience of Argentina's capital city. Draper was enthusiastically received because the ground had been previously fertilised with an account that could trace its pedigree to French Enlightenment and positivism. More or less radicalised views of the conflict thesis were built out of a shuffling of elements picked up from one or the other of these cognate narratives. Variants upon the conflict theme were used in arguments in favour of secularising politics, such as the abandonment of religious teaching in state-supported schools. The conflict idea is apt for the task. The neat bipartite division of history it implies maps onto an analogous separation of the secular and religious spheres whereas the epistemological devaluation of religion inherent to the thesis legitimizes the restriction of religious influence in society. The isomorphism between the thesis operating at the discursive level and the political field of real conflict reinforced one another.

Also, around 1883–1884 in Argentina evolution was synonymous with Darwin (who had died in 1882) and progressive thinkers were proud to call themselves "Darwinists." The early reception of Darwin took place within the framework of a nascent, ebullient secularist ideology which implied the surrendering of social functions so far in the hands of the Church to a religiously neutral state. Much of the spirit of the Enlightenment still floated over the arguments of the secularists, whose understanding of the relationships between science and religion drew upon Draper's "conflict thesis." The responses of the senior Catholic leaders—almost all of them lawyers—were scientifically uninformed; they saw Darwinism as an enemy of a society based on Christian principles and identified evolution with liberalism, positivism, and secular thought—in other words, with the modernity Rome had condemned in the 1864 *Syllabus* of errors.

In his reminiscences of his generation in the 1880s, the Argentine historian and sociologist Juan Agustín García wrote in 1921: "We smiled when we walked in front of a church…our gods, the successors of the old and inept superstitions, inhabited volumes filled with facts and numbers. The mystery of the Universe had been discovered…".[230] This mildly ironic view from a distance captures the right juncture at which the clash of a new and an old social and religious

---

[229] This kind of comparison points to considerable differences between responses in various societies (even in those with a common religion and language) and reinforce once more the importance of local studies at the time of investigating the relationships between science, religion, and secularisation.

[230] Juan Agustín García, *Nuestra incultura* (Buenos Aires: Claridad, 1922), 17.

order and the ever denser atmosphere of scientism and positivism potentiated each other. In was in this politically charged scenario of secularisation that Draper's book, the conflict thesis, and Darwinism deployed the full strength of their power.

Chapter 5
Science and Secularism: Transitional Times

The participants in the Third International Congress of Freethinkers that took place in Buenos Aires in September 1906 were a mixture of Freemasons, socialists, anarchists, spiritualists, and nondescript radicals. The only scientist attending the assembly was Eduardo L. Holmberg. Two years before, in a preparatory meeting for the congress, he had affirmed that there had never been any conflict between science and religion because such a confrontation would have been like a theoretical discussion between a savant and a two-year old child. Holmberg held religion in contempt: "How is it possible for a person who has read and understood Darwin, Spencer, Lamarck, and Haeckel to lose a moment of time with the absurdities of the Bible?" he claimed.[1] The president of the 1906 congress, Juan Balestra, took a slightly different slant on the issue. In his opening speech, he proclaimed that "the creeds founded on the absolute and the supernatural" had been "definitely overthrown;" it was science which stirred the sincere person to recognise the relativity of all things and "to establish unprejudiced criteria of truth."[2] The incompatibility of the ideologies of the motley crew participating in the meeting emerged on occasion of the speech of the young radical activist and future doctor Alicia Moreau, who proposed that the natural sciences should be the basis of all teaching. While the socialist physician Enrique Dickmann took opportunity to declare that "the popularisation of science should have the working class as its target audience," the conservative Holmberg complained about Dickmann's criticism upon the ruling elite in Argentina.[3] The anarchists laughed out at the proposals of the spiritualists who were forced to leave the conference room in protest in the midst of the ensuing racket—one of the troublemakers was arrested by the police.[4] The differences in the political agendas of the various factions highlight the diversity of forces operating in the secularist camp in pre-World War I Argentina.

One of the objectives of this chapter is to explore some aspects of this kaleidoscope of ideas. Two broad directions of thought can be distinguished here: one of them associated in some way to the idea of "spirit," the other to the notion of matter. Catholic culture was caught in a pincer movement from opposite ends of

---

[1] Quoted in *Album biográfico de los Libre-Pensadores de la República Argentina en el Primer Centenario de su Independencia* (Buenos Aires: "El Progreso." Periódico racionalista, 1910), 89.
[2] "Congreso de Libre Pensamiento. Su inauguración," *La Nación*, 21 September 1906.
[3] "Congreso de Libre Pensamiento. El día de ayer," *La Nación*, 22 September 1906.
[4] Ibid.; Cosme Mariño, *El espiritismo en la Argentina* (Buenos Aires: Constancia, 1963), 221–22.

the secularist spectrum: non-Christian spiritualism and scientific materialism. There were other alternatives related to "philosophical 'substitute-religions'" and to the survival of nineteenth-century anti-religious positivism.[5] After looking at two positivist authors who sought to come to grips with the religious, I shall examine the interactions between spiritualism and Catholics focusing on questions of science and religion. The analysis of the reception in Buenos Aires of the French debate over the bankruptcy of science discussed in the third section of this chapter shows that the secular rationalism of the liberal conservative generation of the 1880s was still in good shape in the 1910s. The following sections investigate how the anti-evolutionist Jesuit José M. Blanco denounced the use of faked evidence to support Florentino Ameghino's theory of human origins and how the scientific intelligentsia invented a "secular religion" around the figure of the famous palaeontologist. The chapter concludes with an analysis of the controversy over mind and matter between Father Blanco and the German-Argentinian neurobiologist Christfried Jakob.

Positivism, Religion, and the Unknowable

Historians of ideas have tried to impose some order on the different currents of positivism in Argentina.[6] It is not my intention to reopen this discussion. Here I shall briefly comment on two authors who adhered respectively to Comte's non-transcendental religion and to Spencer's notion of the Unknowable. Comte's philosophy was formally introduced in Argentina through teachers connected with the Normal School of Paraná in the province of Entre Ríos who were also active in the neighbouring province of Corrientes (both of them situated between the Paraná and the Uruguay Rivers): the Italian-Argentinian naturalist Pietro

---

[5] Donald G. Charlton, *Secular Religions in France, 1815–1870* (London: Oxford University Press, 1963), viii.
[6] Ricaurte Soler, *El positivismo argentino* (Buenos Aires: Paidós, 1968); Hugo Biagini, ed., *El movimiento positivista argentino* (Buenos Aires: Editorial de Belgrano, 1985); José L. Romero, *Las ideas políticas en la Argentina*, 2nd ed. (Mexico: FCE, 1956); Luis Farré, *Cincuenta años de filosofía*; Torchia Estrada, *La filosofía en la Argentina*; Diego F. Pró, "Periodización y caracterización de la historia del pensamiento filosófico argentino," in idem, *Historia del pensamiento filosófico argentino. Cuaderno 1* (Mendoza: Universidad Nacional de Cuyo, Facultad de Filosofía y Letras, 1973), 143–84; Oscar Terán, *Historia de las ideas políticas en la Argentina. Diez lecciones iniciales, 1810–1980* (Buenos Aires: Siglo XXI, 2008); Jens R. Hentschke, "Argentina's Escuela Normal de Paraná and Its Disciples: Mergers of Liberalism, Krausism, and Comtean Positivism in Sarmiento's Temple for Civilizing the Nation, 1870–1916," *Iberian and Latin American Studies* 17 no. 1 (2011): 1–31.

[Pedro] J. Scalabrini and José Alfredo Ferreira.[7] The school was founded by Sarmiento in 1870 as part of his program of introducing North-American education in Argentina, in particular the ideas of Horace Mann. Its first director (1871–1876) had been George A. Stearns, a graduate from Harvard; his wife Julia A. Stearns acted as principal.[8] Scalabrini, hired as science teacher by Stearns, was filled with the ideals of the Italian Risorgimento; he would become a proficient palaeontologist and the founder of two provincial museums of natural history in his adoptive country.[9] The Normal School of Paraná evolved into a hotbed of Comtean positivism, Darwinism, and anticlerical feeling. It was the centre of a movement which in unison with a blend of advanced pedagogical principles and positivism "took to the provinces [the idea of] secular elementary teaching together with the feeling of emancipation from the shallow dogmatism of the sacristy."[10] Writing in 1889, Scalabrini set out a cartography of ideas in fin-de siècle Argentina: "Materialism dominates the School of Medicine in the capital city of Buenos Aires... Darwinism finds its most accomplished representatives in the Museum of La Plata, directed by [Francisco P.] Moreno and [Florentino] Ameghino... Positivism has much influence in the press and has conquered the sympathies of the talented youth in the provinces."[11] The Catholic writer Manuel Gálvez narrated that around the turn of the century in Buenos Aires "all the law students were materialists; they saw Spencer as a god and also Comte, although the latter was only known by hearsay."[12] Scalabrini, a Freemason, rejected materialism due to its "contempt for all religions" and granted that the creeds, in particular Catholicism, had "rendered some services" to humanity (Scalabrini's brother was bishop of Piacenza).[13] His adscription to positivism was somewhat free from the anti-religious intolerance of many of his colleagues. The Italian-Argentinian naturalist admitted that there were two ways of accounting for the exis-

---

[7] Ángela M. Fernández, "Pedagogía y positivismo a fines del siglo XIX: Pedro Scalabrini, Alfredo Ferreira y *La Escuela Positiva*," *Estudios Ibero-Americanos* 29, no. 1 (2003): 73–92.
[8] Samuel G. Inman, "Parana, Exponent of North American Education," *Bulletin of the Pan American Union* 53 (1921): 463–74; Hentschke, "Argentina's Escuela Normal de Paraná."
[9] Juan B. Ambrosetti, "Prof. Pedro Scalabrini (1849–1916). Fundador y director de los museos provinciales de Entre Ríos y Corrientes," *Anales del Museo Nacional de Historia Natural de Buenos Aires* 28 (1916): 227–40; Julio R. Contreras Roqué, Bárbara Gasparri, Adrián Giacchino and Yolanda Davies, *Pedro Scalabrini (1848–1916). Educador y naturalista* (Buenos Aires: Fundación Azara-Universidad Maimónides, 2019).
[10] Alejandro Korn, *El pensamiento argentino* (Buenos Aires: Nova, 1961), 207.
[11] Pedro Scalabrini, *Materialismo, darwinismo, positivismo. Diferencias y semejanzas* (Paraná: La Velocidad, 1889), 6.
[12] Manuel Gálvez, *Recuerdos de la vida literaria*, vol. 1 (Buenos Aires: Taurus, 2002), 68.
[13] Ibid., 13.

tence of the world—the "theological and the scientific"—and saw the logic of supporting "either the supernatural explanation based on faith or the natural explanation based on observation" but considered that holding both at the same time was inconsistent.[14]

Alejandro Korn has appropriately observed that while liberals "polemicised aggressively against Catholic religion...positivists accepted the existence of the Church as a historical fact, without hostility toward it."[15] Ferreira's career amply confirms this idea. An influential educationist, he was a student of Scalabrini and a fairly orthodox Comtean who supported the master's Religion of Humanity.[16] *The Positive School* [*La escuela positiva*], a magazine founded and edited by Ferreira in Corrientes during the last decade of the nineteenth century, featured articles from Scalabrini and other local Comteans. One of them claimed that "the contemporaneous God is the scientific God, destined to unify science and human consciousness as the theological God unified them in the Middle Ages."[17] In a lecture given in the city of Mercedes (province of Corrientes) in 1902, Ferreira discussed his own views on religion.[18] He admitted that science, technology, and progress had made obsolete "the dogmas of our old religions" but although the world had reached the "scientific-industrial" stage of evolution, a "profoundly religious epoch" awaited humanity in the future.[19] Ferreira understood "religion" as "feeling" and considered "the supernatural" as superfluous; following in the steps of Comte he recognised that it was futile to fight against religion, becase humanity's "aptitude for religion" is larger than any science or philosophy.[20] In contradistinction to the liberals of the generation of 1880, who were anti-Catholic and saw Protestantism as the lesser evil, Ferreira valued highly Catholicism; he saw it as a "noble religious synthesis of its times" and claimed that "it had contributed to decisive progresses" in the history of human-

---

14 Ibid., 14.
15 Korn, *El pensamiento argentino*, 210.
16 For Comte's Religion of Humanity see, for example, Henri de Lubac, *The Drama of Atheist Humanism*, trans. Anne E. Nash (San Francisco: Ignatius Press, 1995), 131–268; Andrew Wernick, *Auguste Comte and the Religion of Humanity. The Post-Theistic Program of French Social Theory* (Cambridge: Cambridge University Press, 2001); Mary Pickering, *Auguste Comte. An Intellectual Biography*, vols. 2 and 3 (Cambridge: Cambridge University Press, 2009).
17 Ricardo Colbert, "La evolución de Dios," *La escuela positiva* 4, no. 6 (1898): 253–55, quoted in Malvina A. de Gabardini, *Revista "La escuela positiva" (Corrientes, 1895–1899). Introducción e índices* (Resistencia: Editorial Gómez Lestarri, 1995), 35.
18 José A. Ferreira, "Evolución y educación religiosa," *Revista Positiva* (Mexico, D.F.) no. 54 (26 March 1905), 165–89.
19 Ibid., 171, 173.
20 Ibid., 187–88.

kind.²¹ He was not altogether satisfied with secular elementary instruction and prophesied that in the end "education would be religious"—after all, he went on, for all their anticlericalism the great liberal figures of the country used to "send their children to Catholic schools."²² In an article published in 1926, Ferreira commented upon the 1885 controversy between Spencer and the English Comtean Frederic Harrison and cleary took the side of the latter.²³ Although sympathetic to Comte's Religion of Humanity, Ferreira did not practice its liturgy; contrarily to what happened in Brazil, no positivist chapels were ever erected in Argentina.

While Ferreira followed Comte's Religion of Humanity, the sociologist and writer Carlos Octavio Bunge celebrated Spencer's notion of the Unknowable even when the popularity of the British philosopher in Argentina was on the decline.²⁴ It has been claimed that Bunge's interest in the Unknowable was consistent with his desire to conciliate positivism with "those moral and existential questions which connected him with the fin-de siècle spiritual malaise."²⁵ Bunge thought that religion would die a peaceful death and it was not worth combating it. He believed that the confrontation between science and religion was "childish" for each of them had its particular sphere: "science has to do with the intelligence and religion with feelings." In this view, religious statements are devoid of truth value and seen as "conventional lies." Bunge, born into a family which grew very rich from the grain trade, was a conservative who thought that religion was useful "to bring some method to the revolution and to help it reach [its goals] in a peaceful and evolving way."²⁶ But his advo-

---

21 Ibid., 179, 182.
22 Ibid., 174, 184.
23 José A. Ferreira, "La concepción religiosa de Comte y Spencer," in idem, *Ensayos de ética* (Buenos Aires: Imprenta Ferrari, 1944), 135–55; originally published in *El positivismo* 1, no. 2 (1926): 3–21. See Frederic Harrison and Herbert Spencer, *The Nature and Reality of Religion. A Controversy* (New York: Appleton, 1885); Sydney Elsen, "Frederic Harrison and Herbert Spencer: Embattled Unbelievers," *Victorian Studies* 12, no. 1 (1968): 35–56.
24 Carlos O. Bunge, *Principios de psicología individual y social* (Madrid: Daniel Jorro, 1903), 229–38. See Novoa, "The Rise and Fall," who argues that by the first decade of the twentieth century Spencerism was wanig.
25 Oscar Terán, *Vida intelectual en el Buenos Aires fin-de-siglo (1880–1910)* (Buenos Aires: FCE, 2000), 147.
26 Carlos O. Bunge, "Notas de psicología social," in idem, *Estudios filosóficos* (Buenos Aires: La Cultura Argentina, 1919), 149–50 (originally published in 1894–1895). Bunge's brother Alejandro was an influential Catholic economist and his sister Delfina was a Catholic writer. See Eduardo J. Cárdenas and Carlos M. Payá, *La Argentina de los hermanos Bunge, 1901–1907* (Buenos Aires: Sudamericana, 1995); Hernán González Bollo, *La teodicea estadística de Alejandro E.*

cacy of Spencerian evolutionism was a far cry from the socialist endorsement of "the new materialist and evolutionist philosophies" which he saw as "metaphysically absurd" hypotheses.[27] In his imaginative and dreamlike essay "Travel through the [human] lineage," Bunge distinguishes Darwin's theory from its materialist interpreters. The passage awakens echoes of Spencer: "His [Darwin's] theory is not as anti-religious as some superficial critics and commentators pretend. In the Origin of Life, of Matter, of Force, Darwin always found an Unknown Principle which can be called God. And in his most inner and secret being, he adored God, the God of his grandparents."[28] Two worlds coexist: "the world of the phenomena and science, the human world, and the world of myths and religion, the supra-human world."[29] In Bunge's short story, Darwin himself affirms: "I have never attacked God... The human being will never discover the Beginning and the End. And the Beginning and the End are always God... Man [sic] carries God within himself. He carries within himself the aspiration of creating God in the world of his intelligence."[30] In Bunge, a view of science and faith as different spheres is associated to a non-materialist interpretation of evolutionism and a disbelief tolerant of the "social fictions" of religion.[31]

Following in the footsteps of either Comte or Spencer, the authors discussed in this section strove to include some form of the religious into their positivist interpretations of human life leaving out any trace of the transcendent sacred.[32] They were both dissatisfied with materialism (Bunge claimed that it was "absurd to base the solid construccion of the moral sciences upon the study of biology") but neither stepped beyond the boundaries of the secular.[33] It could be said of them what Charlton said of their French contemporaries: "they still believed that religious commitment of one sort or another" was "essential—even when they interpreted 'religion' as little more than an ethical, political, or metaphysical system."[34]

Bunge, 1880–1943 (Buenos Aires: UCA-Imago Mundi, 2012); Lucía Gálvez, *Delfina Bunge. Diarios íntimos de una época brillante* (Buenos Aires: Planeta, 2000).

27 Bunge, *Principios de psicología individual*, 232–33.
28 Carlos O. Bunge, "Viaje a través de la estirpe," in idem, *Viaje a través de la estirpe y otras narraciones* (Buenos Aires: Biblioteca de La Nación, 1908), 28.
29 Ibid., 32.
30 Ibid., 33.
31 Carlos O. Bunge, "La enseñanza de la tradición y la leyenda," *El Monitor de la Educación Común* 36, no. 458 (1911): 269.
32 This has been described as the "spiritualisation" of positivism in Argentina, in Terán, *Vida intelectual*, 148.
33 Bunge, *Principios de psicología individual*, 233.
34 Charlton, *Secular Religions in France*, 4.

Spiritualism, Theosophy, and Science

When Henry Steel Olcott visited Buenos Aires in 1901, there were three theosophical branches operating in it; a local section of the Theosophical Society was established in 1919.[35] In the first decade of the twentieth century the esoteric and the occult proved terribly attractive to some intellectuals in Buenos Aires. As Lynn Sharp said of French nineteenth-century freethinkers and romantic socialists, the "secular spirituality [of spiritualism] ... and the natural supernatural offered a 'third way' between the secular and the Catholic."[36] The psychiatrist Giuseppe Ingegnieri (José Ingenieros) and Alfredo Palacios (the first socialist member of Parliament in 1904) graced the meetings of one of the Theosophical groups in Buenos Aires ("Luz" [Light], founded in 1893) and contributed articles to its journal *Philadelphia*.[37] The modernist poet Leopoldo Lugones, who was distancing himself from his youthful radicalism, was also a member of this circle. The articles of these authors at that time denounced the epistemological limits of positivism and materialism and proclaimed the need to look for something beyond. Alfredo Palacios declared that "science has stopped satisfying the aspirations of human intelligence and strives now for broader horizons where to deploy its activity" (proof of which were Sir William Crookes's spirit photographs). He commended Haeckel's monism, which in his view was far from being a "vulgar materialism" and opened the way to "the innumerable elementary spirits of the cells."[38] Ingenieros thought similarly: to "orthodox" science practiced by "official" savants within the narrow limits of standard methodology he opposed the path-breaking studies of Crookes and Alfred Russel Wallace, who drawing upon occultism and Theosophy had revealed new phenomena in contradiction with the laws of official science.[39] In Ingenieros's

---

35 Henry Steel Olcott, *Old Diary Leaves. The Only Authentic History of the Theosophical Society. Fourth Series, 1887–1892* (London: The Theosophical Publishing Society; Madras: Theosophist Office, 1910), 500.
36 Lynn L. Sharp, *Secular Spirituality. Reincarnation and Spiritism in Nineteenth-Century France* (Lanham: Rowan and Littlefield 2006), 207.
37 Dora Barrancos, "El otro rostro de la modernidad: socialistas y ciencia esotérica (1890–1930)," *Estudios sociales* (Santa Fe) 21, no. 40 (2011): 101–26; Soledad Quereilhac, *Cuando la ciencia despertaba fantasías. Prensa, literatura y ocultismo en la Argentina de entresiglos* (Buenos Aires: Siglo XXI, 2016), 125–59; Pedro L. Barcia, "Lugones y los saberes ocultos," in Leopoldo Lugones, *Estudios esotéricos*, ed. Pedro L. Barcia (Buenos Aires: Docencia, 2018), 9–86.
38 Alfredo Palacios, "Las religiones y las ciencias ante la teosofía," *Philadelphia* 3, nos. 4 and 5 (7 October and 7 November 1900), as quoted in Barrancos, "El otro rostro," 115.
39 José Ingenieros, "La ciencia oficial y la Facultad de Ciencias Herméticas," *La Montaña* 1, no. 11 (1st September 1897): 6–7. For spiritualism and science, see Richard Noakes, *Physics*

view, hermetic knowledge fit well with political radicalism and could be used to combat religious and scientific orthodoxy.

In his 1898 contribution to *Philadelphia*, Lugones initiated his lifelong commitment to theosophical doctrines. At that time he perceived a strong "spiritualist reaction" against "recalcitrant positivism" and the failure of the socialism he had once embraced. Picturing himself as one of those "who once believed in materialism but have now turned to God," he affirms that it is necessary "to believe without explaining the reason for it." Lugones rejects dogmatic religions and demands a mystery religion: "we wish to have religion, we wish to affirm the Absolute." Science should also "take part in this great movement."[40] Born in the city of Córdoba in 1874 and with only two years of high-school learning, Lugones arrived in Buenos Aires in 1896 and soon joined the theosophist circles. As a rebellious youth, he had read about evolutionism, proclaimed himself a socialist and then an anarchist, and marched down the streets of his native city shouting "Death to God!".[41] In 1893, he published his first poem, "The Worlds," an imaginative account of the evolution of the universe in which he refers to the priestly office of the savants, those "divers of darkness" who "dictate the word of science" to the world.[42] Much of Lugones's literary output would rest upon a combination of scientific and occult motives and arguments. Several of the fantastic short stories he published in 1906 (*Strange forces*) have plots dealing with the scientific imagination; the compilation is complemented by the "Essay of cosmogony in twelve lessons."[43] It is generally accepted that "the basic common goal of the fictions and the essay is the articulation of scientific and theosophical conceptions in a harmonious blend."[44] On 14 August 1920, Lugones gave a lecture in the University of Buenos Aires later published in book format as *The*

---

and Psychics. The Occult and the Sciences in Modern Britain (Cambridge: Cambridge University Press, 2019); Courtenay Raia, *The New Prometheans. Faith, Science, and the Supernatural Mind in the Victorian* Fin de Siècle (New York: Columbia University Press, 2019).

40 Leopoldo Lugònes, "Acción de la Teosofía," in idem, *Estudios esotéricos*, 91–101.

41 Leonardo Castellani, *Lugones* (Buenos Aires: Biblioteca Nacional, 2011), 119–20; Efraín U. Bischoff, *Leopoldo Lugones. Un cordobés rebelde* (Córdoba: Editorial Brujas, 2005), 6.

42 "Y el sabio, el sacerdote / que dicta de la Ciencia la palabra; / buzo de las tinieblas." Leopoldo Lugones, "Los Mundos," *Nosotros* 3, no. 7 (1938): 173.

43 Leopoldo Lugones, *Strange Forces*, trans. Gilbert Alter-Gilbert (Pittsburgh: Latin America Literary Review Press, 2011).

44 Pedro L. Barcia, "Composición y temas de *Las fuerzas extrañas*," in Leopoldo Lugones, *Las fuerzas extrañas*, ed. Pedro L. Barcia (Buenos Aires: Ediciones del 80, 1984), 12. Cf. José M. Martínez, "La arquitectura de *Las fuerzas extrañas* de Leopoldo Lugones: analogía universal y taxonomías de lo fantástico," *Bulletin of Hispanic Studies* 96, no. 5 (2019): 533–52.

Size of Space. An Essay on Mathematical Psychology (1921).⁴⁵ In his speech, he argued against the intuitive notion of space and claimed that geometrical notions (e. g. Riemann's non-Euclidian geometry or Minkowski's four-dimensional manifold) should be considered in purely rational terms. In Lugones's *The Size of Space* Einstein's cosmology becomes part of the theosophical doctrine of cosmic evolution. A newspaper article published at that time even celebrated Lugones's anticipation of Einstein's idea of a finite measurable universe.⁴⁶ When in his 1925 visit to Buenos Aires a journalist asked the latter what he thought about this claim, Einstein answered that "Lugones's ideas on the subject were rather...nebulous!".⁴⁷ Lugones was a self-taught pundit with an impressive literary talent which allowed him to get away with conceptual mystifications; a contemporary Catholic critic who knew him personally put it bluntly: "He touched upon all fields, but always incompletely and without any serious foundation—except for his poetry, which was perfect."⁴⁸

Lugones is significant for our inquiry because he was considered a scientific authority even by those who should have known better. A sizable part of scientific discourse in fin-de siècle Argentina was used to convey philosophical ideas. It is significant that Eduardo Holmberg and Eduardo Wilde, who should be counted among the staunchest supporters of secularist science, were also creative persons of letters who gave shape to the new genre of scientific fantasy and through a diversified literary production helped to fashion a certain cultural image of science. Literary critic Sandra Gasparini has convincingly shown how the emergence of this genre in Argentina was associated to the consolidation of modern science in the 1880s and to the diffusion of evolutionist ideas.⁴⁹ The occult sciences and spiritualism allowed those who did not wish to overstep the boundaries of positivistic science to experience the frisson of the occult without losing face.

---

45 Leopoldo Lugones, *El tamaño del espacio. Un ensayo de psicología matemática* (Buenos Aires: El Ateneo, 1921); see the English versión in Miguel de Asúa and Diego Hurtado de Mendoza, trans., "The Size of Space (An Essay on Mathematical Psychology). Leopoldo Lugones," *Science in Context* 18, n° 2 (2005): 317–36; cf. Miguel de Asúa and Diego Hurtado de Mendoza, *Imágenes de Einstein. Relatividad y cultura en el mundo y en la Argentina* (Buenos Aires: Eudeba, 2006), 240–61.
46 "'El tamaño del espacio' por Leopoldo Lugones," *La Nación*, 13 February 1921.
47 Conrado Eggers-Lecour, "Einstein, la inteligencia más madura y el espíritu más joven del siglo," *El Hogar* 21, no. 809 (1925): 7, 64.
48 Castellani, *Lugones*, 76.
49 Sandra Gasparini, *Espectros de la ciencia. Fantasías científicas de la Argentina del siglo XIX* (Buenos Aires: Santiago Arcos, 2012), 44. See also Oscar Hahn, "Estudio," in idem, *El cuento fantástico hispanoamericano en el siglo XIX* (Mexico: Premia, 1978), 9–84.

Whereas Theosophy's clientele pertained to the bourgeoisie and to artistic circles, spiritualism catered to a more popular public. Spiritualism was introduced in Argentina by the Spanish immigrant Justo de Espada in the 1870s. In the 1880s there were two vigorous spiritualist societies, *Constancia* (1877) and *Fraternidad* (1880). The former was founded by Angelo Scarnicchia and engineer Rafael Hernández, the brother of José, author of the *Martín Fierro* (the national saga of gaucho literature). *Constancia*, which published a journal of the same name, was soon led by Cosme Mariño.[50] Eventually, a more scientifically oriented group emerged, the Argentine Magnetological Society presided by the chemist Ovidio Rebaudi, who had been born in Asunción (Paraguay) and ended his days in his native country. The Magnetological Society had laboratories of chemistry, physics, and magnetology and published a journal, the *Revista Magnetológica*.[51] Spiritualists fought against two fronts: positivist scientists and Catholics, who in turn opposed one another. In his memories, Mariño told that "by 1881 the spiritualist 'masonry' began to call the attention of the priests and of a few men of science."[52] Noted positivist psychiatrists such as José María Ramos Mejía and Lucio Meléndez denied the phenomenon of hypnotism and affirmed that spiritualism was a cause of madness.[53] In a speech given by the spiritualist Hernández, a student rose and claimed defiantly that "Comte's positivism is the philosophical school dominating scientific circles; despite [the vogue of] spiritualism, today's science is materialist." The lecturer recommended the student to read Spencer, Tyndall, and Schopenhauer and called Comte's positivism a "mutilated and deficient metaphysics."[54] By the turn of the century, spiritualist leaders sought an alliance with socialists in order to confront their common enemy, Catholicism. Mariño and the above mentioned Alfredo Palacios tried to compete with Father Friedrich (Federico) Grote's Workers' Circles creating an analogous organisation; the experiment did not attract supporters among the workers

---

50 Cosme Mariño, *El espiritismo en la Argentina* (Buenos Aires: Editorial Constancia, 1963); Bianchi, *Historia de las religiones en la Argentina*, 140–49; Gustavo A. Ludueña, "Popular Epistemologies and 'Spiritual Science' in Early Twentieth-Century Buenos Aires," in *Handbook of Religion and the Authority of Science*, ed. James R. Lewis and Olav Hammer (Leiden: Brill, 2010), 609–31.
51 Anna J. Conforte, "Cultura científica, magnetología y espiritismo durante fines del siglo XIX en Buenos Aires: un análisis de Ovidio Rebaudi," *Anuario de Investigaciones de la Facultad de Psicología, UBA* 24 (2018): 25–33.
52 Mariño, *El espiritismo*, 46.
53 Ibid., 48, 51.
54 Ibid., 97.

and by 1916 the Socialist Party hardened its line and expelled anyone who would defend spiritualism.⁵⁵

Catholics reacted against the spread of spiritualism. In 1878, Miguel Navarro Viola translated to Spanish Camille Deban's *Discours contre le spiritisme* (1865).⁵⁶ The peak of the controversy took place in 1883 coinciding with the polemic about the secularisation of elementary schools. When Mariño published a spiritualist cathecism of religion, the Catholic daily *La Unión* proclaimed that spiritualist apparitions were "demonic phenomena."⁵⁷ A week later, the newspaper launched a series of articles reproducing an anti-spiritualist book by the Catalan priest Félix Sardá y Salvany.⁵⁸ Mariño answered to this challenge with an article against *La Unión* in the pages of *Constancia*.⁵⁹ Four years later, in October 1887, both the Catholic newspaper and the liberal *El Nacional* published anonymous notes making fun of Mariño and his spiritualist catechism; the latter described the spiritualist centres in Buenos Aires as "madhouses."⁶⁰ In that month, on the occasion of the Commemoration of the Faithful Departed, Father Grote (at that time the leader of social Catholicism and a relevant public figure) prayed a novena in the Redemptorist church of Our Lady of Victories in Buenos Aires and preached a series of sermons critical of spiritualism. *La Unión* published brief celebratory reports of some of them.⁶¹ In turn, Mariño published in *El Nacional* a series of long notes with his refutation of Father Grote's sermons.⁶² These were answered by two articles in the Catholic newspaper. In the first, the author argues that spiritualism is neither religion, nor philosophy, nor science.⁶³ The second article explains why spiritualism is not and could never be scientific. The author claims

---

55 Juan P. Bubello, "De 'Jesús no es Dios' a 'Jesús...es el verdadero fundador del socialismo.' Ocultismo y política en el espiritismo kardecista argentino (1870–1930)," *Melancolia* 1 (2016): 51–74.
56 Camille Debans, *Discurso sobre el espiritismo...por un médico incrédulo*, trans. Miguel Navarro Viola, in *La Biblioteca Popular de Buenos Aires* 13 (1878): 126–73.
57 "Supersticiones," *La Unión*, 17 June 1883.
58 The first was "¿Qué hay sobre el espiritismo?," *La Unión*, 26 June 1883; the last was published on 1st August 1883. Cf. Félix Sardá y Salvany, *¿Qué hay sobre el espiritismo? Cuatro palabras sobre esta secta* (Barcelona: Tipografía Católica, 1872).
59 [Cosme Mariño], "*La Unión* y el espiritismo," *Constancia* 6, no. 7 (30 July 1883): 193–96.
60 "Producciones espiritistas," *La Unión*, 19 October 1887; [Untitled article on Mariño], *El Nacional*, 25 October 1887; "Un espiritista serio y razonado," *La Unión*, 21 October 1887.
61 The first was "Conferencia sobre el espiritismo," *La Unión*, 26 October 1887. For Grote, see Néstor T. Auza, *Grote y la estrategia social*, vol. 1, *Aciertos y fracasos sociales del catolicismo argentino* (Buenos Aires: Docencia, Don Bosco and Guadalupe, 1987).
62 The first was Cosme Mariño, "Refutación a los sermones del Padre Groote [sic] en la Iglesia de las Victorias," *El Nacional*, 31 October 1887.
63 "Pláticas y refutaciones. El espiritismo carece de base," *La Unión*, 3 November 1887.

that since it came to life in New York by mid-nineteenth century, spiritualism has contributed nothing to technology and progress "while electricity and steam have effected wonders." Moreover, it is rejected by "materialist savants" who attribute spiritualist phenomena to fluids which do not obey any natural law and also by Catholic scientists, who impute them to the agency of an evil intelligent creature.[64] In his 1887 diocesan tour, Msgr Aneiros presented each parish priest with a copy of the Spanish translation of Ambroise Matignon's *Les morts et les vivants. Entretiens sur les communications d'outre-tombe* (Paris, 1862), an antispiritualist book.[65]

The articles against spiritualism published in the Jesuit journal *Estudios* can be taken as a sample of local Catholic opinion after the Holy Office had issued the decree of 24 July 1917 condemning the assistance of Catholics to spiritualist séances.[66] One of them was written from the point of view of "empirical psychology." After reviewing the different phenomena associated with spiritualism (levitation, materialisations, talking tables, spirit rapping, automatic writing) the author concludes that "though natural explanations are so far unsatisfactory, it seems there is no necessity as yet for appealing to preternatural explanations."[67] Most priests thought otherwise. The Catalan Jesuit Juan Planella, a classical scholar who also wrote on ancient religions and taught in the Archdiocesan Seminary in Buenos Aires, criticised as a fraud the case of a Chilean man who could supposedly see through solid objects and thus diagnosed internal affections—he was popularly known as "the man of x-ray vision."[68] Later, Planella published a series of articles commenting on a case which had excited public interest. In 1921, distressing phenomena began to occur in the house of the Mullaly family, members of the Irish colony in the town of San Andrés de Giles, near the capital city. As reported in *The Standard*, one of the English newspapers in Buenos Aires, "the spectre appears to be one of the unpleasant type classified by the German savants (great authorities on Spooklore) as a Poltergeist."[69] An insulting and mocking voice was heard during the night, pieces of furniture

---

64 "Sobre espiritismo. Por qué el espiritismo no es, ni ha sido, ni jamás será ciencia," *La Unión*, 6 November 1887.
65 Ambroise Matignon, *Los muertos y los vivos. Conferencias acerca de las comunicaciones de ultratumba* (Barcelona: Tipográfica Católica, 1872). See [Cosme Mariño], "El arzobispo señor Aneiros y el Espiritismo," *Constancia* 10, no. 130 (30 January 1887): 1–2.
66 "De spiritismo," *Acta Apostolicae Sedis* 9, no. 1 (1917): 268.
67 Vicente Segarra, "La psicología empírica y los fenómenos mediánicos," *Estudios* 18 (January–June 1920): 291.
68 Juan Planella, "¿Impostura o verdad?," *Estudios* 13 (July–December 1917): 44–53.
69 "The Giles Ghost. Violent Visitation from Nether World," *The Standard*, 28 October 1921.

flied round the apartment, and one of the children exhibited signs of snake bite in an arm. The phenomena were attributed to the spirit of a fellow Irishman, Peter Rossiter, who had recently died in the neighbourhood in a ditch full of water. When a Passionist priest said mass at a nearby church, "the crucifix was dashed to the ground, the candles were blown out, and the building shook."[70] Other Catholic priests attended the case, said mass, and performed the rites of exorcism but to no avail. Eventually, the spirit ceased to torment the family. Joaquín Trincado, an anticlerical socialist who in 1911 had created the "Magnetic Spiritualist School of the Universal Commune," interpreted the case in terms of spiritualism and took opportunity to denigrate the clergy.[71] Newspaper reports were written in a tongue-in-cheek style.[72] Father Planella concluded that this was a case of "diabolic influence" and claimed that what happened in spiritualist séances should be attributed to the evil one and not to the spirits of the death; in his view, hypnotism had the same supernatural causation.[73] At long last, Planella visited San Andrés de Giles, talked with the members of the family, and performed some kind of informal exorcism, which was also unsuccessful. His final conclusion was that "all the apparitions of spirits and messages of [dead] souls are the product of the devil."[74]

In the years following the Great War, the occult ceased to attract the attention of the scientifically-minded local intelligentsia (Lugones was the exception). There was also a growing mood of disillusionment with the promises of unending scientific progress. In 1920, the writer and historian Carlos Ibarguren, a conservative nationalist, regretted his youthful faith in science: "Idealism and spiritualism were drowned in the new god: the laboratory which revealed to men [sic] the truth of positive science. The modern scientific spirit, which made us see everything through the prism of matter, taught us also that determinism is the law of the universe."[75] But for many the authority of science remained intact and ca-

---

[70] "Do the Death Return? Strange Tale from Carmen de Areco," *Buenos Aires Herald*, 28 October 1921.
[71] Joaquín Trincado, *El espiritismo estudiado o política del creador y gobierno del espiritismo* (Buenos Aires: Voz Informativa, 1922).
[72] "Los muertos no hablan. Misterioso fenómeno que se viene produciendo en la provincia;" "El temible espíritu del irlandés de Giles;" "El extraño fenómeno de San Andrés de Giles," *La Razón*, 27, 28 and 29 October 1921.
[73] Juan Planella, "Lo de Giles y a propósito de ello," *Estudios* 22 (January–June 1922): 334, 340, 345.
[74] Juan Planella, "Un testimonio más sobre 'lo de Giles'," *Estudios* 23 (July–December 1922): 475.
[75] Carlos Ibarguren, *La literatura y la Gran Guerra* (Buenos Aires: Agencia General de Librería y Publicaciones, 1920), 7–8.

pable of turning its full firepower against religion. This is the theme of the following section.

The Bankruptcy of Science

In 1895, an article of the literary critic Ferdinand Brunetière unleashed a polemic that touched the very nerve of the conflict between Catholicism and science in France. The "bankruptcy of science" announced by Brunetière became a symbol of the rising challenge posed by Catholics to the triumphant scientism of the Third Republic.[76] With a delay of two decades, the question was taken up in Buenos Aires, where it flared up during the crepuscular years of secularist conservative governments. The French debate—a conceptually rich and politically involved affair—was reduced in Argentina to a black and white issue: progressive science against backward religion. Common to both settings was the alliance between science and the secularist state: those who wrapped themselves in the banner of science belonged to the upper echelons of the state bureaucracy and spoke confidently from a position of power. Any challenge to science was perceived as a challenge to the identity of the nation.

Brunetière was a powerful voice in the Parisian literary world. Maître des conferences in the École normale superieure, professor at the Sorbonne, and since 1893 member of the Académie française and director of the *Revue des Deux Mondes*, he was a rationalist who had published essays on Darwin and in his lectures defended an evolutionary view of literary genres.[77] His 1895 article "Après une visite au Vatican" was the result of the private interview he had held the previous year with Pope Leo XIII.[78] Brunetière was not at that time a Catholic (he converted in 1900). The article called for a joint action of rationalists and Catholics against the view (defended by Taine among others) that the moral sciences should be based on the natural sciences. In 1891, Leo XIII had issued the encyclical *Rerum novarum* on social Catholicism; a year later he proclaimed his *Inter Sollicitudines*, a call to the French Catholics to collaborate with the secula-

---

[76] Harry W. Paul, "The Debate over the Bankruptcy of Science in 1895," *French Historical Studies* 5, no. 3 (1968): 299–327.
[77] Ferdinand Brunetière, *L'évolution des genres dans l'histoire de la littérature*, 2 vols. (Paris: Hachette, 1890–1892); idem, "Charles Darwin," *Revue politique et littéraire*, 3rd series, 2, no. 17 (29 April 1882): 518–24. See John Clark, *La pensée de Ferdinand Brunetière* (Paris: Nizet, 1954).
[78] Ferdinand Brunetière, "Après une visite au Vatican," *Revue des Deux Mondes*, 4th period, 127, no. 1 (1st January 1895): 97–118.

rist Third Republic (the *ralliement*). The relationships between Catholics and republicans were strained to the maximum with the Dreyfus affair (1894–1906), in which Brunetière played the role of a moderate anti-Dreyfusard.[79] In his critique of science as the basis of moral knowledge, Brunetière underscores the boundaries among the natural, the philological, and the historical sciences. He claims that the physical sciences "are impotent...to address the only important questions":[80] where do we come from? What is our destiny? What are we? Evolutionary science can provide an account of our animal origins, but not of human origins. With Renan as his main target, Brunetière postulates the limits of textual critique applied to the Bible and comments on the incongruences resulting from applying the comparative study of ancient religions to Christian origins. The central idea of the article is that if science has aspired to replace religion, for the time being it has lost the game.[81] Two chemists, the Nobel-prize winner Charles Richet and Marcellin Berthelot—who saw himself as a token of the alliance between science and the Third Republic—answered to Brunetière's challenge.[82] On 4 April 1895, a large banquet was celebrated in honour of Berthelot, a social event seen by the public as a glorification of science and a devastating counter-attack of the Parisian illuminati against Catholic reaction (the minister of public instruction and the president of the Chamber of deputies held places of honour and big names such as Auguste Rodin and Émile Zola were among the attendees).[83] It is time to return to Buenos Aires and survey the local reactions to Brunetière's article.

Antonio F. Piñero was a psychiatrist, professor of clinical psychiatry, and director of the Women's Asylum between 1890 and 1905. He graduated from the University of Buenos Aires with a thesis on brain localisation of vision and auditory disorders and then studied medicine in Paris.[84] As director of the Women's Asylum, Piñero was a progressive reformer who wrestled from the state support

---

[79] Thomas Lue, "L'apologétique de Ferdinand Brunetière et le positivisme: un bricolage idéologique 'génereux et acueillant'," *Revue des sciences philosophiques et théologiques* 87, no. 1 (2003): 101–26.
[80] Brunetière, "Après une visite au Vatican," 99.
[81] Ibid., 104.
[82] Charles Richet, "La science a-t-elle fait banqueroute?," *Revue scientifique* 3 (12 January 1895): 33–39; Marcellin Berthelot, "La science et la morale," *Revue de Paris* (1st February 1895): 449–69.
[83] Jacqueline Lalouette, "La querelle de la foi et de la science et le banquet Berthelot," *Revue historique* 300, no. 608 (1998): 825–44.
[84] Antonio F. Piñero, *Localizaciones cerebrales en las alteraciones auditivas y de la visión* (Buenos Aires: La Nación, 1883); "Antonio F. Piñero," in *Argentines of Today*, ed. William Belmont Parker, vol. 2 (Buenos Aires and New York: The Hispanic Society of America, 1920), 865–66.

to build impressive new facilities for the institution (designed by the Swedish architect who had reformed the Salpetrière) and also planned a rural colony for the patients.[85] In 1906, he was elected to the Chamber of Deputies. In an address to the Chamber on a bill regulating women's and child labour introduced by Alfredo Palacios, Piñero defended the benefits of science "in the economic, political, and moral order of things," as had been shown by Berthelot on the occasion of the debate about the bankruptcy of science. Piñero diagnosed the idea of science going bankrupt as "the mystical manifestation of an unbalanced hysterical individual [i.e. Brunetière], a delusion of those alien to the scientific spirit."[86]

Pedro Arata, a chemist, physician and the organizer of the School of Agronomy and Veterinary of the University of Buenos Aires, confronted the ideas of Spencer and Brunetière. In his speech during a banquet celebrated in 1910 in honour of the university, Arata started with Spencer's notion of science as a sphere of ever increasing radius expanding into the infinite unknowable and dismissed as "backward spirits" those who believed that the unfathomable depths of the unknown announced the bankrupt of science.[87] In 1915, Martín Gil, a Cordobese writer and politician well-known in his time as a populariser of astronomy, called Brunetière a "person sick with chronic backwardness." He asserted that the problem of those who like Brunetière decry science on the grounds that it cannot reach the ultimate mysteries is that they throw away the baby with the bathwater: science does not aim at absolute truth, which "the human brain cannot hold." An enthusiast of progress, Gil looked patronizingly upon those weak souls unable to bear "the cold and penetrating gaze of the great enigma" of the universe and who took comfort in the non-scientific pronouncements of the "sacred books."[88] "Peace through science" was the title of the 1914 commencement speech in which the rector of the University of La Plata, Joaquín V. González, intertwined the themes of nationalism and science. In it, he declares that science constitutes the "social core of our national identity." Science, he goes on, substitutes "the authoritarian dicta of inherited religious and philosophical dogmas." Only science is capable of satisfying the spirit of international social justice, longed for by "those saints of an undogmatic profane religion" such as Franklin,

---

85 Jonathan P. Ablard, *Madness in Buenos Aires. Patients, Psychiatrists, and the Argentine State, 1880–1983* (Calgary: University of Calgary Press, 2008), 35–37.
86 Congreso Nacional, Cámara de Diputados, *Diario de sesiones...Año 1906*, vol. 1 (Buenos Aires: Talleres Gráficos de la Penitenciaría Nacional, 1907), 876–78.
87 Pedro Arata, "Banquete de la Universidad de Buenos Aires," *Revista de la Universidad de Buenos Aires* 13 (1910): 135–38. Cf. Diego Medan, *Pedro Arata. Un viaje a los cimientos de la Argentina* (Buenos Aires: Eudeba, 2017).
88 Martín Gil, "La bancarrota de la ciencia," *Caras y caretas* no. 792 (6 December 1913), 106–7.

Jefferson, and Washington. Anyone who declares "the bankruptcy of science" is wrong for the only problem with science is that those in power disregard its liberating potential from traditions and dogmatic religion.[89]

In a 1915 article criticising intuitionism and Madame Blavatsky's Theosophy (which a decade before he had warmly recommended), Ingenieros identified Bergson's followers as heirs to Brunetière's vision. Both the intuitionist philosophers and those who enthused about the prospect of the bankruptcy of science, he said, deprecated the disciplines they were unable to learn and found consolation in "the door opened into the path of Paradise."[90] In 1919, almost twenty-five years after Brunetière published his momentous article, Armando Donoso, a Chilean literary critic, published an essay on him in the journal edited by Ingenieros. Donoso departs from "the insoluble antinomy between science and [Christian] faith" which in true Draperian fashion he sees as "ever growing". He addresses the main question of Brunetière's argument, i.e. that science is unable to pierce the mystery of reality and impotent to answer to the really fundamental questions of existence. The critic is convinced that those ultimate questions *are* answered by science: every naturalist knows that life is "a passing form of cosmic evolution, a conglomerate of cells" and that our destiny is "to continue the transformation of matter, of which we are an insignificant part."[91]

Catholic reactions were few. In 1907, one year after Brunetière's death, Gustavo Franceschi, a Paris-born young priest of the Archdiocese of Buenos Aires (at that time active on the local Christian social movement and who in 1917 would publish a book on spirituality in French contemporary literature), published an article about the French Catholic convert pointing out how he had sought to achieve a blend of positivism and Catholicism.[92] Franceschi recalled the "fierce hate" awakened by Brunetière's article on the bankruptcy of science and was well aware that the central issue being discussed was the reduction of morals to science. Franceschi's article is the most complete and informed discussion on Brunetière published at that time in Argentina. The First World War provided

---

[89] Joaquín V. González, *La paz por la ciencia. Discurso en la colación de grados y títulos de la Universidad Nacional de La Plata, el 15 de agosto de 1914* (La Plata: Christmann and Crespo, 1914).

[90] José Ingenieros, "Los fundamentos de la psicología biológica," *Revista de Filosofía. Cultura, ciencias, educación* 1, no. 1 (1915): 442–71.

[91] Armando Donoso, "La conversión de Brunetière," *Revista de Filosofía. Cultura, ciencias, educación* 9 (1919): 206–22.

[92] Gustavo J. Franceschi, *El espiritualismo en la literatura francesa contemporánea* (Buenos Aires: Agencia General de Librería y Publicaciones, 1917); idem, "Ferdinand Brunetière," *Revista Eclesiástica del Arzobispado de Buenos Aires* 7 (1907): 147–56.

the occasion for harping on the theme of the bankruptcy of science. The Catholic Uruguayan poet Juan Zorrilla de San Martín observed that at a time where science reigned supreme, "a stream of blood flows down from the arteries of humanity." But he warned his audience from interpreting this as the final bankruptcy of science. Admittedly, there were "unbalances and voids" in science, but Catholics had always defended the truth and could not deny the truth of scientific knowledge.[93]

In Buenos Aires, the "bankruptcy of science" became little more than a catchy slogan and Brunetière the symbol of anti-scientific backwardness (he was even diagnosed as delusional). This controversy shows the extent to which the cultural atmosphere in Argentina in the years of the Great War was still permeated by the cult of science and secularism. The authors examined here, a representative sample of educated opinion, took the conflict view of the relations between science and religion as an unquestionable truth. Yet their violent rhetoric betrays some uneasiness, some cracks in the fortress of science threatened by new currents of thought. In the following section, we shall see how the confrontation between members of the scientific establishment and a Jesuit amateur anthropologist over the empirical evidence supporting a local theory of human origins were enmeshed with attempts of the socialists to create something approaching a secular cult of science.

Shifting Meanings

In 1884, the year of the great debate on religious education in Argentina, the 31-year-old palaeontologist Florentino Ameghino published in Buenos Aires his theoretical work *Filogenia*. During his lifetime and even more after his death, Ameghino would embody the kind of evolutionism which was an essential part of the secularist agenda put forth by the liberals and the new rising socialist movement. His image as the personification of evolutionary science was utilised by the positivistic establishment as a battering ram to storm what they saw as the crumbling fortress of metaphysical and religious deceptions. Paradoxically, it was the representatives of this broad swathe of progressive opinion who were unable or unwilling to recognise the scientific fraud favouring their views, and it was an anti-evolutionist Jesuit who acted as the whistle-blower.

---

[93] Juan Zorrilla de San Martín, "Discurso [21 July 1915, addressed to the Marian Congregation of Colegio del Salvador]," *Estudios* 9 (July–December 1915): 8–19.

George G. Simpson, one of the founders of the evolutionary modern synthesis, affirmed that the palaeontological achievement of Florentino and his brother Carlos "was one of the most remarkable in scientific history."[94] Simpson might have been paying a courteous tribute to the country in which he carried out his own research, but there is more than a kernel of truth in his statement. Ameghino's palaeontological contributions were ground-breaking. The same cannot be said of his anthropological theories: he never abandoned his juvenile conviction about the Tertiary origin of humanity in the Pampas, even in the face of the opposition of the international scientific community. Ameghino was a talented autodidact without more formal education than elementary schooling and a year in a Teachers' School (to which he was admitted thanks to the gracious intervention of the Catholic José Manuel Estrada).[95] He was raised in the rural town of Luján, located 60 km west of Buenos Aires, in an area rich in fossils (it was there that the skeleton of a *megatherium* had been excavated by a Dominican friar in 1776). Since colonial times, the town has been a centre of popular Catholic devotion to Our Lady of Luján, whose seventeenth-century terracotta image is venerated in the neo-gothic basilica of the same name, with construction work beginning in 1890. The image arrived in Argentina from Brazil in 1630 and, according to pious legend, the oxen which drew the cart carrying it through the fields at one point stopped near Luján and stubbornly refused to continue, which was interpreted as a divine sign that a sanctuary should be erected upon that location. Ameghino's newspaper article published in that crucial year of 1884, mocking the sacred image, illustrates his scornful anticlericalism.[96] Luján came to represent a disputed symbolic space which the spokespersons of religion and scientific secularism strove to appropriate for themselves.[97] In the 1920s the socialists would try to turn the town into a place of secular worship, with Ameghino as the object of a civic cult of patriotic science.[98]

---

[94] George G. Simpson, *The Beginnings of the Age of Mammals in South America. Part 1* (New York: American Museum of Natural History, 1948), 19.
[95] Fernando Márquez Miranda, *Ameghino. Una vida heroica* (Buenos Aires: Nova, 1951), 34.
[96] Florentino Ameghino, "Una virgen falsificada," in idem, *Obras completas*, ed. Alfredo J. Torcelli, vol. 23 (La Plata: Taller de Impresiones Oficiales, 1936), 505–9.
[97] This has also been observed and discussed in Roberto Di Stefano and Diego Mauro, "Our Lady of Luján: Mass Mobilization and Identity in Argentina," in *Marian Devotions, Political Mobilizations, and Nationalism in Europe and America*, ed. Roberto Di Stefano and Ramón Solans (Switzerland: Palgrave Macmillan, 2016), 279–313.
[98] This episode and the patriotic cult of Ameghino has been discussed in Máximo Farro and Irina Podgorny, "Frente a la tumba del sabio," *Ciencia Hoy* 8, no. 47 (1998): 28–37; Carolyne R. Larson, "'Argentine Man': Human Evolution and Cultural Citizenship in Argentina, 1911–1940," in *Making Citizens in Argentina*, ed. Benjamin Bryce and David M. K. Sheinin (Pittsburgh:

Ameghino's philosophical leanings took public form in a speech pronounced late in his life, in 1906. From the title onwards ("My Creed") his "synthetic exposition of the universe" can be seen as a mimesis of religious cosmological discourse.[99] The core of Ameghino's world view is the existence of four infinites, one of them material (matter) and three immaterial (time, space, and movement). Two kinds of movement, concentration and expansion, rule the whole evolution of the cosmos. The mass of matter is limited, which explains the struggle for life, for a number of creatures should perish for others to live. Future humans will enjoy a longer span of life and have superior intellects. This grand evolutionary vision of nature is an exercise of the imagination on the theme of a materialistic universe; the whole thing is written in the style of a speculative, aprioristic philosophy of nature, leaving aside sizable chapters of nineteenth-century chemistry and physics.

Ameghino's technical acquaintance with Darwin seems to have been slight, which is not surprising since his rite of passage into the academic world took place in France between 1878 and 1880. In an 1882 lecture in honour of the author of the *Origin*, he talked mostly about his own palaeontological findings within the framework of his phylogenetic research program.[100] It has been claimed (correctly, in my view) that he was a Lamarckian.[101] In *Filogenia*, he used dental and digital formulae to infer the morphology of a hypothetical ancestor from those of its descendants in order to build phylogenetic trees in the style of Haeckel—although the author declares that he "did not use" the works of the famous German evolutionist.[102] (In this respect, he might have also been influenced by the Catholic palaeontologist Jean-Albert Gaudry, a proponent of theistic evolu-

---

The University of Pittsburgh Press, 2017), 43–61. This literature concentrates on the nationalistic exploitation of Ameghino's figure. See Sociedad Luz, *Ameghino. Homenaje de la Sociedad Luz en el XXV aniversario de su muerte. 1911-Agosto 6–1936* (Buenos Aires: Federación Gráfica Bonaerense, 1936).

99 Florentino Ameghino, "Mi Credo," *Anales de la Sociedad Científica Argentina* 52, no. 2 (1906): 64–95. This brief exposition is related to a longer essay, written in French: Florentino Ameghino, "Origine et persistence de la vie (la matière, la vie, la mort et l'immortalité)," in idem, *Obras completas*, ed. Alfredo J. Torcelli, vol. 19 (La Plata: Taller de Impresiones Oficiales, 1935), 184–581.

100 Florentino Ameghino, "Un recuerdo a la memoria de Darwin. El transformismo como ciencia exacta," *Boletín del Instituto Geográfico Argentino* 3 (1882): 213–25.

101 Ángel Cabrera, *El pensamiento vivo de Ameghino* (Buenos Aires: Losada, 1944), 31; Julio Orione, "Florentino Ameghino y la influencia de Lamarck en la paleontología argentina del siglo XIX," *Quipu* 4, no. 3 (1987): 447–71.

102 Florentino Ameghino, *Filogenia. Principios de clasificación transformista basados sobre leyes naturales y proporciones matemáticas* (Buenos Aires: La Cultura Argentina, 1915), 13.

tion under whom he had studied at the Paris Museum.¹⁰³) One of Ameghino's phylogenetic series corresponded to the origin of human beings. He was convinced beyond the shadow of a doubt that the proliferating intermediary forms he postulated had existed as a matter of fact.¹⁰⁴ Links in the series leading to the emergence of *Homo sapiens* such as the hominids *Tetraprothomo, Triprothomo, Diprothomo, Prothomo,* and *Homo Pampaeus* were predictions awaiting confirmation.¹⁰⁵ Such a confidence was not entirely unfounded if we take into account that in 1891, moved by a blind faith in Haeckel's predictions, the Dutch physician Eugène Dubois had found the rests of *Pithecanthropus erectus* in Java.¹⁰⁶

The above mentioned Simpson tersely summarised Ameghino's ideas: "all the mammals in the world had as their ultimate ancestors Patagonian marsupials of various sorts...man also arose...from Patagonian ancestry."¹⁰⁷ This theory, which depended on the sequence and dating of beds in which the fossils were found, brought as a consequence several foreign expeditions to Patagonia in the 1890s and 1900s: Richard Lydekker from the British Museum, the young André Tournouër sent by Gaudry, and John Bell Hatcher's three Princeton expeditions on behalf of William B. Scott.¹⁰⁸ While the theory of South America as centre of origin and distribution of mammals through Africa and Australia through land bridges foundered upon growing international consensus against it, Ameghino turned his attention to the related question of human origins. He first developed his ideas about the birth of the human lineage in the territory of Argentina in the two volumes of his first important work on the paleo-anthropology of La Plata

---

[103] Jean Gaudant, "Albert Gaudry (1827–1908) et les 'Enchaînements du Monde animal'," *Revue d'histoire des sciences* 44, no. 1 (1991): 117–28; Cédric Grimoult, "Albert Gaudry dans l'histoire de la paléontologie évolutioniste," *Physis. Rivista internazionale di storia della scienza* 36, no. 1 (1999): 73–97.
[104] Elvira I. Baffi and María F. Torres, "Ameghino, Florentino (1854–1911)," in *History of Physical Anthropology*, ed. Frank Spencer, vol. 1 (New York: Garland Publishing, 1997), 54–56.
[105] Ameghino, *Filogenia*, 495–97.
[106] Pat Shipman, *The Man Who Found the Missing Link. Eugène Dubois and his Lifelong Quest to Prove Darwin Right* (Cambridge, MA: Harvard University Press, 2002), 57–61.
[107] George Gaylord Simpson, *Discoverers of the Lost World* (New Haven: Yale University Press, 1984), 92.
[108] Ronald Rainger, *An Agenda for Antiquity. Henry Fairfield Osborn and Vertebrate Palaeontology at the American Museum of Natural History, 1890–1935* (Tuscaloosa: The University of Alabama Press, 1991), 189–200; Peter J. Bowler, *Life's Splendid Drama* (Chicago: University of Chicago Press, 1996), 409–10; Irina Podgorny, "Bones and Devices in the Constitution of Paleontology in Argentina at the End of the Nineteenth Century," *Science in Context* 18 (2005): 249–83.

region.[109] His opinions were strengthened in 1907 and 1909 by certain anthropological fossil findings in the city of Buenos Aires and along the southern Atlantic coast of the province of the same name, in terrain that Ameghino thought corresponded to the Upper Miocene or Lower Pliocene (Tertiary).[110] He believed these findings were remains of the hominid species he had first postulated in *Filogenia* in 1884; only a handful of scientists in the international community accepted this view. With the Smithsonian expedition of Aleš Hrdlička and Bailey Willis to Buenos Aires in 1910 and their negative judgement, at least for the anglophone scientific community the issue was over.[111] Things were not so clear-cut before that. Writing in the first decade of the twentieth century about the findings of a "primeval Adam," the British Arthur Stuart Pennington, an amateur naturalist who had lived more than two decades in Argentina, affirmed that although their value was a question to be decided, "his [Ameghino's] deductions are exceedingly interesting and worthy of careful perusal."[112]

The process of investing the figure of Ameghino with religious meaning immediately followed his death in the city of La Plata, where he spent his last years. One and a half months after his death in August 1911, a "civic funeral" was set up in the local university, a recently founded house of higher learning and a bulwark of secular positivism. The staging of this event, attended by the famous French socialist politician Jean Jaurès—by chance in the country at that time—was evocative of the pageantry of Republican Rome, with allegorical images of the virtues and science and large Greek columns decorated with wreathes.[113] Against this backdrop, José Ingenieros gave a speech, reproduced in the newspapers as "The Modern Sanctity."[114] Ingenieros reinterpreted several Christian notions in a secular key while retaining their symbolic aura. Modern sanctity was "knowledge." The new "saints," of which Ameghino was a prime example, enjoyed a "serene mysticism" free of superstitions; they believed in "doctrines" which were not exposed as dogmas, had "faith" but were not fanatics, and although they could not "make miracles," they knew how to search for the

---

109 Florentino Ameghino, *La antigüedad del hombre en el Plata*, 2 vols. (Paris: Masson; Buenos Aires: Igon, 1880–81).
110 Gustavo G. Politis and Mariano Bonomo, "Nuevos datos sobre el hombre fósil," in Asociación Paleontológica Argentina, *Vida y obra de Florentino Ameghino*, special issue, 12 (2011): 101–19.
111 Aleš Hrdlička, W. H. Holmes, Bailey Willis, Frederick E. Wright and Clarence N. Fenner, eds., *Early Man in South America* (Washington: Government Printing Office, 1912).
112 A. Stuart Pennington, *The Argentine Republic* (London: Stanley Paul, 1910), 241.
113 *Funeral cívico de homenaje a la memoria del sabio naturalista Dr. D. Florentino Ameghino en La Plata* (La Plata: Taller de Impresiones Oficiales, 1911).
114 José Ingenieros, "La santidad moderna," *La Nación*, 19 September 1911.

truth. The irony of this rhetorical exercise comes to light when we consider that two decades earlier, the speaker had contributed a ribald piece on the pilgrimages to the sanctuary of Luján to a radical newspaper he had founded.[115] In a slightly later article, Ingenieros reformulated his ideas about a "secular religion." He denied any conflict between religion and science on the grounds that religion (more precisely, "religious feeling") was experiencing a transformation, it was "adapting itself to the new truths." In this view, "faith" is not a way of knowing, it is a "heat embracing those ravished by an ideal." The future religion, Ingenieros claimed, "will be a moral aspiration" rather than "a concrete dogmatism;" it will not impose rules of any kind, "its temple will be the school...its only rite the admiration for the great characters of the race."[116] He was thinking, of course, of his compatriot Ameghino.

The Uncomfortable Truth

The year 1916 marked the end of thirty-four years of conservative rule in Argentina and the ascent to power of the Radical Civic Union (a middle-class reformist party, hereinafter UCR, after its name in Spanish). It was also the year in which Father Blanco came to the fore as one of the most notable critics of Ameghino. In his *History of Argentine Literature*, the secularist intellectual Ricardo Rojas and rector of the University of Buenos Aires between 1926 and 1930, affirmed that "the learned men of the church, such as Father Blanco, have retorted to Ameghino's doctrines and even denied that the Argentine savant has been born in our country" (the Jesuit was right on both accounts: Ameghino's anthropological theories were refuted and he was born in Genoa).[117]

Father José María Blanco was born in Galicia (Spain) and arrived in Buenos Aires as a child.[118] He joined the Society of Jesus in Argentina and studied philosophy in Tortosa (Catalonia). Blanco taught sciences and spent a period of training in the recently established biological laboratory of Javier Pujiula, a Jesuit embryologist who had studied in Austria and Berlin and was a declared enemy of evolution, in particular in relation to human origins and the origin

---

115 José Ingenieros, "Los reptiles burgueses. I. Los que van al santuario," *La Montaña* 1, no. 2 (15 April 1897): 5.
116 José Ingenieros, "La religión de la raza," *Caras y caretas* no. 839 (31 October 1914): 47.
117 Ricardo Rojas, *La literatura argentina. Los modernos*, 2nd ed., vol. 1 (Buenos Aires: La Facultad, 1925), 78.
118 "El padre José María Blanco († 9 de agosto de 1957)," *Estudios* 47, no. 487 (1957): 53–58.

of life.[119] In 1910, Pujiula gave a series of lectures in Valencia in which he criticised Ameghino's hypothesis of Tertiary humans in the pampas, which had gained followers in Spain.[120] Blanco returned to Río de la Plata and after a brief stint in Montevideo took a position as professor of natural sciences in the Archdiocesan Seminary of Buenos Aires. In 1924, he began teaching at the College of El Salvador where he organised a small anthropological museum with the materials he collected in the course of his expeditions and his lecture tours all over the country. Due to his journalistic and educational activities in Jesuit secondary schools, Blanco became a relatively well-known figure in his time. His obstinate anti-evolutionistic views never changed: in 1950, the year in which Pious XII published the encyclical *Humani generis* which opened a small door to evolution theory among Catholics, he considered that the finding of Peking Man, in which his more famous confrere Teilhard de Chardin was much involved, was "sheer mystification."[121]

In Buenos Aires, popularisation of science was at the core of the socialist programs for workers education. Biological evolution, the materialist conception of life, and the origin of the universe were the staple subjects.[122] Astronomy was also a favourite theme of this agenda. The inaugural lecture of *Sociedad Luz*, a nascent socialist cultural centre, was on "The planetary system and the Earth"—delivered in June 1899 by the engineer student Maurice Klimann, it was hampered neither by the heavy Russian accent of the lecturer nor by the asphyxiating smoke generated by the kerosene lamp of the magic lantern he used.[123] Between 1912 and 1913 the socialist French physicist and mathematician Camille Meyer, who introduced Planck's quantum theory to the country, gave a series of lectures on the universe published in a fine volume of three hundred pages.[124] Ameghino became a cult figure among the socialists (although he was not one of them).

---

[119] Juli Peretó and Jesús I. Català Gorgues, "A Reconciliation with Darwin. Erich Wasmann and Jaime Pujiula's Divergent Views on Evolutionism: Biologists and Jesuits," *Mètode. Science Studies Journal* 7 (2017): 87–93.
[120] Jesús I. Català Gorgues, "La polémica sobre el hombre terciario y su expresión en la Valencia de comienzos del siglo XX," *Asclepio* 64, no. 1 (2012): 63–96.
[121] "El padre José María Blanco," 57.
[122] As can be seen in the papers published in the socialist magazine *La Universidad Popular* between 1905 and 1907.
[123] Ángel Giménez, *Páginas de historia del movimiento social en la República Argentina* (Buenos Aires: Imprenta La Vanguardia, 1927), 66–68. Cf. Dora Barrancos, *La escena iluminada. Ciencias para trabajadores (1890–1930)* (Buenos Aires: Plus Ultra, 1996).
[124] Camilo Meyer, *Conferencias de astronomía popular dadas en 1912 y 1913* (Buenos Aires: Kidd, 1916); Carlos Galles, "La obra de Camilo Meyer por la cultura científica argentina," in

Blanco lectured on the anthropological theories of the Argentine savant in the Jesuit College of El Salvador and he also published two papers criticising Ameghino's philosophical ramblings.[125] This triggered a series of answers in the socialist daily *La Vanguardia*. The cause of outrage was not that Ameghino's scientific competence had been called into question—many had done that before. What the critics could not tolerate was that a member of a "corporation" famous for its obscurantism, was using scientific arguments: "it would have been tolerable," protested one of the authors, "if he had used the usual arguments of religion."[126] Another complained of "the meddling of a priest with the natural sciences," which meant that the Church had "entered surreptitiously into the fortress of science" and presented itself as a "dicephalus monster: at one time assailant and defender of scientific truth."[127] The message was clear: science is a strictly secular business.

At the same time that Blanco was being accused of intruding on the turf of science, in the Chamber of Deputies of the Province of Buenos Aires the simmering dispute over the scientific and religious meaning of the town of Luján eventually boiled over. While discussing the annual budget of the province, the socialist deputy Adolfo Dickmann proposed to assign 5,000 *pesos* to buy Ameghino's natal house in that city.[128] Contrasting the unedifying spectacle of "the magnificent basilica" rising to the sky with "the healthy poverty of Ameghino's house," he proposed that the latter should be the centre toward which Argentine children "should peregrinate to obtain inspiration" from the "spirit of humility" of its former dweller. The manoeuvre to take over the religious symbolism attached to the centre of popular piety and substitute secular scientific meanings for the original sacred ones foundered upon the uncomfortable fact that Ameghino's precise birthplace was a matter of speculation. At that point, the conservative Catholic deputy Nicanor de Nevares reminded the chamber that no parochial local register of the savant's baptism had been found in Argentina, but one had popped up in the small Genoese town of Moneglia, from

---

La ciencia en la Argentina. Perspectivas históricas, ed. Miguel de Asúa (Buenos Aires: CEAL, 1993), 134–44.

125 José M. Blanco, *La evolución antropológica y Ameghino* (Buenos Aires: Alfa y Omega, 1916); idem, *El Credo de Ameghino* (Buenos Aires: R. Herrando, 1916).

126 Esteban Dagnino, "El 'Credo' de Ameghino juzgado por un fraile," *La Vanguardia*, 27 October 1916.

127 Ricardo Calatroni, "Malabarismo católico. La Iglesia y la obra de Ameghino," *La Vanguardia*, 17 November 1916.

128 Provincia de Buenos Aires, Cámara de Diputados, *Diario de sesiones 1916* (La Plata: Taller de Impresiones Oficiales, 1917), 1049–59 (session of 25 August 1916).

whence the family had emigrated (the name entered there was not Florentino, but Fiorino). In the end, a commission was formed to look into the case. The question of Ameghino's birthplace turned into a local cause célèbre, which lasted for decades. In the succeeding months, the issue became inextricably enmeshed with the discussion of Ameghino's anthropological theories and the unceasing archaeological and anthropological findings that seemed to support his ideas and were duly reported by the newspapers with patriotic pride. During 1916, the Catholic newspaper *El Pueblo* and the socialist *La Vanguardia* crossed swords with each other on this account. The articles in both periodicals were a sounding board of the ever louder sound and fury of the political and ideological clash between the two groups. The Catholic newspaper did not mentioned Blanco; instead, it resurrected, with notable prolixity, Burmeister's anti-evolutionistic critiques against Ameghino. The goal was twofold: to deprive the latter of any scientific credibility and to invalidate evolutionism.[129] The anonymous journalist denounced his socialist antagonists, stating that, in trying to present Ameghino "as the highest exponent of Argentine science," they were waging war against "the traditional notions of God as creator of the orb."[130]

At the same time, Blanco and his students from the Seminary of Buenos Aires went on writing articles critical of Haeckel's evolutionism. That made sense, since cheap editions of Haeckel were regularly brought out by the socialist printing houses as part of a sustained effort to educate the working class. The seminarians used as one of his sources the works of Erich Wasmann, the Jesuit entomologist with a moderate theistic version of evolutionism who had famously contended with Haeckel's monism.[131] In 1917, Blanco published an article which opened with an account of Wasmann's 1907 Berlin lectures, which the Argentine Jesuit interpreted as a vindication of the inductive methodology he favoured.[132] It is not unlikely that Blanco could have intended to model his own dispute with the followers of Ameghino onto the more exalted exemplar of Wasmann versus Haeckel. In any case, what was at stake in the local interchanges was the assignation of spheres of authority for science and religion. In one of a series of articles published in a respected newspaper in 1916, Lucas Kraglievich, a paleontologist at the Museum of Natural History in Buenos Aires, had affirmed that

---

129 The main series of articles was collected and edited as: *La nacionalidad y la obra de Ameghino*, 2nd ed. (Buenos Aires: Imprenta de El Pueblo, 1917).
130 Ibid., 5.
131 Robert J. Richards, *The Tragic Sense of Life. Ernst Haeckel and the Struggle over Evolutionary Thought* (Chicago: The University of Chicago Press, 2008), 350–71; Andrew J. Lustig, "Erich Wasmann, Ernst Haeckel, and the Limits of Science," *Theory in Bioscience* 121, no. 3 (2002): 252–59.
132 José M. Blanco, "Las ideas preconcebidas y la ciencia," *Estudios* 13 (1917): 423–35.

Blanco was an amateur aspiring to handle "without any serious knowledge of the matter the delicate problems of naturalist philosophy, the understanding of which presupposes...the profound and methodical analysis of the materials deposited in cabinets and museums."[133] Blanco inverted the terms of the discussion: it was the naturalist Ameghino who had erred grossly when he had believed himself worthy of the title of philosopher. In September 1917, a popular daily published a brief manuscript by Ameghino entitled "Notion of God and notion of space," which affirmed that the "primitive...infantile" notion of a God, born out of fear, will eventually be dissolved into the "much more grandiose" idea of infinite matter moving infinitely into infinite space.[134] Blanco's response contended that it was not so easy to solve the fundamentals problems of philosophy. For that purpose, he said, "it is not enough to be a PALEONTOLOGIST; something more is needed...which Ameghino lacked" (small caps in the original).[135]

The Piltdown Skull, as it is known, was found by Charles Dawson in East Sussex in 1912.[136] In that same year, an analogous and less illustrious hoax took place in far-away Argentina. The story of the fraudulent archaeological and anthropological findings in the ravines of a coastal town in Buenos Aires by Lorenzo Parodi has been told several times.[137] Here I shall highlight only those aspects related to Blanco's role in those events. In the summer of 1912–1913, little more than a year after Ameghino's death, several stone objects—among them a bola and a flint knife—were found in the locality of Miramar. This was taken as evidence of human habitation of the region from two million years in the past, which amounted to a new confirmation of Ameghino's hypothesis regarding the Argentine origin of the human lineage.

Parodi, an almost illiterate Genoese immigrant with a great talent for paleontological surveying, had been commissioned to explore the site by Carlos Ameghino, at that time the chief of the palaeontology section of the Museum of Buenos Aires. These findings raised two issues. Firstly, had the objects been

---

133 Lucas Kraglievich "Las teorías de Ameghino. Sobre una titulada réplica," in idem, *Obras de Geología y Paleontología*, ed. Alfredo J. Torcelli and Carlos Marelli, vol. 1 (La Plata: Taller de Impresiones Oficiales, 1940), 65 (originally published in *La Nación* in November 1916).
134 Florentino Ameghino, "Noción de Dios y noción de espacio," *La Razón*, 16 September 1917.
135 Graco Nebel (José M. Blanco), "¡Ameghino filósofo!," *Estudios* 13 (1917): 277–79.
136 Frank Spencer, *Piltdown: A Scientific Forgery* (Oxford: Oxford University Press, 1990).
137 Leonardo Daino, "Exégesis histórica de los hallazgos arqueológicos de la costa bonaerense," *Prehistoria Bonaerense* (1979): 95–145; Eduardo P. Tonni, Ricardo C. Pasquali and Mariano Bond, "Ciencia y fraude: el hombre de Miramar," *Ciencia Hoy* 11, no. 62 (2001): 58–62; Mariano Bonomo, "El hombre fósil de Miramar," *Intersecciones en Antropología* 3 (2002): 69–87.

buried at the time of formation of the sediments (primary site) or had they been interred afterwards as a result of dispersal (secondary site)? Then, there was the thorny question of geological dating.[138] In 1914, a commission of Argentine scientific notables surveyed the area and concluded that the sites were primary. The findings continued. In December 1914, Carlos Ameghino and Parodi found a toxodont femur pierced by an arrow.[139] The next year, the amateur archaeologist Major Antonio A. Romero published a booklet with a harsh critique of the findings followed four years later (1919) by a paper in which he claimed that the site was not primary.[140] In 1920, the young student Milcíades Vignati, who would later become a noted anthropologist, fired off a contentious pamphlet against Romero, which the military man took as a personal affront. He sent his seconds to Vignati, asking him to make a public retraction or face a duel. Vignatti opted for the safer alternative.[141] In 1921, Father Blanco enthusiastically joined the fray with a short paper.[142] He referred to the publication in one of the Buenos Aires newspapers of the discovery of several bolas in Miramar by a scientific commission on 22 November 1920 and, quoting freely from Romero's articles, he denounced the findings as a sham, orchestrated by Parodi with the connivance of Carlos Ameghino, bent on supporting his brother's notion of Tertiary Man.

In 1919 and 1921, the Swedish Eric Boman, at that time director of the archaeological section of the Museum of Buenos Aires, published two papers offering a sceptical perspective on the Miramar findings.[143] The Catholic daily *El Pueblo* reproduced Boman's articles under the title "Ameghinism—anti-scientific charlatanism and Machiavellian industry. The word of the men of science."[144] In his unceasing campaign, Blanco reproduced Boman's article and also the fragments criticising Ameghino's anthropological theories in Marcellin Boule's *Les*

---

138 Tonni et al., "Ciencia y fraude," 59.
139 Daino, "Exégesis histórica," 26–29.
140 Ibid., 29–31, 38–42.
141 Manuel J. Samperio, "Cuestión de paleoantropología argentina," *Estudios* 18 (1920): 351; Milcíades A. Vignati, *Los restos de industria humana de Miramar. A propósito de los despropósitos del comandante Romero* (Buenos Aires: Oceana, 1919).
142 José M. Blanco, "Las bolas de Parodi ¿serán bolas?," *Estudios* 20 (1921): 31–35.
143 Eric Boman, "Encore l'homme tertiaire dans l'Amérique du Sud," *Journal de la Société des Américanistes*, new series, 11 (1919): 657–64; Eric Boman, "Los vestigios de industria humana encontrados en Miramar (República Argentina) y atribuidos a la época terciaria," *Revista Chilena de Historia y Geografía* 39, no. 42 (1921): 330–52.
144 "El ameghinismo. Charlatanismo anticientífico e industrialización maquiavélica. La palabra de los hombres de ciencia," *El Pueblo*, 8, 9–10, and 11 January 1922.

hommes fossils (Paris: Masson, 1921).[145] The discussion among the Argentine specialists went on for years to come.[146] Whereas some of the protagonists changed sides in the course of the polemic, Ameghino's most faithful disciples defended the validity of the Miramar findings to the end. As late as 1941, Vignati still claimed that scepticism regarding the Miramar findings was the result of "ignorance and bad faith."[147] Only recently have Argentine members of the palaeo-anthropological community gallantly recognised that Father Blanco's criticisms were muffled on account of the anti-religious bias impregnating Argentine science at that time.[148]

Blanco went on writing on human evolution. In the second semester of 1923 he published two articles on the antiquity of human beings on Earth in which he affirmed, on the one hand, that the Bible could not be utilised as a chronological guide to establish the origin of humans, and on the other, that scientific results about that issue were unreliable, although it was clear that our ancestors could not be dated beyond the Quaternary.[149] In further articles he stated that "biology does not favour in any way the hypothesis of evolution as applied to man."[150] In his copious writings, he never addressed specifically the question of evolution per se, restricting himself to human evolution and the criticism of Ameghino's ideas and hypothesis about the Tertiary origin of the human being in the pampas.[151]

The Spanish Jesuits who worked as science teachers at the College of El Salvador and the Archdiocesan Seminary shared anti-evolutionistic views. In 1917, a group of students directed by Father Ubach made a series of presentations on human evolution whose contents reproduced Father Blanco's arguments against Ameghino's theories and the origin of "Tertiary Man" in Argentina. They were also in tune with Blanco's rotund affirmation of the human "spiritual, i.e. imma-

---

[145] José M. Blanco, "Hablan los hombres de ciencia del país sobre las asendereadas teorías de Ameghino," *Estudios* 22 (1922): 428–45; "Ameghino juzgado por Boule. *Les hommes fossiles*," *Estudios* 20 (1921): 419–26.

[146] Bonomo, "El hombre fósil," 74–81.

[147] Milcíades A. Vignati, "Descripción de los molares humanos fósiles de Miramar," *Revista del Museo de La Plata. Nueva Serie. Sección Antropología* 1, no. 8 (1941): 292.

[148] Eduardo P. Tonni and Laura H. Zampatti, "El 'hombre fósil' de Miramar. Comentarios sobre la correspondencia de Carlos Ameghino a Lorenzo Parodi'," *Revista de la Asociación Geológica Argentina* 68, no. 3 (2011): 436–44.

[149] José M. Blanco, "La antigüedad del hombre," *Estudios* 25 (1923): 91–103, 161–74.

[150] José M. Blanco, "¿Es posible la evolución del hombre?," *Estudios* 29 (1925): 51.

[151] For a detailed analysis of all the articles by Blanco, see Miguel de Asúa, "Los artículos del P. José María Blanco S.I. en la revista *Estudios* sobre la evolución y las teorías antropológicas de Ameghino," *Stromata* 65, no. 3–4 (2009): 313–35.

terial soul" which could not be the result of the evolution of animal psychism, but "is essentially the work of God's creative intervention in each individual" (to be discussed in the next section).[152] Five years later, Ubach gave a lecture for the general public in which after a rather exhaustive review of the literature on genetics, heredity, and hybrids, concluded that "evolution remains contained within the narrow limits of systematic species" (variation within species).[153] He further argued that in what concerns human origins, we should abide by the "obvious and literal sense" of the Bible, which should be taken as "the only available historical source" for the understanding of the emergence of the human being.[154]

The great debate regarding Ameghino's nationality was intertwined with the controversy over his anthropological theories on "Tertiary Man" and the authenticity of the Miramar findings. While in 1883–84 the protagonists on either side of the science-versus-religion divide were members of the patrician elite, the main characters in this story were now recent immigrants, in tune with the shift towards a more open and inclusive society. Evolutionism was identified with Ameghino, the local hero of humble origins, patriotic mien, and world achievements, and in the second place, with Haeckel. The avant-garde of the anti-clerical sectors were the socialists, whose combative secularism had sharp ideological edges. Their ultimate goal was the suppression of religion, but the old positivist dream of a "secular religion" lingered on in the attempts to invest secular science with religious trappings. By the second decade of the twentieth century, there were Catholic specialists of different degrees of expertise and representing a spectrum of opinion: from the Jesuit Blanco, an anti-evolutionistic serious amateur, to Ángel Gallardo, a professional naturalist who accepted biological evolution in theistic terms.

It has been claimed that evolutionism gave unbelievers the opportunity that Newtonianism could hardly grant, that is a scientific justification "for their materialism and their atheism."[155] Just as Holmberg had prophesied in 1882, the kind of evolutionism that eventually took root in Río de la Plata was not Darwin's but Haeckel's, which lent itself more readily to being a vehicle for anti-metaphysical doctrines. Symbols play a prominent role in this story. The surname "Ameghino" was made into an icon which stood for evolutionism, patriotic science,

---

[152] José Ubach, *El origen del hombre* (Buenos Aires: Talleres Gráficos Rodríguez Giles, 1917), 6.
[153] José Ubach, "El pasado y presente del problema de la evolución," *Estudios* 22 (January–June 1922): 259.
[154] Ibid., 250.
[155] Bernard Lightman, "Does the History of Science and Religion Change Depending on the Narrator? Some Atheist and Agnostic Perspectives," *Science and Christian Belief* 24 (2012): 151.

and secularism, and as such was venerated and execrated in equal measure by the representatives of the opposing factions. The toponymic "Luján" has attached to it sacred and secular meanings in an unstable and conflictive balance. A journalistic and political battle was fought over its significance: would it be the symbol of the Catholic character of the country or would it be the cipher of a secular society built upon the twin columns of positivism and socialism? Thomas Lessl has utilised Northrop Frye's theory of "displacement" to discuss evolutionism as discourse that considers "some contemporary or at least secular subject matter but orders it in accordance with a narrative form that descends from myth."[156] He devotes a substantial chapter of his book to the discussion of French positivism. We have seen that Comte's Religion of Humanity never entirely acclimatised itself in Argentina; it was the "scientific Comte" who was received, together with Spencer, and generous doses of both of them were added to the concoction of an indigenous version of positivism *cum* evolutionism. Nonetheless, from the 1910s onwards there were ever-renewed attempts at replacing sacred contents with the secular in religiously significant places, persons, and discourse. This strategy sought to accomplish the appropriation of religious symbolism, while at the same time strictly delimiting the boundaries between science and religion and proscribing any incursion of the latter upon the territory of the former.

There is also the question of historical evidence, of the legitimacy of a truth claim. In two instances—the documents demonstrating Ameghino's birth in Genoa and the dating of the supposedly Tertiary lithic objects in Miramar—a majority of the local palaeontological and archaeological circles closed ranks behind what was eventually shown to be the wrong position. The paradox is that it was a representative of Church religion with anti-evolutionistic views who was on the side of scientific probity.

Jesuits were not the only ones impatient with the cult of Ameghino. Writing in 1927, the Italian-Argentinian philosopher Coriolano Alberini, author of *Die deutsche Philosophie in Argentinien* prologued by Einstein, made a caustic comment on this issue:[157]

> Some of those attached to positivism...have devoted themselves to the metaphysical interpretation of Darwinian biology in the manner of Haeckel. One result of this was the brief *Credo* of Ameghino... A noisy and decrepit Haeckelian senility found in its ranks educational scientists and literary psychiatrists [he is referring to Ingenieros] whose pamphlets on

---

156 Thomas M. Lessl, *Rhetorical Darwinism. Religion, Evolution and the Scientific Identity* (Waco: Baylor University Press, 2012), 38.
157 Coriolano Alberini, *Die deutsche Philosophie in Argentinien* (Berlin: Hendriock, 1930).

"the future of philosophy" will be valuable documents for the future historian of philosophical dilettantism on the coast of [Río de la] Plata.[158]

Soul versus Brain

Evolutionism was not the only target of Father Blanco's criticisms. He also polemicised against the views of Christfried (Christofredo) Jakob, the German neuroscientist who arrived in Argentina in 1898 hired by Domingo Cabred, the progressive director of the Men's Psychiatric Asylum (Hospicio de las Mercedes), in order to organise a laboratory of neuropathology. Born in Bavaria, Jakob studied in Erlangen and became an associate of the famous neurologist Adolf von Strümpell.[159] At the time of his emigration to Argentina, he had already published two important works which had been translated to several languages and re-edited.[160] After a decade of intensive work and not entirely smooth relationships with Cabred, Jakob returned to Germany. By 1912, he was back in Buenos Aires, this time as director of the laboratory of neuropathology of the Women's Asylum (an amphiteatrum reproducing the one in Erlangen had been built for him); he was also nominated professor of biology in the Faculty of Philosophy and Letters of the University of Buenos Aires and from 1922 he held an analogous position in the University of La Plata.[161]

Jakob collaborated with Ingenieros and Onelli. The Italian naturalist Clemente Onelli was director of the Buenos Aires Zoo from 1904 to 1924; he had been preceded in this position by Holmberg, the first director of the Zoo opened to the public in 1888. Ingenieros was in his way to become director of the Police Psychiatric Observation Unit; in 1908, he became professor of experimental psychology in the University of Buenos Aires and was a member of the staff of the Chair of Neurology in the School of Medicine of the same institution, in charge of the neurologist and psychiatrist José María Ramos Mejía; in that same year, Ingenieros founded the *Archives of Psychiatry and Criminology* [*Archivos de Psiquiatría y Criminología*]. Jakob contributed many papers to this journal

---

158 Coriolano Alberini "Contemporary Philosophic Tendencies in South America, with Special Reference to Argentina," *The Monist* 37, no. 3 (1927): 328–34.
159 Lazaros C. Triarhou and Manuel del Cerro, "Semicentennial Tribute to the Ingenious Neurobiologist Christfried Jakob," *European Neurology* 56 (2006): 176–88, 189–98.
160 Christfried Jakob, *Atlas des gesunden und kranken Nervensystems* (Munich: Lehmann, 1895); idem, *Das Menschenhirn* (Munich: Lehmann, 1911).
161 For the development of psychiatry at that time, see Ablard, *Madness in Buenos* Aires, 17–50.

as well as to the *Journal of the Zoological Garden* [*Revista del Jardín Zoológico*], directed by Onelli. Jakob and Onelli co-authored an atlas of mammalian brains, with a first German edition and a second augmented Spanish edition.[162] Jakob also published in Buenos Aires his *Folia neurobiológica argentina*, an atlas of the human brain in three volumes, with five accompanying volumes of text.[163]

Psychiatrists, psychologists, and neurologists in pre-World War I Argentina held as a matter of fact materialistic views of science, the human being, and mental pathology (the case of Korn, discussed in the previous chapter, is illustrative in this respect). For the most, they were secularists through and through and not a few were anti-Catholic. Ingenieros was one of the nodes of an extensive network of experts, institutions, academic production, and state policies dealing with mental health.[164] Explosive urban population growth due to massive immigration brought with it an increasing rate of mental pathology and anti-social behaviour. The state answered the challenge developing policies of public health, improved medical facilities, the reformulation of the penal system, and measures of social control allegedly predicated on the ideology of science and modernity.[165]

In 1920, Ingenieros published his *Madness in Argentina*, an account of the history of mental pathology and psychiatry in the country.[166] His is a story of progress and enlightenment which begins in colonial times and concludes in the second decade of the twentieth century. It is also a story in which religious superstition, supernatural causation of mental illness, and belief in the "dogma" of the rational soul are superseded by rational explanations, natural causality, and scientifically-based treatments of mental disease. At the time the author wrote, doctors could resort to "histology in order to understand the physiopathological processes and to biochemistry in order to analyse the causal elements [of mental diseases]."[167] Ingenieros's teacher, José María Ramos Mejía, had paved the way for this kind of discourse. In his book on the neurosis of famous persons in Argentine history, published in 1878–82, he affirmed: "Voltaire's prophecy

---

[162] Christfried Jakob and Clemente Onelli, *Vom Tierhirn zum Menschenhirn* (Munich: Lehmann, 1911); idem, *Atlas del cerebro de los mamíferos de la República Argentina* (Buenos Aires: Kraft, 1913).

[163] Christfried Jakob, *Folia neurobiológica argentina. Atlas*, 3 vols. (Buenos Aires: Aniceto López, 1939–1941); *Folia neurobiológica argentina*, 5 vols. (Buenos Aires: Aniceto López-López y Etchegoyen, 1941–1946).

[164] Ablard, *Madness in Buenos Aires*, 33.

[165] Rodríguez, *Civilizing Argentina*.

[166] José Ingenieros, *La locura en Argentina* (Buenos Aires: Editorial Buenos Aires, 1920).

[167] Ibid., 7.

has been realised. It was not possible to solve the problem of the soul until anatomy could penetrate into the inner constitution of that divine pulp that beats under the skull vault."[168] True to positivism, he goes on: "the 'theological' and 'metaphysical' epochs are happily over... The revelations of [today's savants] have inaugurated the 'positive age' of science."[169] Jakob's views were not substantially different from these, but as a representative of the brilliant tradition of German neuroanatomy and a missionary of Wilhelmine science, he felt never quite at ease in his adoptive country.

Jakob's views on the origin of conscience, dispersed among his many articles and books, are hard to summarise. He thought in terms of dynamism and equilibrium and held fast to a double approach, onto- and phylogenetic. His writings bristle with ever-changing neologisms and at times the sober anatomical descriptions are interrupted by bouts of wild speculation. When boiled to the bones, his idea is that self-awareness and freedom are the last stage in a process of complexification of the basic reflex arc. His position might have been close to some kind of materialistic emergentism.

The article which attracted Blanco's attention was published in 1918.[170] In it, Jakob described thought as "an energetic process entirely analogous to the assimilation of organic material."[171] To the question of what makes a person think and feel, cry and laugh, his answer is: "something dynamogenic neuroplasmatic." Certainly neither a soul nor some kind of nervous fluid different from "vile matter." Psychic phenomena were "the flower of vital trophisms, in the strict sense of the word."[172] It seems that Jakob conceived the nervous impulse in terms of a wave signal of some kind; he also talked of undetermined "energies." Consciousness would emerge from the impressions received by the nervous cells from stimuli, through some kind of harmonious interplay of longitudinal and transversal vibrations ("combinatorial resonance"). Jakob had experimented with the new technique of tissue culture trying to grow tissue taken from foetal animal brains. In the report of his experiments, he asked himself: "Would it be possible to think about an artificial culture of thought in the long-distance future?".[173] The article concludes with an unambiguous definition:

---

168 Ramos Mejía, *Las neurosis*, 91.
169 Ibid., 92.
170 Christfried Jakob, "Del mecanismo al dinamismo del pensamiento. Estudio histórico-crítico de psicología orgánica," *Anales de la Facultad de Derecho y Ciencias Sociales* 18 (1898): 195–238.
171 Ibid., 197.
172 Ibid., 235–37.
173 Christfried Jakob, "El cultivo artificial del órgano del pensamiento," *Revista del Jardín Zoológico*, 2nd epoch, 10, no. 37 (1914): 9–22.

"—'Give me protoplasm, says experimental biology, and I shall give you thought'."[174]

Much as he had done in the case of the "Miramar Man," Father Blanco was able to spot the weak points in Jakob's articles. But also as in the former case, his critical acumen was stronger than his capacity of advancing a consistent thesis.[175] The Jesuit anthropologist explicitly identifies his position with soul-body dualism on the grounds of doctrinal religious orthodoxy. He complained about Jakob's "exotic terminology" and the latter's advocacy of "the most economic hypothesis," i.e. the one which did not involve metaphysical entities. In the *Atlas* he co-authored with Onelli, Jakob affirms that the causes of the emergence of the cortex are twofold, exogenous and endogenous. The cortex would result from an adaptation of vertebrates "superior to fish" to the environment of terrestrial life, more variable than the environment of marine life. When it comes to the endogenous point of view, Jakob rejects the Darwinian notions of variations and natural selection; instead, he talks of "a continuous endogenous process of maturation [of the germ] from an imperfect to a more perfect state" and of a "law of immanent differentiation of the germ-protoplasm."[176] Blanco interprets this as meaning that "progressive intellectual evolution [in the history of humankind] is immanent and is stored in the germs with the result that it can be transmitted through biological heredity."[177] The Jesuit contends that that view leaves cultural heritage unexplained, because biological heredity cannot account for culture. Briefly, materialism does not explain "the giant work of the collective intellect."[178]

On the basis of the increasing number of cortical neurons and of the complexity of the neural networks as we ascend the phylogenetic scale, Jakob claimed that "the richness of our psychic life and that of the animals is undoubtedly in direct proportion with the number of cortical cells" (he compares the neurones to "ants of thought").[179] Blanco's second objection has to do with Jakob's implicit correlation between the size and weight of the brain along the phylogenetic scale and the degree of intelligence: "If the psyche is proportional to the number of cortical neurones, then a larger brain mass would imply a more complex psy-

---

[174] Ibid., 22.
[175] José María Blanco, "Dinámica cerebral," *Estudios* 24 (1923): 5–16, 83–89, 161–67 and 246–54.
[176] Jakob and Onelli, *Atlas*, 23–26.
[177] Blanco, "Dinámica cerebral," 84.
[178] Ibid., 253.
[179] Jakob and Onelli, *Atlas*, 29.

che."¹⁸⁰ But this correlation does not hold, as shown by social insects, which are more intelligent than many mammals. The Jesuit's position was dualistic. Not only did he affirm that the brain was not the primordial basis of the mental world in human beings, it was "not even the basis of the inferior psyche of animals."¹⁸¹ (It is plausible that Blanco was here thinking in terms of a dualistic interpretation of Aristotles's doctrine of the soul.)

Father Blanco challenged Jakob's theories from the philosophical and from the scientific point of view. Jakob was far from being unknowledgeable about philosophy. His lectures on philosophy of nature were an articulate exposition of the field with a reasonable command of the state of the question in Germany at that time.¹⁸² He was also well acquainted with Kantian philosophy. If we add to this his scientific credentials, we have to conclude that Blanco had come up against a formidable opponent. The Jesuit anthropologist was smart enough to focus his criticisms on some evident difficulties in Jakob's theories and the hazy character of many of his arguments. But Blanco's unnegotiable anti-evolutionistic stance put him at a disadvantage with respect to the German neurologist. Jakob conceived the origin of religion in terms of a biological system of protection evolved in primitive humans in the face of dangers and an uncertain future; in his idiosyncratic jargon, "a biological system of extra-empirical prospective psychophylaxis." Primitive humans developed "imaginative constructions" in the form of anthropomorphic supernatural powers as an answer to "a dark and unknown future." Jakob defined "natural religion" in terms of a vitalist philosophy as the expression of the instinctive optimism of the child, which manifests itself in a "will to live, to produce, to fight, and to triumph."¹⁸³ His framework of though was materialistic and evolutionistic, but unlike what was the case with his Argentinian colleagues, his writings are free from secularist pronouncements and anticlerical remarks.

Conclusions

As a reflection of the implicit and uneasy *pax* between conservative liberals and the hierarchy of the Church, in pre-World War I Argentina the banner of aggres-

---

180 Blanco, "Dinámica cerebral," 166.
181 Ibid., 249.
182 Christfried Jakob, *Filosofía de la naturaleza. Un curso de conferencias* (Buenos Aires: Kraft, 1920).
183 Christfried Jakob, "La religión de la naturaleza y el porvenir del hombre," *Humanidades* (La Plata) 22 (1930): 107–19.

sive secularism was passed from the liberal elite to socialism. It has been observed that with the first waves of immigrant workers (many of which were Freemasons and Italian liberals), a new actor stepped on a so far two actors stage (liberals and Catholics).[184] But the more genteel secularism of the liberal elite of the 1880s was still alive and kicking, so that the rapprochement between liberals and socialists in the field of social reformism "was backed by a 'scientific' foundation which eased the overcoming of political or ideological divergences."[185] Positivism and evolutionism were still very influential in this regard. It was in these years when the winds of non-Christian spiritualism began to blow over Argentina, adding complexity to the already diverse landscape of secularist opinion. The "secular cult" of Ameghino and the "secular spiritualities" which combined science and the occult were attempts at redefining the religious in terms of immanent worldviews. As awkward as they were, Father Blanco's polemical interventions in defence of the compatibility between science and Christianity announced that the times were changing. In the next chapter I shall show how his activities were part of a broader network and how these early attempts at disengaging science from secularism were furthered by the appearance of prestigious Catholic scientists, a phenomenon which paved the way for a new configuration of the relationships between science and religion in Argentina.

---

[184] Gustavo Morello, "'It Takes Two to Tango': The Religious and Secular in Argentina's Political Dance, 1860–1960," in *Laicidad and Religious Diversity in Latin America*, ed. Juan M. Vaggione and José M. Morán Faúndes (Cham: Springer, 2017), 61.
[185] Eduardo A. Zimmermann, *Los liberales reformistas. La cuestión social en la Argentina, 1890–1916* (Buenos Aires: Sudamericana, 1995), 217.

Chapter 6
Science, Catholicism, and Politics in Argentina, 1910–1935

On 16 April 1910, just before dawn, Father Fortunato Devoto pointed his binoculars towards the eastern skies from the top of the episcopal palace in the capital city of the province of Buenos Aires. As the acting director of the La Plata Observatory, he was the first person to see Halley's Comet from Buenos Aires.[1] The reporter who interviewed him for a popular illustrated weekly, puzzled perhaps by the seeming contradiction of finding a Catholic priest in a secular temple of science, compared the premises of the observatory to "a cemetery."[2] The cometary appearance which captured the imagination of the world was shadowed in Argentina by the fêtes of the first centenary of the revolution of independence from Spain. Among the several learned meetings convened in Buenos Aires to celebrate the patriotic anniversary, the first International American Scientific Congress stood as a reliable mirror of the country's scientific culture in the twilight of the belle époque.[3] The only priest to present a communication to the congress was Fr Henri Sisson, a French Dominican who that same year had published a comprehensive account of his adoptive country.[4] His contribution ("Humanitarianism in Argentine civilisation") would have passed unnoticed were it not that his passing reference to "the dangers of immigration and anarchism" provoked an irate reaction from the socialist Sara Justo.[5] The paltry visible Catholic representation at the meeting reflected the wide gap between the institutional Church and a strongly secularised scientific world. But for those who could seem them, there were signs that this state of affairs was about to change:

---

[1] "El cometa Halley. Su observación en La Plata," *La Prensa*, 17 April 1910. The comet had been followed from Córdoba with the 12-inch equatorial telescope of the National Observatory since 30 November 1909, see Charles D. Perrine, "Résumé of Observations of Halley's Comet at Córdoba," *Publications of the Astronomical Society of the Pacific* 22 (1910): 211–13. La Plata is the capital city of the province of Buenos Aires, situated about 60 km south of the city of Buenos Aires, capital city of the country.
[2] "Una visita al Observatorio de La Plata," *Caras y Caretas* 13, no. 606 (1910): 77–79.
[3] Miguel de Asúa, "El Congreso Científico Internacional Americano," *Ciencia Hoy* 21, no. 125 (2011): 18–24; 22, no. 126 (2012): 14–20.
[4] Henri-Dominique Sisson, *La République Argentine. Description, étude sociale et histoire* (Paris: Plon, 1910).
[5] Sociedad Científica Argentina, *Congreso Científico Internacional Americano*, vol. 1 (Buenos Aires: Coni, 1910), 452.

the president of the section of biological sciences was the prestigious Catholic engineer and naturalist Ángel Gallardo and the secretary of the astronomy section was Father Devoto, who delivered an informal talk about the solar and lunar tables in the astronomical calendars he was editing at that time.[6] Both of them would play an important role in the building of scientific institutions in the 1930s in Argentina.

In this chapter, I shall show how the culturally ingrained assumptions of a secular science promoted by liberals and socialists were challenged by the rise of politically active Catholic scientists in a period of a Church expansion. After providing some necessary background information about the political, ecclesiastical, and scientific history of Argentina between 1910 and 1935, we shall follow the scientific careers and analyse the views on science and religion of the naturalist Ángel Gallardo and the astronomer Fortunato Devoto, in order to examine the relationships between the Catholic Church, science, politics, and secularisation during that period.

Church and Politics in Transition

The year 1916 marked the end of the long conservative era in Argentina, which had opened in 1880.[7] The Radical Civic Union (UCR) guided the country for the next fourteen years during the presidencies of its leader Hipólito Yrigoyen (1916–1922) and his more conservative successor, Marcelo T. de Alvear (1922–1928). Against the background of the beginning of the Great Depression, Yrigoyen's second term of office (1928–1930) was interrupted by the military coup led by General Félix Uriburu. Uriburu's short rightist and nationalistic rule (September 1930–February 1932)—a failed attempt at establishing a corporatist state—was followed by a return to a more politically liberal orientation with General Agustín P. Justo (1932–1938).[8]

---

[6] Devoto was substituting Francesco Porro de Somenzi, the former director of the Observatory, see Sociedad Científica Argentina, *Congreso*, 1:265; Fortunato Devoto, "Introducción," in Observatorio Astronómico de la Universidad Nacional de La Plata, *Calendario astronómico para el año 1909* (Buenos Aires: Coni, 1908), i–xxxv. As a result of the 1910 meeting, an agreement was reached between the four countries of the southern cone for the joint publication of an astronomical calendar series, prepared by Friedrich W. Ristenpart, director of the Observatory of Santiago de Chile and Devoto: *Calendario astronómico para la parte austral de América del Sur. Año 1911* (Buenos Aires: Coni, 1910).
[7] Ezequiel Gallo, "Society and Politics, 1880–1916."
[8] Rock, *Argentina 1516–1987*, 191–213.

By the beginning of the First World War Argentina had gone through two decades of breakneck economic and demographic expansion; by 1930 growth was keeping pace with that of Canada and Australia.[9] Between 1904 and 1913 the net flow of immigration to the country was never less than 100,000 persons per year and in some years it reached 200,000.[10] In 1914, 1,576,000 people lived in Buenos Aires; about half of them were foreigners.[11] Immigration came mostly from Italy and Spain, but the flow from non-Catholic lands was not insignificant. While the foundational thinkers of the country, such as Sarmiento and Alberdi, had seen immigration as the only available answer to the challenge of creating a modern nation out of the empty wilderness of "the desert," by the turn of the century the ruling elites showed increasing signs of anxiety vis-à-vis the explosive transformation of society; as early as the 1880s literature gave expression to these fears.[12] Answers to the challenge ranged from the law that allowed the executive power to expel any immigrant considered "undesirable" (1902) to efforts at integration and nationalisation through the patriotic rituals implemented in public elementary schools designed to turn children fresh off-the boat into fully fledged Argentine citizens.[13] The visit to Buenos Aires of the Infanta Isabel, the heir to the Spanish throne, during the centennial celebrations of independence was a symbol of the reconciliation with Spain and of the felt need to recover the Hispanic heritage, systematically despised by Francophile fin-de siècle intellectuals. This change in the atmosphere was given expression in the work of a group of influential writers and thinkers such as Manuel Gálvez, Ricardo Rojas, Carlos Ibarguren, and Leopoldo Lugones, who fostered a type of nationalism predicated upon traditional Hispanic values. While some of them remained faithful to the ideals of conservative political liberalism (Ángel Gallardo was among them), others would eventually shift to an overtly authoritarian and militaristic style of nationalism, verging on fascism.[14] Many of the immigrants who huddled in the

---

[9] Ibid., 162–99.
[10] Ernesto Tornquist & Co., *The Economic Development of the Argentine Republic in the Last Fifty Years* (Buenos Aires, 1919), 15.
[11] República Argentina, *Tercer Censo Nacional levantado el 1º de junio de 1914*, vol. 2 (Buenos Aires: Rosso, 1916), 3.
[12] Eduardo Gargurevich, "La reacción anti-inmigrante en la literatura argentina de los ochenta," *Revista de Crítica Literaria Latinoamericana* 20, no. 39 (1994): 91–107.
[13] Lilia A. Bertoni, *Patriotas, cosmopolitas y nacionalistas. La construcción de la nacionalidad argentina a fines del siglo XIX* (Buenos Aires: FCE, 2001).
[14] David Rock, *Authoritarian Argentina. The Nationalist Movement, its History and its Impact* (Berkeley: University of California Press, 1993), 37–54. See in particular Fernando J. Devoto, *Nacionalismo, fascismo y tradicionalismo en la Argentina moderna* (Buenos Aires: Siglo XXI, 2005), 47–119.

slums of Buenos Aires were anarchists and socialists who led the organisation of labour unions and stirred social unrest.[15] The wave of strikes and uprisings and the consequent repression reached its peak in January 1919 in Buenos Aires ("Tragic Week") and in the labourers' revolt in Patagonia (1920–1922), both under governments of the UCR. If during the last decades of the nineteenth century, secular feeling had been the hallmark of the Europeanised elite, by the first decade of the twentieth century anticlericalism was not uncommon among the immigrant masses from rural areas of Mediterranean Europe.

It has already been discussed how the secularising laws of the 1880s, a victory of the liberal conservative elite of cosmopolitan Buenos Aires then in power, had further restricted the social influence of an already weak Catholic Church, isolated from Rome and dependent upon the state. By the beginning of the 1890s, Estrada, Goyena and other lay leaders who had fought to stop the tide of secularisation by involving Catholics in politics had died and with them the ideal of a Catholic party. Catholic lay and clerical activism shifted from political to social concerns, as shown by the organisation of the Workers' Circles by Father Grote and other initiatives aimed at the creation of a Catholic labour movement strong enough to withstand the challenge of socialism and anarchism.[16] While membership in Catholic unions during the 1915–1930 period amounted to 40–45,000 affiliates, by 1920 the socialists could boast around 100,000 supporters.[17] Lay Catholic intellectuals such as Santiago O'Farrell, Alejandro Bunge, Arturo M. Bas, Juan F. Cafferata, Joaquín Cullen, or Emilio Lamarca were social reformers preoccupied by questions of the law, economy, and society. There were no Catholic natural scientists in Argentina before the twentieth century. While by the turn of the century the energies of the Church were spent in the social question, the hierarchy reached some kind of accommodation with the liberal elites in the face of the threat of radicalism and anticlerical sentiment; this implicit settlement was at the back of the lavish donations for charities and church buildings and the sumptuous diplomatic display on the world scene of the belle époque—a stage upon which Gallardo would move at ease.[18]

Catholic social action also found expression in the several unstable organisations of the laity which succeeded each other in the course of the first two decades of the twentieth century and were inspired by the ideas of European Chris-

---

[15] Jeremy Adelman, ed., *Essays in Argentine Labour History, 1870–1930* (Basingstoke: Macmillan, 1992).
[16] Auza, *Grote y la estrategia social*, 22.
[17] Ivereigh, *Catholicism and Politics in Argentina*, 64.
[18] José M. Ghio, *La iglesia católica en la política argentina* (Buenos Aires: Prometeo, 2007), 37; Lida, *Historia del catolicismo en la Argentina*, 41–44, 55–57.

tian democracy.¹⁹ In the 1920s, Catholic culture flourished with new institutions such as the Centre of Religious Studies for Ladies (1919, hereinafter CER, after its name in Spanish), the Courses of Catholic Culture (1922, hereinafter CCC), and the journal *Criterio* (1928). As had been the case with the Catholic social and labour movement, this effervescence of lay cultural enterprises was in the event bottled up by the hierarchy, who tried to bring it under its control.²⁰

The strong crisis concerning the nomination of the Archbishop of Buenos Aires that took place during Alvear's presidency was a signal that the paradigm of a weak church at the service of a strong state was changing. The Vatican repeatedly rejected the nomination of Msgr Miguel de Andrea, Bishop of Temnos (1920), the government's candidate, who was basically friendly to liberal democracy (in line with the *patronato*, the government sent a shortlist of nominees to Rome to be approved).²¹ Much of the negotiations was conducted by Ángel Gallardo, by then Alvear's minister of foreign affairs.²² After three years with a vacant see, the government was compelled to accept De Andrea's resignation of his candidacy and the nomination of a compromise candidate, the Franciscan Fr José M. Bottaro; in return, the papal nuncio was declared *persona non grata*. The protracted conflict was a serious challenge to the prevailing regalist settlement. The new assertiveness of the Argentine Catholic Church was tied to its acquiescence with the increasing Romanisation of the world church, and locally with the growing enthusiasm of laity and clergy alike for the model of integral Catholicism tied with authoritarian versions of nationalism which would mark the relationships between church and state in the interwar years and beyond.²³

Ivereigh has aptly described the "Catholic Renaissance" of the Argentine Catholic Church in the first decades of the twentieth century as a consequence of the transition from the almost Gallican church of the liberal state consolidated

---

19 Di Stefano and Zanatta, *Historia de la Iglesia en la Argentina*, 367–77.
20 Carlos A. Floria and Marcelo Montserrat, "La política desde *Criterio* (1928–1977)," *Criterio* 50, no. 1777–1778 (24 December 1977): 762–89; Fernando J. Devoto, "Atilio Dell'Oro Maini," *Prismas* 9 (2005): 187–204; Raúl Rivero de Olazábal, *Por una cultura católica. El compromiso de una generación argentina* (Buenos Aires: Claretiana, 1986); Miranda Lida and Mariano Fabris, eds., *La revista* Criterio *y el siglo XX argentino. Religión, cultura y política* (Rosario: Prohistoria, 2019).
21 Ambrosio Romero Carranza, *Itinerario de Monseñor de Andrea* (Buenos Aires, 1957), 213–23; Miranda Lida, *Monseñor Miguel de Andrea. Obispo y hombre de mundo (1877–1960)* (Buenos Aires: Edhasa, 2013), 91–108.
22 Jorge E. Gallardo, *Conflicto con Roma (1923–1926). La polémica por Monseñor de Andrea* (Buenos Aires: El Elefante Blanco, 2004)
23 Fortunato Mallimaci, *El catolicismo integral en la Argentina (1930–1945)* (Buenos Aires: Biblos, 1988); Devoto, *Nacionalismo, fascismo y tradicionalismo*, 169–262.

in 1880 to the "integral Catholicism" of the 1930s and early 1940s.[24] Interpretations focused on the development of social Catholicism have seen the changes in the three first three decades of the twentieth century as the transition from a "federative" plural church with leadership of the laity, modelled upon German social Christianity, to an Italian-type of Church in which clericalism, Romanisation, unification, and obedience to the hierarchy were the characteristic traits.[25]

The secularist impulse that characterised the 1880s in Argentina began to grind to a halt in the 1890s and barely survived in the 1930s. In this transitional period, the Catholic Church experienced a renovation in all aspects of its life. For the first time in the history of the country, a number of Catholic scientists became publicly visible. This phenomenon is of peculiar interest, because it seemed to contradict what had become a common-sense assumption resulting from the hegemonic influence of secularising positivism, that the very idea of religion having anything to do with science was inconceivable.

Catholics and Science

Whereas socialists devoted serious efforts to the popularisation of science, Catholics cultivated science in the context of a network of teaching religious congregations (most of them of French origin), the Jesuit schools, and the Archdiocesan seminary in Buenos Aires, at that time in charge of the Society of Jesus. The cases of astronomy and entomology are illustrative of the achievements and the limits of this kind of "Catholic" science, a distinctive characteristics of which was to convey the message that science and religion need not be at odds. In 1904, the Jesuits of the province of Aragón established a research complex in a locality near Tortosa (Catalonia), with laboratories of biology and chemistry and the Ebro Observatory.[26] This scientific complex was influential in the development of Jesuit science in Argentina (Blanco, Ubach, and also Argentinian Jesuits were trained there; until 1918 Argentina was part of the Jesuit province of Aragón).[27]

In 1902, Julio A. Roca (then president of the country for a second time) visited the Jesuit College of the Immaculate Conception in the city of Santa Fe,

---

24 Ivereigh, *Catholicism and Politics in Argentina*, 18, 84–91.
25 Néstor T. Auza, *El proyecto episcopal y lo social*, vol. 3, *Aciertos y fracasos sociales del catolicismo argentino* (Buenos Aires: Docencia, 1988), 11–20; Di Stefano and Zanatta, *Historia de la Iglesia en la Argentina*, 377–81.
26 Agustín Udías, *Searching the Heavens and the Earth. The History of Jesuit Observatories* (Dordrecht: Springer Science, 2003), 213–18.
27 Hugo Storni, "Argentina," in *Diccionario histórico de la Compañía de Jesús*, 1:227–30.

founded in 1862 upon the readmission of the Society of Jesus to the country. He was astonished to see the quality of the physics laboratory equipment, imported at great cost from France and Germany. It seems that "it bothered him that a college he considered 'dogmatic' and antiquated should have a physics cabinet superior to those in the state colleges." While showing the physics demonstration equipment, Father Pablo Hernández "seized the opportunity to tell him [Roca] that [at the college] the secular sciences were as appreciated and cultivated as much as the ecclesiastical sciences."[28] Despite the prejudices of the secularising elite, Catholic secondary schools in Argentina offered first-class scientific education. Most of them had well-equipped laboratories, school museums of natural history, and observatories or at least good telescopes; the teachers were as a rule well qualified and devoted. Besides a museum of natural history, the Jesuit college in Santa Fe had a meteorological observatory which began recording weather registers in 1882 under the direction of Father Enrique M. Cappelletti; some of these observations were published. In 1919, the instruments were put in a room on the first floor over the entrance to the new building of the college. Also in that year, a system of radiotelegraphy was put into function which allowed to receive a time signal from Buenos Aires marking the official national time; this was retransmitted to the whole town through a spotlight situated on the church tower. Cappelletti made observations of the Great Comet of 1882 from 13 September to 12 February 1883 with a German refracting telescope of 89 mm of aperture and published them in letters to the Catholic daily *La Unión* in Buenos Aires.[29] He was well aware of similar observations made in the Córdoba observatory and acquainted with the then current literature on comets. The report of his observations is framed within the context of the relationships between science and religion. He begins extolling "the Providence of God, author of nature, who has sowed the infinity of space with...phenomena so varied and extraordinary" that they should motivate all the peoples on Earth "to admire the greatness of His omnipotence" and to give Him praise.[30] Whereas Cappelletti's account opens with a reference to the physico-theological argument, it ends with a reminder of the limits of human reason. After commenting on the uncertainties regarding the course and nature of the comet, the author

---

28 Guillermo Furlong, *Historia del Colegio de la Inmaculada Concepción*, vol. 5 (Santa Fe: Colegio de la Inmaculada Concepción, 2011), 91.
29 Enrique M. Cappelletti, "Ligeros estudios sobre el gran cometa de 1882," *La Unión*, 29, 30 and 31 May 1883; 1st, 2 and 3 June 1883. Published as a pamphlet in idem, *El gran cometa de 1882. Lijeros estudios sobre sus apariencias y curso hechos en el Colegio de Santa Fe (R. A.)* (Buenos Aires: Biedma, 1883).
30 Cappelletti, *El gran cometa de 1882*, 3–4.

remarks that the intellect "should not aspire to go beyond the limits imposed on it by God."[31]

In Buenos Aires, Ubach also made observations from the Archdiocesan Seminary. He had a recently acquired equatorial Zeiss astrograph of 2.35 m of focal length and an aperture of 130 mm, which had been presented to the Seminary by Msgr Luis Duprat, the vicar general of the Archdiocese between 1900 and 1923 and in 1903 director of the Christian Democratic League.[32] Ubach observed and photographed the 1918 annular solar eclipse, published his data, and also gave public lectures on the theme at the College of El Salvador.[33] He also made observations of the solar eclipses of 1921 and 1927; in the latter case, he took 94 astrophotographs.[34] A short unsigned article about the sun eclipse of 1925 was published in the section of Ecclesiastical Ephemerides of the journal of the Archdiocese of Buenos Aires; it was written by Ubach in all probability.[35] By 1927, an observatory had been set up on the roof of the main entrance of the Archdiocesan Seminary.[36] The significance of this comes to light if we take into account that the observatory of the National College of Buenos Aires —mostly aligned with secularism since in 1883 its rector, the Catholic Estrada, was dismissed by the government—was inaugurated only after 1935 with a refracting telescope of 80 mm of aperture and a focal length of 141 cm.[37]

In 1858, The French-Basque congregation of Betharramites founded the San José College in Buenos Aires. Catering to the large number of Basque immigrants, it soon expanded and became one of the best Catholic private schools. In 1914, a telescope, donated by a student's father, was housed in a small observatory in a tower of the main building. It was an equatorial double astrographic Mailhat re-

---

[31] Cappelletti, *El gran cometa de 1882*, 39.
[32] José Ubach, *Observaciones del eclipse anular del 3 de diciembre de 1918 en Buenos Aires (República Argentina)* (Tortosa: Algueró y Baiges, 1919), 7.
[33] "Eclipse solar del 3 de diciembre. Conferencias de divulgación científica," *La Prensa*, 1st September 1918. See also José Ubach, *El eclipse anular del 3 de Diciembre de 1918 en la República Argentina, Chile y Uruguay* (Buenos Aires: R. Herrando, 1918).
[34] José Ubach, "Observaciones astronómicas del eclipse del 1 de octubre de 1921," *Estudios* 22 (January–June 1922): 19–29; idem, *Observaciones astronómicas del eclipse del 3 de enero de 1927* (Tortosa: Algueró y Baiges, 1927).
[35] José Ubach, "El eclipse del día 3 de enero," *Revista Eclesiástica del Arzobispado de Buenos Aires* 27 (3 January 1927): 105–106.
[36] All that remains of it are the walls and a ruined dome. See Mario A. Poli, "El Seminario Metropolitano de Buenos Aires en la Facultad de Teología (1622–2015)," in *Facultad de Teología, 100 años de la Facultad de Teología: memoria, presente, futuro* (Buenos Aires: UCA, 2015), 169–93.
[37] Colegio Nacional de Buenos Aires, Departamento de astronomía, "El Observatorio," accessed 18 September 2020, https://astro.cnba.uba.ar/?page_id=24.

fractor of 127 mm of aperture and 1880 mm of focal length.[38] By the end of the nineteenth century, a school museum of natural history was organised, with the advice of Karl Berg and Hermann Burmeister, who catalogued some of the collections and once in a while dropped by to have tea and talk in French with the brothers.[39] Francisco P. Moreno, an alumnus of the institution, also contributed some specimens—it seems that his unorthodox religious beliefs did not perturb his former teachers.[40] In 1914, the San José school museum was reorganised and enriched with natural history specimens imported from France. The school periodical, *F.V.D.*, used to publish short articles on science and religion, most of which were intended as arguments against the conflict thesis. One of them reproduced a 1926 note in *Le Figaro* (Paris) with a survey among famous French scientists about the compatibility of science and religion.[41] A short note by a student drew rather incongruously on Thomas Huxley's writings.[42] There was a series of notes on Christian savants such as Augustin-Louis Cauchy, Alessandro Volta, Édouard Brauly, and Charles Tellier.[43]

The law of 7 July 1904 pushed by Émile Combes in France at the height of the conflict between the state and the Catholic Church (a year later, they separated) ordered the exclusion from elementary, secondary, and higher teaching of the members (female and male) of religious congregations.[44] As a result of the law, around 30,000 brothers and nuns abandoned France to teach elsewhere.[45] Many of these highly qualified teaching personnel travelled to Argentina, where some of the teaching orders had secondary colleges. For example, the first school of De La Salle Brothers (Brothers of the Christian Schools) in Buenos Aires was founded in 1891; its laboratories of physics and chemistry were magnificently equipped with French instruments. The Marist Brothers arrived in Argentina in

---

38 Basilio Sarthou, *Historia centenaria del Colegio San José de Buenos Aires (1858–1958)* (Buenos Aires, 1960), 160–61.
39 Sarthou, *Historia centenaria*, 234–36.
40 "Como se funda un museo. Una de las glorias del Colegio San José: el doctor Francisco P. Moreno," *F.V.D.* 5, no. 55 (September 1955).
41 "Ciencia y religión," *F.V.D.* 6, no. 66 (6 August 1926).
42 Edgardo C. Eyzaguirre, "Ciencia y religión," *F.V.D.* 12, no. 135 (May 1932): 126.
43 "La fe y los sabios [Cauchy]," *F.V.D.* 7, no. 77 (July 1927); "Los sabios cristianos: Charles Tellier," *F.V.D.* 12, no. 139 (September 1932): 250; "Los sabios cristianos: Eduardo Brauly," *F.V.D.* 12, no. 140 (October 1932): 288; "Los sabios cristianos: Alejandro Volta," *F.V.D.* 12, no. 141 (November 1932): 325.
44 Patrick Cabanel and Jean-Dominique Durand, eds., *Le grand exil des congrégations religieuses françaises, 1901–1914* (Paris: Le Cerf, 2005).
45 Patrick Cabanel, "Le grand exil des congrégations enseignantes au début du XXe siècle. L'exemple des Jesuites," *Revue d'histoire de l'Église de France* 81, no. 206 (1995): 207–17.

1903; the new building of the Manuel Belgrano College in Buenos Aires, erected in the 1940s, was provided with a small astronomical observatory.

Members of religious congregations and orders also contributed to the growth of entomological studies in Argentina. Jean Brèthes was a French Lasallian Brother who arrived in Argentina in 1890, when he was nineteen years old. He taught at the newly established La Salle School in Buenos Aires and was the organiser of its museum of natural history; Brèthes published his first articles under the pseudonym of "Frère Judulien" and as such is mentioned in the *Souvenirs entomologiques* by Jean-Henri Fabre, with whom he corresponded. Brèthes soon entered into contact with the local entomological community—in particular with the Lynch Arribálzaga brothers—and on account of his talent and his tireless work was named head of the entomological section of the National Museum of Natural History by Florentino Ameghino, at that time director of the institution; he continued working in that position for 26 years and also worked in other research institutions. At one time he quit religious life. Since 1921 until his death in 1928, he was professor of Agricultural Entomology in the School of Agriculture of the University of La Plata; he also taught natural history at the Jesuit College of El Salvador in Buenos Aires and was chief of the entomological section at the Biological Institute of the Argentine Rural Society. Brèthes took charge of the study of coccinellids (ladybugs) of the British Museum and of South American Hymenoptera (wasps, bees, ants, and sawflies) of the Deutsche Entomologische National-Museum; he described more than 1100 new species and published around 200 papers.[46] He was nominated corresponding member of the Museum d'Histoire Naturelle in Paris.[47] During his lifetime, Brèthes was a rather well-known naturalist in the country (there is a bronze bust of him in the MACN) and his death received ample press coverage.[48]

The Jesuits were able to establish an important entomological tradition. Juan B. Mühn was one of six brothers who joined the Society of Jesus in Argentina; two sisters also entered into religious life. They were the sons and daughters

---

[46] Enrique Herrero Ducloux, "Profesor Juan Brèthes, 1871–1928," *Revista de la Facultad de Agronomía* 18, no. 1 (1928): 5–12; Ernesto Dallas, "Dr. Juan Brèthes. Bio-bibliografía," *Revista de la Sociedad Entomológica Argentina* 2 (1928–1929): 103–12; Alejandro Rossi Belgrano and Mariana Rossi Belgrano, *Juan Brèthes (Frère Judulien Marie): primer entomólogo del Museo Nacional* (Buenos Aires, 2018).

[47] "Un naturalista argentino fue nombrado miembro del Museo de París," *La Razón*, 19 January 1925.

[48] "Falleció el entomólogo profesor Juan Brèthes," *La Razón*, 2 July 1928; "Se efectuó el sepelio de los restos del professor Juan Brèthes," *La Prensa*, 4 July 1928; "En la metrópoli falleció ayer el entomólogo Juan Brèthes," *El Diario* (La Plata), 4 July 1928.

of a family of German settlers in the province of Santa Fe.⁴⁹ Mühn studied in Tortosa, where he trained under the Catalan Jesuit naturalist and entomologist Longinos Navàs.⁵⁰ Mühn then spent a year and a half studying biology and doing research in Munich and returned to Argentina in 1924, where he taught courses of science at the Seminary of Buenos Aires. In 1923, seminary students gave a lecture on Mendel and the laws of heredity, which means that these matters were taught there at the time.⁵¹ Mühn organised an entomological collection which was exhibited in the First Entomological Exposition in Argentina (1928) and also recruited two Jesuit student who would continue the entomological tradition he started, Albino Bridarolli and Gregorio Williner.⁵² Bridarolli, the brightest of these Jesuit entomologists, studied philosophy at the Ignatiuskolleg (Valkenburg), where he met the German entomologist Hermann Schmitz, a specialist in Phoridae who had been a student of the more famous Jesuit entomologist Erich Wasmann. Bridarolli obtained his doctor's degree from the University of La Plata but died tragically in a car accident. Williner went on building the entomological collection in the Colegio Máximo of San Miguel; in the end, an entomological institute was established there (in the 1970s, the collection had around 100,000 specimens).⁵³

Until the 1930s, there were no Catholic scientific institutions in Argentina. In 1910, there was a failed attempt at organising a Catholic University around a Law School; the government did not allow it to grant professional diplomas and after a decade it closed.⁵⁴ The scientific activities of Jesuits from the Iberian Peninsula trained in Tortosa and the Ebro Observatory and others born in Argentina from immigrant families from Bavaria and Tyrol showed that science could be cultivated within a Catholic context. The high tide of French secularism, which was still the model for many local thinkers, had as an unexpected consequence

---

49 Gregorio J. Williner, "P. Jorge Mühn," *Noticias de la Provincia Argentina* 22, no. 159 (1952): 19–23.
50 Juan-Jesús Bastero Montserrat, "Longinos Navás S.J. An approach to his Life and Entomological Work," in *Overview and Strategies of Ephemeroptera and Plecoptera*, ed. J. Alba-Tercedor y A. Sánchez Ortega (Gainesville, FL: Sandhill Crane Press, 1991), 581–84.
51 Juan Isern, *La formación del clero secular de Buenos Aires y la Compañía de Jesús* (Buenos Aires: San Miguel, 1936), 487.
52 Juan B. Mühn, "Primera exposición entomológica argentina, 19 al 25 de septiembre de 1928," *Estudios* 36 (1928): 151–56.
53 Miguel de Asúa, "Los entomólogos de la Compañía de Jesús en Argentina," *Stromata* 73, no. 2 (2017): 231–43.
54 Juan C. Zuretti, "La fundación de la primera Universidad Católica," *Universitas* 9, no. 38 (July–September 1975): 89–101; Hebe C. Pelosi and Geraldine Mackintosh, "El ensayo de Universidad Católica (1909–1920)," *Archivum* 16 (1994): 185–95.

the improvement of Catholic scientific education in Argentina. But as discussed in the preceeding chapter when treating the teaching of evolution at the College of El Salvador, there were clear limits imposed by Catholic doctrine which the teachers would not trespass. In a way, the situation was analogous to Early Modern Jesuit science, which excelled in empirical accuracy and thoroughness but floundered when addressing questions of natural philosophy.[55] Observational astronomy, experimental physics, and taxonomical natural history were fields which could be cultivated safely.

Gallardo

It is against the background depicted in the previous sections that Gallardo, born into a wealthy family of Spanish origin, must be understood. A lifelong Catholic, Ángel Gallardo (1867–1934) participated as a youth in the 1890 armed revolt that gave birth to the Civic Union (later UCR), and though he later formally quit the party, he nevertheless remained within its orbit. When, in 1919, president Yrigoyen called him to preside the National Council of Education (hereinafter CNE, after its name in Spanish), he was already a distinguished personality. He went on to serve as representative of his country in Italy for a year (1921) and then as minister of foreign affairs during Alvear's presidency (1922–1928) when Argentina "glided gently in years of prosperity, without shocks or difficulties."[56] Between September 1927 and January 1928, Gallardo made a diplomatic farewell tour in the course of which he was entertained by almost all the heads of state of Western Europe. During the presidency of Justo, he served for two years as rector of the University of Buenos Aires (1932–1934), at that moment shaken by student unrest. At the time of his death, Gallardo belonged to most scientific institutions and learned academies in Argentina; he had crossed the ocean back and forth seven times, walked with kings and, as expressed by one chronicler, experienced "the spiritual peace of a savant and a believer."[57] His funeral was attended by President Justo, his ministers, the Archbishop of Buenos Aires, and other civil and ecclesiastical personalities.[58]

Gallardo never showed any enthusiasm for integral Catholicism and in his public career acted as if the Church should be subordinate to the state. But

---

55 William B. Ashworth, "Catholicism and Early Modern Science," in *God and Nature*, 136–66.
56 Félix Luna, *Alvear* (Buenos Aires: Hyspamerica, 1986), 66.
57 *La Prensa*, 16 May 1934, cited in Guillermo Furlong, *Ángel Gallardo* (Buenos Aires: ECA, 1964), 160.
58 Furlong, *Ángel Gallardo*, 151–55.

the conflict built around the nomination of De Andrea as Archbishop of Buenos Aires put his allegiances to test.[59] Later, he declared that at that time he had "intimately prayed to God to allow a solution with detriment neither to the Fatherland nor to the Church."[60] Although he seems to have made every effort to reach a negotiated outcome between the state and the Catholic Church, at no time did he hesitate to loyally follow Alvear's policy.[61] A few years earlier, in 1918, Gallardo had also showed his prioritization of the civil sphere over the ecclesiastical. As director of the CNE, he supported a regulation (in the end relaxed) that prescribed that teachers in elementary and secondary schools should be graduates from Argentine institutions, thus excluding from those positions the foreign priests who taught in private Catholic schools.[62]

Throughout his life, Gallardo was a democratic conservative and a nationalist. Nationalism seems to have been in him a sentiment closely related to traditions, duty, and the love of the land and its creatures—a different thing from the kind cultivated in the 1930s and 1940s by far-right Catholics, some of them transatlantic camp followers of Charles Maurras's Action Française (to be discussed in the next chapter). Gallardo was certainly situated towards the right of the political spectrum and he did not hide his strong anti-communist convictions, but he always remained within the bounds of liberal democracy.[63] If we are to take him at face value, his passing admiration for Mussolini, whom he met twice, and his press declarations about fascism (which he saw as "sympathetic for its patriotism, its nationalist ideal, and its disinterested spirit") were those of a hidebound conservative in the 1920s.[64] In August 1926, a socialist representative in the Chamber of Deputies accused him of "a profound admiration for the political system" dominating Italy at that time. Gallardo candidly admitted that his spirit "might be reactionary," but that his "democratic faith was absolute, frank, and

---

[59] Ángel Gallardo, *Memorias para mis hijos y mis nietos* (Buenos Aires: Academia Nacional de la Historia, 1982), 347–86.
[60] Gallardo, *Memorias*, 357.
[61] Gallardo, *Memorias*, 383–386; Gallardo, *Conflicto con Roma*, 58–62.
[62] Dimitri P. Papanikas, "La iglesia de la raza. La iglesia católica española y la construcción de la identidad nacional en Argentina" (PhD diss., Universidad Autónoma de Madrid, 2012), 125–26, https://repositorio.uam.es/handle/10486/10264.
[63] For De Andrea on this, see John J. Kennedy, *Catholicism, Nationalism, and Democracy in Argentina* (Notre Dame: University of Notre Dame Press, 1958), 139–46.
[64] "Declaraciones del ministro argentino de relaciones exteriores," *La Nación*, 24 December 1922; Gallardo, *Memorias*, 316–19, 438–40; Alfredo L. Palacios, "El ideal de las democracias ibero-americanas," *Nosotros* 17, no. 172 (1923): 35.

incontrovertible," and he considered any "anti-democratic suggestion" a "true crime."⁶⁵

Gallardo's speech at the moment of his nomination as rector of the University of Buenos Aires on 11 May 1932 was a manifesto of nationalistic faith, in tune with the spirit of the age. Though he decried "aggressive and xenophobic nationalism," he declared that the very reason of existence of the university was "the patriotic and national ideal" and "the shaping of an enlightened national conscience," besides the cultivation and progress of abstract science.⁶⁶ Two years after these unambiguous declarations, Gallardo accepted heading the local board of the Buenos Aires Pacific Railway.⁶⁷ As a result of the global depression and local circumstances, the British-owned railway companies were experiencing large losses. In order to reach an agreement acceptable to them and the government, President Justo set up an official commission in which Gallardo should sit as representative of the companies—an arrangement that did not materialise for he died shortly after.⁶⁸ The growing wave of anti-British feeling that swept the country in those years (partly as a result of the 1933 commercial treaty with Britain, urged by Argentina to maintain the quota of meat exports and generally seen as favouring the British) was propelled by far-right nationalist intellectuals and integral Catholics.⁶⁹ One of them, Ibarguren, at one time a good friend of Gallardo's, considered that the treaty "reinforced the old submission of Argentine economy to the British empire."⁷⁰ Gallardo's position was subtly but crucially different from this; his nationalism was nearer to Justo's conservative liberal state than to a fascist conception of society.

Biologist and Naturalist

Gallardo obtained his degree in engineering in the University of Buenos Aires (1894) and subsequently got a doctor's degree in the natural sciences from the

---

65 Congreso Nacional, Cámara de Diputados, *Diario de sesiones...Año 1926*, vol. 3 (Buenos Aires: Imprenta de la Cámara de Diputados, 1926), 551–55.
66 "El nuevo rector de la Universidad," *Archivos de la Universidad de Buenos Aires* 7 (April–May 1932): 11–13.
67 Furlong, *Ángel Gallardo*, 143; Buenos Aires & Pacific Railway Company, *History and Characteristics, 1882–1933* (Buenos Aires, 1933).
68 Julian S. Duncan, "British Railways in Argentina," *Political Science Quarterly* 52, no. 4 (1937): 570–72; Winthrop R. Wright, *British-Owned Railways in Argentina. Their Effect on Economic Nationalism, 1854–1948* (Austin: University of Texas Press, 1974), 136–57.
69 Rock, *The British in Argentina*, 276–86.
70 Carlos Ibarguren, *La historia que he vivido* (Buenos Aires: Dictio, 1977), 602.

same institution (1902). Much of his published work rests on the intersection of these fields. In the communication he read to the 1900 International Congress of Mathematics in Paris, he argued for the application of statistics to the problems of biological variation, inheritance, and evolution.[71] Gallardo taught botany and zoology at the schools of Exact and Natural Sciences, of Agronomy, and of Pharmacy of the University of Buenos Aires and presided the SCA and the Argentine Society of Natural History.[72] He organised the first Latin American Scientific Congress (1900) and was vice-president of the 1910 International Congress of Americanists in Buenos Aires. Undoubtedly, his most durable institutional commitment was as director of the Museum of Natural Sciences in Buenos Aires (1911–1918). Although his efforts to move the museum from its cramped old quarters to a new location that the institution badly needed were unsuccessful, he was able to recruit qualified naturalists and promoted the growth of the collections.[73]

Gallardo studied natural history in Buenos Aires under Karl Berg, a naturalist born in Courland who succeeded Burmeister as director of the Public Museum in Buenos Aires, but he learnt his biology in Paris, where, during his first and second European voyages (April 1895–March 1896, November 1899–June 1901) he became acquainted with the zoologist Yves Delage and took courses with Gustave Loisel, Léon Guignard, Alfred Giard, and Félix Le Dantec (Delage, Giard and Le Dantec were anticlerical supporters of the Third Republic.) [74] The kind of biological problems he tackled would be those discussed in this circle of natural scientists, who for the most sustained a neo-Lamarckian view of evolution.[75] Gallardo gained a certain notoriety with his "dynamic" theory of cell division,

---

[71] Ángel Gallardo, "Les mathématiques et la biologie," in *Compte rendu du deuxième Congrès international des mathématiciens*, ed. Ernest Duporcq (Paris: Gauthier-Villars, 1902), 395–403.
[72] "Datos biográficos de don Ángel Gallardo," *Boletín de la Academia Argentina de Letras* 2 (1934): 105–13.
[73] Martín Doello-Jurado, "Ángel Gallardo (1867–1934). Su actuación en el Museo de Buenos Aires," *Anales del Museo Argentino de Ciencias Naturales* 38 (1934–1936): ix–xliv; Miguel de Asúa, "Dos siglos y un museo," in *El Museo Argentino de Ciencias Naturales. 200 años*, ed. Pablo Penchaszadeh (Buenos Aires: MACN-Conicet, 2012), 13–69. Cf. Sheets-Pyenson, *Cathedrals of Science*.
[74] Gallardo, *Memorias*, 79–82, 95–111.
[75] For Giard, Delage, and Le Dantec, see Peter J. Bowler, *The Eclipse of Darwinism. Anti-Darwinian Evolution Theories in the Decades around 1900* (Baltimore: The Johns Hopkins University Press, 1983), 110–15; Christophe Charle and Eva Telkes, *Les professeurs de la Faculté des sciences de Paris. Dictionnaire biographique (1901–1939)* (Paris: Institut National de la Recherche Pédagogique, 1989), 104–107, 136–137 and the articles in *Dictionary of the History of Science*, ed. Charles C. Gillispie, 16 vols. (New York: Scribner's Sons, 1981), 4:11–13 (Delage), 5:385–86 (Giard), and 5:581–82 (Guignard).

according to which the mitotic apparatus was an expression of a force field generated by the centrosomes charged with polar charges, analogous to Faraday's magnetic fields.[76] His dissertation on this matter was enthusiastically reviewed in *Nature* by the Irish natural historian Marcus Hartog, who began experimental work along the same lines.[77] The theory was not entirely original, having been proposed by Hermann Fol in 1879 and in 1895 by H. E. Ziegler, but Gallardo worked it out independently.[78] In his dissertation, the latter allowed himself to depart from his experimental results and freely postulate a "cariocynetic force," a Newtonian force of unknown nature which could also be used to account for the phenomena of heredity and fecundation not in terms of matter but of energy.[79] Gallardo's neo-vitalism drew upon the natural philosophical speculations of Johann Reinke, Lord Kelvin's theory of vortex atoms (both of whom he knew at second hand), and the book of the French physiologist Louis Bard, which he had reviewed.[80]

Gallardo introduced genetics in Argentina in a series of works for specialists and also for the general public published between 1908 and 1910.[81] At the time of the polemics between Mendelians and the followers of the biometric school of Karl Pearson, he insisted upon the lack of contradiction between the ap-

---

[76] Ángel Gallardo, "Essai d'interprétation des figures karyokinétiques," *Anales del Museo Nacional de Buenos Aires* 5 (1896): 11–22; Horacio Damianovich, "Ángel Gallardo y su teoría de la cariocinesis," *Anales de la Sociedad Científica Argentina* 133 (1942): 102–31.

[77] Ángel Gallardo, *La interpretación dinámica de la división celular* (Buenos Aires: Coni, 1902); Marcus Hartog, "Dynamic Interpretation of Cell Division," *Nature* 67, no. 1724 (13 November 1902): 42–43.

[78] See the discussion in Edward B. Wilson, *The Cell in Development and Inheritance*, 2nd ed. (New York: Macmillan, 1911), 108–11.

[79] Gallardo, *La interpretación* dinámica, 84–95.

[80] Johann Reinke, *Die Welt als That. Umrisse einer Weltansicht auf wissenschaftlicher Grundlage* (Berlin: Gebrüder Paetel, 1899); Helge Kragh, "The Vortex Atom: A Victorian Theory of Everything," *Centaurus* 44 (2002): 32–114; Ángel Gallardo, "Problemas biológicos. Algunas reflexiones sobre la especificidad celular y la teoría física de la vida, de Bard," *Revista de Derecho, Historia y Letras* 4 (1899): 540–65 (review of *La spécificité cellulaire*, by Louis Bard [Paris: Georges, Carré et Naud, 1899]).

[81] The main publicaciones are Ángel Gallardo, *Las investigaciones modernas sobre la herencia* (Buenos Aires: La ciencia médica, 1908); idem, "Sur l'épreuve statistique de la loi de Mendel," *Comptes rendus hebdomadaires des séances de l'Académie des sciences* 146 (1908): 361–62; idem, "Recientes contribuciones matemáticas al estudio de las leyes de la herencia biológica," *Anales de la Sociedad Científica Argentina* 68 (1909): 185–208. See also Nancy Stepan, *"The Hour of Eugenics." Race, Gender, and Nation in Latin America* (Ithaca: Cornell University Press, 1991); Novoa and Levine, *From Man to Ape*, 107–109.

proaches.[82] He also devoted a number of papers to botanical subjects, in particular to teratology, perhaps a result of his having being in charge of the government's Division of Agriculture (1904–1905).[83] But his most lasting interest, and one which accompanied him from childhood to mature years, was ants, a subject to which he made enduring contributions.[84]

Science, Religion, and Evolution

The traditional notion of God's manifestation in nature is not absent from Gallardo's writings: the value of learning the laws that rule the stars is that "we can discern in them the sublime harmony established by God in Creation;" fungus-growing ants "show as in lightning God's supreme intelligence, reflected in the narrow aspect of their instinct."[85] These were articles for the general public; his academic papers were free from any philosophical or religious allusions. As shown by his 1902 dissertation, he did entertain broad natural philosophical schemes and advanced hypotheses (which he calls *Arbeitshypothesen*) that could explain the phenomena of the living beings in terms he saw as congruent with a theistic world view, but he was conscious that indulging in this kind of speculation was "to overstep the strictly positivistic scientific terrain."[86] Gallardo was educated in a cultural atmosphere oversaturated with the positivism of fin-de-siècle Argentina. "Positivism" was for Gallardo just standard scientific methodology.

---

[82] See Karl Pearson, "On the Ancestral Gametic Correlations of a Mendelian Population Mating at Random," *Proceedings of the Royal Society of London. Series B, Containing Papers of a Biological Character* 81 (1909): 225–29. For background of this issue, see Lyndsay A. Farrall, "Controversy and Conflict in Science: A Case Study. The English Biometric School and Mendel's Laws," *Social Studies of Science* 5 (1975): 269–301; William B. Provine, *The Origins of Theoretical Population Genetics* (Chicago: The University of Chicago Press, 2001), 56–89; Nicholas Gillham, "The Battle between the Biometricians and the Mendelians: How Sir Francis Galton Caused his Disciples to Reach Conflicting Conclusions about the Hereditary Mechanism," *Science & Education* 24 (2013): 61–75.
[83] "Bibliografía del Dr. Gallardo," *Anales de la Sociedad Científica Argentina* 133 (1942): 172–73.
[84] Carlos Bruch, "La obra entomológica del doctor Ángel Gallardo," *Revista de la Sociedad Entomológica Argentina* 6 (1934): 234–42.
[85] Ángel Gallardo, "Prologue," in Martín Gil, *Cosas de arriba* (Córdoba: La Italia, 1909), iv; Ángel Gallardo, "El instinto de las hormigas," *Revista de Filosofía. Cultura, ciencias, educación* 2 (1915): 20.
[86] Gallardo, *La interpretación* dinámica, 63, 82.

In his 1916 conference on science and belief before university students, Gallardo argued that there should be no conflict between religion and science because "they correspond to different spheres of the human spirit, between which there could be no interference."[87] Religion is "supra-rational"—it concerns "mysteries which reason can neither demonstrate, nor even conceive"—while the scientific method is "positivistic" since "departing from the data offered to the senses, reason deduces more or less general principles." "Absolute" truth is like the limit of a variable function, which can never be attained. In the lab, one should be "entirely positivist and objective" and leave aside any qualms about research results conflicting with the faith because scientific truth could never be in contradiction with absolute truth. In 1939, Nobel-prize winner Bernardo Houssay (1947), himself a liberal, saw Gallardo as a "fervent believer" and "a savant of deeply religious spirit." On the very sensitive issue of evolution, he pointed out that "since his beginnings in science [Gallardo] adopted evolutionistic ideas which he held during all his life, without ever experiencing conflict with his ingrained religious beliefs."[88]

Certainly Gallardo did not doubt the fact of biological evolution, but he was less sure about its mechanism—in this, he concurred with his French teachers. In a brief article in a 1914 issue of the *Illinois State Register*, he calls Darwin's evolution "one of the great progressive movements of the last century" but at the same time remarks that "the mechanism is debated" and goes into hazy speculation about objections to the theory "from a philosophical point of view."[89] He opens the note remarking that he does not believe "that there could be any conflict between religion and science" and proclaims his accord with Father John A. Zahm CSC, the author of a work which essayed a reconciliation between evolution and Catholic doctrine (*Evolution and Dogma*, 1896; Zahm had visited Argentina in 1916 accompanying Theodore Roosevelt on his Latin American tour).[90] If

---

87 Ángel Gallardo, "Creencia y ciencia," *Tribuna Universitaria* 3 (1916): 103–7.
88 Bernardo Houssay, "Ángel Gallardo y el porvenir de las ciencias en la Argentina," in idem, *Escritos y discursos* (Buenos Aires: El Ateneo, 1942), 497.
89 Ángel Gallardo, [Note on Darwin], in "Darwin, Science, and the Church," ed. Elmer J. Kneale, *Illinois State Register*, 23 August 1914.
90 John A. Zahm, *Through South America's Southlands* (New York: Appleton, 1916). In his memoirs, Gallardo mentions that he met him at the International Congress of Americanists in Buenos Aires, but Father Zahm is not mentioned in the proceedings of the meeting. Gallardo, *Memorias*, 135; cf. Thomas F. O'Connor, "John A. Zahm, C.S.C.: Scientist and Americanist," *The Americas* 7, no. 4 (1951): 435–62; Philip R. Sloan, "Bringing Evolution to Notre Dame: Father John Zahm, C.S.C. and Theistic Evolutionism," *The American Midland Naturalist* 161 (2009): 189–205. There is a Spanish version of Zahm's book: *La evolución y el dogma*, trans. Miguel Asúa (Madrid: Sociedad Editorial Española, 1905).

Gallardo had any reservations about Darwinian evolution, they were motivated by scientific, not religious, questions. In his 1916 address as president of the Argentine Society of Natural History, he claimed that evolutionary theory, "initiated by Lamarck and...established upon more solid bases by Darwin," showed the importance of "a dynamical point of view" in biological problems (see above). He recalls that it was in the Argentine Pampas where Darwin adumbrated his new theory "while galloping along the immense and lonely plains," as a result of the impression upon his spirit of the South-American natural world; Gallardo then goes on to expand on the paleontological contributions made by Ameghino, former director of the museum Gallardo was directing.[91] It is telling that in the homage paid to Ameghino in 1915, Gallardo warmly remembered his predecessor and even extolled his debatable theories about the Tertiary origin of the human being in the Pampas.[92] Moreover, in his handbook of zoology intended for preparatory courses for the university and also used in secondary schools, Gallardo included a brief section on the variability of species in which he provides sketches of evolutionary ideas (Lamarck, Darwin, and De Vries).[93]

The Bishop Astronomer

Fortunato Devoto (1872–1941) was born in Buenos Aires soon after the arrival in Argentina of his middle-class Genoese immigrant parents.[94] At fifteen years old, he entered the Pontifical Latin American College in Rome and studied in the Gregorian University, where he obtained degrees in philosophy, theology, and canon law while also studying astronomy under Gaspar S. Ferrrari SJ. He was ordained at the end of 1895 and returned to his country, where he enjoyed the patronage of Msgr Mariano Espinosa, the Archbishop of Buenos Aires, eager to promote bright young priests who had been trained in Europe. Devoto became chaplain of nun

---

91 Ángel Gallardo, "Los estudios biológicos en la República Argentina," in *Primera Reunión Nacional de la Sociedad Argentina de Ciencias Naturales. Tucumán, 1916* (Buenos Aires: Coni, 1919), 4.
92 "Florentino Ameghino. Homenaje a su memoria," *La Nación*, 7 August 1915.
93 Ángel Gallardo, *Zoología*, 8th ed. (Buenos Aires: Estrada, [1917]), 80–83.
94 "Datos biográficos del Iltmo. y Rvmo. Sr. Dr. D. Fortunato Devoto," *Revista Eclesiástica del Arzobispado de Buenos Aires* 28 (1928): 109–13; "Excmo. y Revmo. Monseñor Fortunato J. Devoto. Falleció en Buenos Aires el 29 de junio de 1941," *Revista Eclesiástica del Arzobispado de Buenos Aires* 41 (1941): 457–60; "Falleció esta madrugada Mons. Fortunato J. Devoto," *El Pueblo*, 29 June 1941; "Profundo pesar produjo el fallecimiento del Excmo. Mons. Dr. Fortunato J. Devoto, *El Pueblo*, 30 June–1st July 1941; Félix Aguilar, "Monseñor Fortunato Devoto," *Revista Astronómica* 13 (1941): 259–63.

congregations who ran elite secondary schools for girls; he was named secretary of the cathedral chapter of Buenos Aires and in 1900 took charge of the yearly publication of the archbishopric of Buenos Aires. Never abandoning his astronomical studies, in November 1907 he joined the Observatory of La Plata, was named its director in March 1910, and resigned after little more than a year as a result of an institutional conflict.[95] Two months later, he travelled to Paris on a fellowship granted by the national government.[96] There he obtained the *licenciature* in sciences, and from 1914 to May 1917 worked at the Paris Observatory with much application ("beaucoup de zèle"), if we are to trust his supervisor, Guillaume Bigourdan.[97] It seems that at the end of his first year of study in Paris he toyed with the idea of devoting himself entirely to his priestly work, but Pius X urged him not to abandon his astronomical career.[98] His training included equatorial reductions, attending the observatory's time service, use of the meridian circle, and astronomical photography.[99] An asteroid (1328 Devota) was named in his honour by its discoverer Benjamin Jekhowsky, who had been Devoto's fellow student at the Paris Observatory.[100]

Devoto published his observations of the Delavan comet and was offered the directorship of the observatory of the Castle of Abbadia (Hendaye), a dependency of the Académie des sciences; since the war demanded that the institution should be headed by a native in the event the arrangement did not work.[101] Devoto nevertheless worked in that small observatory for a year and kept a life lasting connection with France: in 1933 he was decorated an Officer of the Legion of Honour for "service rendered to France during the war."[102] The school magazine of the San José secondary school run by the Betharramites in Buenos Aires took

---

95 "Observatorio de La Plata. Renuncia del director," *La Prensa*, 8 June 1910; "Renuncia Fortunato Devoto," *Revista Eclesiástica del Arzobispado de Buenos Aires* 11 (1911): 508–12.
96 [Decree of President Roque Sáenz Peña, 14 July 1911], *Revista Eclesiástica del Arzobispado de Buenos Aires* 11 (1911): 593.
97 *Rapport annuel sur l'état de l'Observatoire de Paris pour l'année 1914* (Paris: Imprimerie Nationale, 1915), 22.
98 José Canovai, "Oración fúnebre," *Estudios* 67 (1941): 70–77.
99 *Rapport annuel sur l'état de l'Observatoire de Paris pour l'année 1916* (Paris: Imprimerie Nationale, 1917), 17.
100 *Dictionary of Minor Planets*, 5th ed., ed. Lutz D. Schmadel (Berlin: Springer, 2007), 108.
101 Fortunato Devoto, "Observations de la comète 1913f (Delavan), faites à l'Observatoire de Paris (équatorial tour de l'Ouest de 0^m, 305 d'ouverture)," *Comptes rendus hebdomadaires des séances de l'Académie des sciences* 160 (January–June 1915): 128–29.
102 "Argentine Prelate Honoured by France," *National Catholic Welfare Conference News Service* no. 33–2640 (18 March 1933).

opportunity to show the pupils how Devoto's career was "a new eloquent testimony of the beautiful accord between science and faith."[103]

Devoto returned to his country in 1918, after seven years in Europe; the press organ of the archbishopric of Buenos Aires dubbed him "one of the most renowned Argentine savants," while extolling his example as a rebuff to "those who declare that science and faith cannot walk hand in hand."[104] During the 1920s, he was a chief protagonist of the Catholic cultural renaissance in Buenos Aires as counsellor of the CER, an active figure in the CCC, and also through his participation in *Criterio*. During the years the government was enmeshed with Rome over the issue of the new archbishop, Devoto ascended several echelons in the hierarchy of the archdiocese of Buenos Aires: Archbishop Espinosa named him canon (1920) and at his death in 1923 Devoto was appointed vicar general and afterwards delegate of the apostolic administrator Msgr Juan A. Boneo.[105] When the consecration of Fr José M. Bottaro (1926) solved the impasse between Rome and the government, Devoto again became vicar general. In 1927 he was made Bishop of Attea and auxiliary bishop of the archdiocese by Pius XI.[106]

An incidental remark by Devoto in the first issue of the review of the archbishopric of Buenos Aires that he himself edited throws some light on his view of the relationships between science and religion. Commenting upon the Fifth International Catholic Scientific Congress (Munich, 1900), Devoto claimed that "it was necessary to instruct the Catholics into the procedures of the rigorous scientific method while vindicating them from the charge of dilettantism in the sciences." He thought that this was the only way to show that the alleged "incompatibility between the scientific spirit and the Catholic spirit" was unfounded.[107] His career would show that he took at heart his program of neutralising the conflict view by showing that Catholics could do serious and prestigious science.

---

103 "Ciencia y fe," *F.V.D.* 13, no. 149 (September 1933): 357–58.
104 [Untitled], *Revista Eclesiástica del Arzobispado de Buenos Aires* 18 (1918): 809.
105 "Nota al Exmo. Sr. Ministro de R. E. y Culto sobre nombramientos en la curia," *Revista Eclesiástica del Arzobispado de Buenos Aires* 23 (1923): 379.
106 "Monseñor Dr. Fortunato J. Devoto. Nuevo Obispo Auxiliar de la Arquidiócesis," *Revista Eclesiástica del Arzobispado de Buenos Aires* 27 (1927): 711; "Sagrada Congregación Consistorial. S.E. Revma. Mons. Dr. Fortunato J. Devoto, Obispo Auxiliar del Excmo. Sr. Arzobispo," *Revista Eclesiástica del Arzobispado de Buenos Aires* 33 (1933): 345.
107 [Fortunato Devoto], "El V Congreso Científico Internacional de Católicos," *Revista Eclesiástica del Arzobispado de Buenos Aires* 1 (1901): 50.

Observatories

We have seen that the National Observatory in Córdoba had been created in the 1870s by President Domingo F. Sarmiento and put in the charge of American astronomers.[108] From 1909, it was headed by Charles D. Perrine, who despite many efforts had not been able to configure the mirror for the sixty-inch reflector he was supposed to install. By the beginnings of the 1930s, the institution was in disarray and the mounting climate of chauvinistic nationalism blamed the situation on its American staff.[109] The government appointed two experts to investigate the issue, who reported that the observatory was "a foreign mission in our territory" instead of "the really national institution" it was meant to be.[110] One of the members of the commission was Félix Aguilar, a geodesic engineer graduated from La Plata at the time Devoto was director there, who, after studying in Europe, returned to take charge of that observatory (1916–1920). He held teaching positions in the major institutions of the army and was head of the geodesic division in the Military Geographical Institute, where Devoto was asked to head the astronomy division, which he declined (1925). In June 1933, Perrine's Córdoba observatory was put under the authority of a National Council for Observatories (hereinafter CNO after its name in Spanish), created by law, with Devoto as its director and Aguilar as one of the members.[111] In turn, Aguilar presided over a newly created commission for the measure of a segment of a meridian arch, in which Devoto participated.[112]

The creation of the CNO was seen as an important institutional advance and the Catholic press took opportunity to promote the figure of the bishop astronomer.[113] The CNO was significant for two actions. First, it supervised the handing over of the Córdoba observatory to an Argentine director, Enrique Gaviola, at that

---

108 Hodge, "Benjamin Apthorp Gould."
109 John E. Hodge, "Charles Dillon Perrine and the Transformation of the Argentine National Observatory," *Journal of the History of Astronomy* 8 (1977): 12–25.
110 Jorge Landi Dessy, "Charles Dillon Perrine y el desarrollo de la astrofísica en la República Argentina," *Boletín de la Academia Nacional de Ciencias* 48 (1970): 235–39.
111 *Boletín Oficial de la República Argentina* 41, no. 11714 (June 1933): 741.
112 Félix Aguilar, "La medición de un arco de meridiano en la República Argentina," *Anales del Instituto Popular de Conferencias* 22 (1937): 199–214; Congreso Nacional, Cámara de Diputados, *Diario de sesiones... Año 1936*, vol. 5 (Buenos Aires: Imprenta del Congreso Nacional, 1937), 527–31. See Eduardo L. Ortiz, "La Comisión del Arco de Meridiano. Astronomía, geodesia, oceanografía y geofísica en la Argentina de 1935–1945," *Saber y Tiempo* 19 (2005): 127–87.
113 "S. E. R. Monseñor Fortunato Devoto fue nombrado presidente del Consejo Nacional de Observatorios," *El Pueblo*, 3 June 1933; "El progreso de la astronomía en la Argentina," *La Razón. Anuario 1934* (Buenos Aires: La Razón, 1934), 63.

time the country's foremost physicist, who was able to finally supervise the shaping of the mirror for a new sixty-inch reflector in Pittsburgh (he had been responsible for theoretical and technical advances on mirror shaping working with John D. Strong at Mount Wilson).[114] Later, Gaviola duly recognised Devoto's role in restructuring of the observatory.[115] Second, the CNO, presided over by Devoto, founded the Jesuit Observatory of San Miguel, in the vicinity of Buenos Aires. This new institution was modelled upon the Jesuit Ebro Observatory; its first director was the Jesuit astronomer Ignacio Puig.[116] The new observatory would study the effects of solar and cosmic activity on the Earth. Originally, it should consist of three sections: geophysics, electro-meteorology, and solar physics. Funding came from several corporations and individual donors, a novel scheme in the country.[117] This was underlined by Devoto in his lecture at the inauguration of the first building of the observatory in December 1935. In his brief intervention, he refers three times to the advantages of encouraging private individuals to establish scientific institutes, which "could become official [institutions] as far as they are willing to subordinate to the national authorities."[118] The creation of San Miguel Observatory did not aim to establish a Catholic system of scientific research parallel to that of the government; it was instead an example of the principle of subsidiarity, of "collaboration between the state and its subjects." Devoto's speech evokes that of a civil servant vigilant regarding the prerogatives of the government. The final invocation to raise the hearts of the audience to the Redeemer of this world in memory of the 1934 International Eucharistic Congress which had taken place in Buenos Aires (see next chapter) caps a discourse whose register sounds more civic than religious. Devoto and Aguilar were the embodiment of a scientific policy marked by state intervention, nationalism, and collaborative relationships between the military and the church within a conservative but politically liberal order.

---

114 Roscoe F. Sandford, "The Seventy-fifth Anniversary of the Córdoba Observatory," *Publications of the Astronomical Society of the Pacific* 58, no. 345 (1946): 341–48; Omar A. Bernaola, *Enrique Gaviola y el Observatorio Astronómico de Córdoba* (Buenos Aires: Saber y Tiempo, 2001).
115 "Nuevas autoridades en observatorios nacionales," *Revista Astronómica* 29 (1956): 85.
116 Udías, *Searching the Heavens*, 142–45; Ignacio Puig, *El Observatorio de San Miguel, República Argentina* (Buenos Aires 1935).
117 "El país contará próximamente con otro observatorio," *La Nación*, 8 September 1934.
118 "Con la asistencia del Gral. Justo inauguróse el Observatorio de Física Cósmica," *El Pueblo*, 13 December 1935.

The Rise of Catholic Scientists

On the grounds that Ángel Gallardo had decided to go on financing the building of the "Argentine church" in Rome (Santa Maria Addolorata, originally donated by his brother Msgr José León Gallardo, who died in 1924), a radical student referred to him in the magazine of the Buenos Aires Centre of Medical Students as "the illustrious biologist who believes in God, the manufacturer of Adam and Eve."[119] The irony, predicated on the assumption that any reader would endorse the conflict thesis, suggests that a large segment of educated public opinion saw science as an essentially anti-religious pursuit. In an article published in 1910, the Catholic social leader Emilio Lamarca merged in a single contentious argument spontaneous generation, Haeckel's evolutionism, and Rousseau's social contract.[120] For him, as for Argentine Catholic social thinkers in the last two decades of the nineteenth century, evolution and science in general were part of the liberal and socialist onslaught on Christianity.[121] The letter to Father Devoto of a conservative Catholic activist blaming the CER for teaching the young ladies too much Plato instead of instructing them against "transformist geology" and "transformist anthropology" is also an expression of this mentality of a church besieged by the spread of evolutionary ideas.[122] Science had been one of the main instruments in secularising public discourse and positivists sought to resignify religious symbols and rituals into a secular cult of science, the founding cultural hero of which was Ameghino. Although the scientific activity of the Iberian Jesuits (Blanco and Ubach) was far from negligible, it was considered by contemporary standards to be backward and defensive.[123] It is against this somewhat dull backdrop that the full extent of the novelty of Gallardo and Devoto should be measured. Both had faultless scientific credentials and a solid French training. They had conducted locally prestigious scientific institutions and by the end of their careers had joined the highest circles of political power and shaped the development of science in their country. By the early 1930s, those who took Catholic scientists as a joke did so at their own risk. Among the science textbooks recommended by an official commission named by the government in

---

119 Edgardo Casella, "Nuestros 'hombres de ciencia'." *Revista del Círculo Médico Argentino y Centro de Estudiantes de Medicina* 27 (1927): 230.
120 Emilio Lamarca, "Necesidad de la acción social," *Revista Eclesiástica del Arzobispado de Buenos Aires* 10 (1910): 3–30.
121 For these authors, see the second section of this chapter.
122 Fortunato Devoto, "A 'Patria y Hogar.' En defensa del Centro de Estudios Religiosos," *Ichtys* 2, no. 25 (1923): 343–46.
123 "R. P. José Ubach Medir S.J.," *Revista Astronómica* 7 (1935): 337.

1927 were Gallardo's *Zoology*, Eduardo Holmberg's *Botany*, and a manual of cosmography by Eduard Brugier, a German Jesuit who taught at the Instituto Politécnico (Quito, Ecuador), Colegio San Ignacio (Santiago de Chile) and the College of El Salvador (Buenos Aires).[124] When in the course of the parliamentary debate on the issue, someone objected to Gallardo's textbook, the socialist deputy and physician Enrique Dickmann replied that, although he disagreed with Gallardo in political matters, he felt compelled to testify that his book was "an excellent text long in use in state schools...[written by] an expert on the matter."[125]

Conclusions

This chapter focuses on an inflexion point in the course of the relationships between science and religion in Argentina. Science and scientific imagery were certainly a main component in the self-fulfilling prophecies of secularising discourse of the liberal elite of the 1880s and later of socialism. We have taken a close look at the moment at which the secularistic monopoly of science was broken by the arrival upon the stage of scientists with strong confessional commitments. This coincided with the first inklings of the local Catholic renaissance and the shift from a model of a moderately "Gallican" church to another closer to integral Catholicism (Gallardo and Devoto could be seen as transitional characters in this respect). This process was not unrelated to the resurgence of different stripes of nationalism. In the 1930s and 1940s, a growing number of Catholic scientists branched into two groups: some of them favoured political authoritarian regimes and neo-Thomist philosophy, while the others defended liberal democracies and were less preoccupied about questions of doctrinal orthodoxy (the latter were among the best Argentine science was able to offer in the years after the Second World War). This is the theme to be developed in the following two chapters.

The irruption of scientists identified with the majority church of the country, as examined here, refutes any simplistic interpretation of science as an agent of secularisation. What was crucial for the transactions between science and reli-

---

124 Eduardo Brugier, *Elementos de cosmografía*, 7th ed. (Buenos Aires: Estrada, 1933); Congreso Nacional, Cámara de Diputados, *Diario de sesiones... Año 1927*, vol. 2 (Buenos Aires: Imprenta y Encuadernadora de la Cámara de Diputados, 1927), 35. Despite being an actualised compendium, Brugier's text mentions the three "systems of the universe" (Ptolemaic, Tychonic, and Copernican) in the long established tradition of eighteenth-century Jesuit commentaries of natural philosophy, see 86–87.
125 Congreso Nacional, Cámara de Diputados, *Diario de sesiones... Año 1927*, 2:29.

gion was the rise and fall of science-saturated ideologies legitimazing secularist agendas tied to local politics. Ascribing science any kind of "causal" role in this process would be taking the effect for the cause. Conversely, it is the particular type of dynamics of secularisation which seems to have modulated and even shaped the kind of relationships between science and religion obtained in this society.

Chapter 7
Science and Integral Catholicism in Interwar Argentina

The 32th International Eucharistic Congress in Buenos Aires (1934) was a turning point in the history of Argentine Catholicism: mass concentrations of at least half a million people and a record of 200,000 male communicants on the night of 11 October signalled a shift in the social and political role of the Catholic Church.[1] New technology came into play. The science journal *Nature* narrated how radio telephone had made possible the live transmission of Pius XI's closing message and the broadcasting of the main speeches to three continents.[2] Communications had been set up by Engineer Ricardo T. Mulleady, a graduate of the University of Birmingham.[3] On the day of the largest mass concentration, 27 telephones connected to the medical emergency station allowed the efficient management of around 500 cases of heat stroke; this operation was controlled by Miguel A. Petty, FRCS, a graduate of Downing College, Cambridge trained in St Thomas' Hospital (London). Petty was the president of the Anglo-American Gentlemen's Commission of the Congress and the first president of the Consortium of Catholic Physicians created in Buenos Aires in 1929 (see below).[4] The latest technology had assured the success of a massive event with global repercussions.

The 1934 Eucharistic Congress was the living embodiment of what has been called "integral Catholicism," a particular type of relationship between Church and society which was attained at certain stages in Taylor's "paleo-Durkheimian" societies.[5] Poulat has extensively discussed the history, scope, and nuances of notions such as "integral Catholicism," "integralism," and "integrism," fo-

---

[1] Jesús Méndez, "Church-State Relations in Argentina in the Twentieth Century: A Case Study of the Thirty-second International Eucharistic Congress," *Journal of Church and State* 27 (1985): 223–43; Miranda Lida, "Los orígenes del catolicismo de masas en la Argentina, 1900–1934," *Jahrbuch für Geschichte Lateinamerikas* 46 (2009): 345–70.
[2] "Religious Broadcasting at the Eucharistic Congress," *Nature* 136, no. 3438 (21 September 1935): 471.
[3] Ricardo T. Mulleady, *Breve historia de la telefonía argentina, 1886–1956* (Buenos Aires: Kraft, 1957), 5.
[4] "La radio en el Congreso," in *XXXII Congreso Eucarístico Internacional*, vol. 1 (Buenos Aires: Comité ejecutivo del Congreso, 1935), 100–106; "Obituary. Mr. M. J. Petty," *British Medical Journal* no. 4810 (10 March 1953): 623.
[5] Taylor, *A Secular Age*, 454–56, 486–87.

cusing on French Catholicism during Pius X's anti-modernist reaction.⁶ For integral Catholicism, "the Church, a perfect society, the sole holder of integral truth" is a closed system which "does not leave any domain free from its apprehension."⁷ As such, it is the reverse of secularisation in the three senses described by Taylor: it aspires to subordinate political structures to religion (sense 1); it promotes church attendance (sense 2); and it turns unbelief into a socially questionable issue (sense 3).⁸ Drawing upon Poulat's analysis, Mallimaci characterised Argentina's integral Catholicism (1930–1946) as a period in which "the Catholic Apostolic Church is the supreme norm and only warrant of transcendent unity."⁹ Ivereigh has interpreted the Argentine "Catholic revival" of the first decades of the twentieth century as a result of the transition from a regalist quasi-Gallican church associated with the liberal state consolidated in 1880 to the integral Catholic church of the 1930s and early 1940s.¹⁰ This chapter addresses the question of how the relationship between science and religion was conceptualised in interwar Argentina, a historical setting in which secularism was temporarily in retreat. For that purpose, I shall examine the views of Catholic scientists about "science and faith" in the period of dominance of the ideal of integral Catholicism. In the next chapter, I shall look at those Catholic scientists with a more liberal view of the relationships between Church and society, in a slightly later period.

Historical Background

The Catholic revival of the twenties in Argentina was a cultural phenomenon consisting of the creation by the laity of teaching institutions, magazines, and other expressions of a vital Catholic culture. Its two main manifestations were the Courses of Catholic Culture (1922, from now on CCC) and the journal *Criterio* (1928), both initiatives led by the jurist, intellectual, and future politician Atilio

---

6 Émile Poulat, *Intégrisme et catholicisme intégral. Un réseau secret international antimoderniste: "La Sapinière" (1909–1921)* (Paris: Casterman, 1969); idem, "La querelle de l'intégrisme en France," *Social Compass* 32 (1985): 343–51.
7 René Remond, "L'intégrisme catholique. Portrait intellectual," *Études* 370, no. 1 (1989): 99–100.
8 Taylor, *A Secular Age*, 1–3.
9 Mallimaci, *El catolicismo integral en la* Argentina, 5.
10 Ivereigh, *Catholicism and Politics in Argentina*, 18, 84–91.

Dell'Oro Maini.¹¹ The decade was marked by economic bonanza and political stability under the presidency of Marcelo T. de Alvear (1922–1928), the leader of the conservative wing of the UCR, which had risen to power in 1916, when its popular leader Hipólito Yrigoyen was elected president (1916–1922). Against the background of the World Great Depression, Yrigoyen's second term of office, which began in 1928, was interrupted by the military coup of September 1930, led by General Félix Uriburu, who called general elections fourteen months after his assumption.¹² Uriburu's short rightist rule was followed by a shift to a more liberal conservative orientation under General Agustín P. Justo (1932–1938), resulting from fraudulent elections under a coalition of conservatives, Radicals followers of Alvear, and a fraction of the socialists. The attempt at a gradual return to a real democracy by Justo's successor, the Radical Roberto M. Ortiz (1938–1942), was thwarted by his death in June 1942. Ortiz was succeeded by Ramón S. Castillo, his conservative vice president who was unable to control the economic crisis and the international pressures resulting from the Second World War. He was overthrown by a coup in June 1943, which brought to power General Pedro P. Ramírez, whose regime (June 1943–March 1944) included many elements of right-wing Catholic nationalism. In less than a year, Ramírez was displaced by General Edelmiro Farrell (March 1944–June 1946), whose government was dominated from the beginning by the rising star of Colonel Juan D. Perón. This was the high point of Catholic nationalism, integral Catholicism, and the Catholic right. It was also the acme of the public activity of Dr César E. Pico and of the naturalist Emiliano J. Mac Donagh, two right-wing nationalists and accomplished exponents of the Catholic cultural scene of the previous two decades.

During the thirties, Argentina's institutional Church went through momentous structural transformations. In the years near to the Eucharistic Congress, ten new dioceses were created, six others were promoted to archdioceses, and the Archbishop of Buenos Aires, Msgr Santiago Copello, was appointed as the first Latin American cardinal.¹³ The Argentine Catholic Action, inspired by the analogous movement created by Pius XI and conceived as a militant lay branch of the Church, amply fulfilled its assigned role. The attempt at a construction of a new Christendom was associated with cultural nationalism of Hispanic stamp and a corporatist view of society.¹⁴ Argentina's past was rewritten as the story

---

11 Lida, *Historia del catolicismo en la Argentina*, 91. See also Floria and Montserrat, "La política desde *Criterio*;" Devoto, "Atilio Dell'Oro Maini."
12 Rock, *Argentina, 1516–1987*, 191–213.
13 Zuretti, *Nueva Historia Eclesiástica*, 412–16.
14 Di Stefano and Zanatta, *Historia de la Iglesia en la Argentina*, 426–29.

of an essentially "Catholic nation" which had lost its bearings disoriented by the mermaid's song of liberalism and was coming home again.[15] A significant part (though not all) of Catholic educated opinion was in this period never far off from nationalism, which expressed itself forcibly in the military take-overs of 1930 and 1943; many intellectuals who had begun their careers as admirers of Charles Maurras ended up in the Catholic fold.[16] A small but vocal group of democratic (sometimes called "liberal") Catholics, who sided with the Allies and rejected Franco's regime, lived uneasily side by side with their confreres (more on them in the following chapter).

Catholics and Einstein

César E. Pico (1895–1966), a graduate of the Jesuit College of El Salvador, joined as a youth the Catholic Student Centre while studying medicine at the University of Buenos Aires. The spiritual leader of *Convivio* (a gathering of literary and artsy people around the CCC), Pico contributed articles of political opinion and philosophical essays to the same nationalist and Catholic magazines which brought Mac Donagh's essays on natural history.[17] Dazzled by Maritain and other French neo-scholastic philosophers, Pico regarded himself as a local representative of that tradition. Amongst his students and followers in the courses of the CCC, he enjoyed a reputation of profound philosopher and theologian enhanced by an unconventional personality.[18] Pico was a hardliner of integral Catholicism and Hispanic tradition, and eventually a declared fascist. Perhaps incongruously, he was also an admirer of the liberal Spanish philosopher José Ortega y Gasset. Throughout his professional life, he held a research position in the prestigious National Bacteriological Institute, where he ascended the bureaucratic echelons until with the de facto regime of 1943 he became acting director of the institution. He published around 60 research papers and technical memoirs, most of them on serology and almost all of them in local journals.[19]

---

15 Loris Zanatta, *Del estado liberal a la nación católica. Iglesia y ejército en los orígenes del peronismo, 1930–1943* (Buenos Aires: Universidad Nacional de Quilmes, 1996).
16 Devoto, *Nacionalismo, fascismo y tradicionalismo*, 169–262.
17 *La Nueva República, Número, Criterio, Signo, Sol y Luna*.
18 Rivero de Olazábal, *Por una cultura católica*, 108.
19 Card index of the Library of the School of Medicine of the University of Buenos Aires, s. v. "Pico, César."

When Albert Einstein visited Argentina (March–April 1925), Pico already had a reputation as a philosopher of science in Catholic circles.[20] In 1916, he had published a youthful article distinguishing between "positive science," which he approved of, from "positivism," the application of the observational and experimental method beyond the empirical sciences, which he condemned.[21] For the most, educated Catholics in Argentina looked askance at Einstein's relativity theory. The Jesuit astronomer José Ubach gave two lectures on relativity in September 1920, probably as a response to the success of the conference by the influential anti-Catholic intellectual Leopoldo Lugones, discussed in chapter 5. Ubach doubted the experimental confirmation of relativity theory and contended that it did not offer anything new; he was alarmed by its subversion of the standard (Newtonian) philosophy of nature and concluded that it "granted objective character to what was purely mathematical and formal."[22] Two years later, the Jesuit cultural magazine (*Estudios*) published a lecture by the Spanish mathematician Julio Rey Pastor, then in Buenos Aires, which was a competent and sympathetic exposition of the theory for the general public.[23] The Catholic daily *El Pueblo* reflected the spectrum of reactions to Einstein's visit in a series of articles. The first was celebratory.[24] After that, the paper reproduced approvingly fragments of the anti-relativist book by the Spanish Augustinian meteorologist Ángel Rodríguez (undistinguished director of the Vatican Observatory between 1898 and 1905), who qualified the theory as "incoherent, vague, and full of shocking affirmations."[25] Around the days of Einstein's departure, *El Pueblo* came back with a well-informed fragment by the Jesuit astronomer in Havana

---

20 Eduardo L. Ortiz, "A Convergence of Interests: Einstein's Visit to Argentina in 1925," *Ibero-Amerikanisches Archiv* 21 (1995): 67–126; Asúa and Hurtado de Mendoza, *Imágenes de Einstein*.
21 César E. Pico, "Discurso sobre el positivismo contemporáneo," *Estudios* 6 (1916): 425–36.
22 José Ubach, *La teoría de la relatividad en la física moderna. Lorentz, Minkowski, Einstein* (Buenos Aires: Amorrortu, 1920), 43–44. For criticism of relativity in the Spanish-speaking world, see Thomas Glick, *Einstein y los españoles. Ciencia y sociedad en la España de entreguerras* (Madrid: CSIC, 1986), 206–15.
23 "En la Academia Literaria del Plata. Conferencia del doctor Rey Pastor. La teoría de la relatividad," *Estudios* 23 (1922): 219–24.
24 "Albert Einstein y la celebración de su obra," *El Pueblo*, 25 March 1925.
25 "Sobre la teoría relativista del Dr. Alberto Einstein juzgada por el sabio Agustín [sic] Ángel Rodríguez," *El Pueblo*, 26 March 1925; cf. Ángel Rodríguez, *Sobre la teoría de la relatividad propuesta por el Dr. A. Einstein* (Madrid: Imprenta del Asilo de Huérfanos del Sagrado Corazón de Jesús, 1924).

Mariano Gutiérrez Lanza, who claimed that although "admirable," relativity theory was no more than "a castle in the air."[26]

In the course of Einstein's visit, Pico gave a conference on theory of science in the Archdiocesan Seminary (at that time in the hands of the Jesuits). He claimed that relativity was not "applicable to the essence of things," because it was no more than "an algebraic expression of the measurements of physical phenomena."[27] This approach was inspired by Maritain's chapter on relativity in *Théonas* ("La mathématisation du temps"), which Pico duly quotes.[28] In a later article, Pico embellished his anti-relativist arguments with a condensed account of the relationships between science and the higher orders of knowledge: "the metaphysical heavens of essences and highest causes," he says, reflects itself analogically upon empirical reality, illuminating it with its intelligibility.[29]

Pico saw the relationships between science and religion as necessarily involving the mediation of philosophy. In a review of the Spanish translation of Johannes Hessen's *Erkenntnistheorie* (1926), he criticises the Catholic German epistemologist for the "irreducible dualism" he posits between faith and knowledge and "for denying any possibility of relationship between both" as a result of attributing an autonomous realm to religion.[30] It is because knowledge of reality through mathematised science "boils down to a poor thing, even to nothing" that we are compelled to affirm "the supereminence of Aristotelian metaphysics over all the disciplines of natural knowledge."[31] In a 1939 paper, Pico belittled science as "a mechanical style of thought" and disparaged the unilateral approach to things characteristic of scientists.[32]

Pico never addresses the specific question of the relationships between science and religion, but the issue crops up in his essays on epistemology and philosophy of science, which amount to little more than variations upon Maritain's theme of the degrees of knowledge. Science was for Pico a result of Modernity and as such a subaltern route to knowledge: "Modern science suffers from... the sickness which devours the body of our culture." Moreover, he sets the "stu-

---

[26] Mariano Gutiérrez Lanza, "La teoría de la relatividad," *El Pueblo*, 19 April 1925; cf. idem, *La teoría de la relatividad* (Havana: Imprenta Molina, 1929).
[27] César E. Pico, "Las reglas del método experimental y las consecuencias filosóficas de su aplicación incorrecta," *Estudios* 29 (1925): 120.
[28] Jacques Maritain, *Théonas* (Paris: Nouvelle librairie nationale, 1921), 73–102. See also, idem, "De la métaphysique des physiciens," *La Revue universelle* no. 10 (15 August 1922): 426–45.
[29] César Pico, "El hombre de ciencia y la filosofía," *Criterio* no. 52 (28 February 1929): 268.
[30] César Pico, "Teorías del conocimiento," *Número* no. 2 (February 1930): 14–15.
[31] Pico, "El hombre de ciencia."
[32] César Pico, "Absurdos del especialismo," *Sol y Luna* 2 (1939): 32–46.

pid dogma of progress" against "the spiritual regulation known to Christendom."[33] The integral view of things is aptly expressed in his pronouncement that "only the Church of Christ sustains with powerful arm...the rights of human intelligence sheltered in the impregnable strongholds of traditional philosophy."[34] He qualifies "modern epistemology" as a "sin against light, a diabolic attitude against the image of God as revealed in the things [of this world]."[35]

In his nationalistic political articles, Pico saw a fitting congruence between the hierarchical view of society he endorsed (with the Catholic Church as the ultimate instance of values and legitimation of political authority) and the stratified approach of the Aristotelian-Thomist epistemology, in which science remained subjected to natural philosophy, in turn ordained to metaphysics, and ultimately to theology. In 1931, during Uriburu's rule, Pico published in the rightist nationalist magazine *Número* a barrage of articles with provocative titles: "Anti-democracy," "Self-destruction of Democracy," and "Syllabus."[36] In the latter, he denies (among other things): (a) that the phenomenal sciences could ever provide an explanation of the material universe; (b) that traditional Biblical exegesis should be subordinated to the results of the positive sciences; (c) that the spirit of modern democracy is compatible with the doctrines of the Catholic Church.[37] In an article published previously in another rightist nationalist political magazine, the parallelism between politics, metaphysics, and philosophy of science is also evident. Departing from Nikolái Berdiáyev's *Le Nouveau Moyen Âge* (1914), Pico bemoans "the disorder of the hierarchies," rallies against "old-fashioned scientism," and hails the impending "vindication of the Church and Thomism."[38]

---

33 César Pico, "Materialismo. Ciencia y religión," *Iatría* 8 (October 1937–May 1938): 14.
34 Ibid., 9.
35 César Pico, "Las ciencias. Del conocimiento en general," *Iatría* 8 (September 1938): 9.
36 César Pico, "Antidemocracia," *Número* no. 17 (May 1931): 44; idem, "Autodestrucción de la democracia," *Número* no. 20 (August 1931): 63; idem, "Syllabus," *Número* no. 23–24 (December 1931): 78. *Número* was founded by a group of Catholic nationalists who quit *Criterio*.
37 Pico, "Syllabus."
38 César Pico, "Inteligencia y revolución," in *Los nacionalistas*, ed. María I. Barbero and Fernando Devoto (Buenos Aires: CEAL, 1983), 96–98; first published in *La Nueva República*, 1st January 1928.

Catholics, University Reform, and Hans Driesch

Born into a middle-class family in a town in the province of Buenos Aires, Emiliano Mac Donagh (1896–1961) was one of many immigrants to Argentina who without ever losing touch with his Hibernian roots became attached to the local culture; all through his life he felt at home with those anglophone naturalists, such as William Hudson and Charles Darwin, who had projected the pampas into the world scientific and literary scene.[39] Mac Donagh studied in the Museum of La Plata, where he got his PhD in natural sciences. Later, he became professor of zoology, director of the Department of Zoology, and eventually director of that institution (1946–1949); he also held positions in the Agronomical School of La Plata and the General Direction of Hygiene of Buenos Aires. Mac Donagh began working in entomology, but ended up as a specialist in ichthyology, and as such had many disciples. Although he fought for the professionalisation of the natural sciences, he also cultivated the broad range of interests of a field naturalist. Unlike Pico, in his published articles he never discussed politics. With the coup of 1943, Mac Donagh was nominated Director General of [State] Schools in the province of Buenos Aires. At that time, the government reintroduced religious teaching in state-supported elementary schools, a symbolic restorative measure which annulled the law voted by Parliament in 1880, which secularists considered as a historical triumph.

When the biologist and philosopher Hans Driesch visited Argentina in 1928, his neo-vitalism had already attracted the attention of materialist scientists and Catholic intellectuals in the country. Just as local controversies over relativity theory had brought into focus questions of secularism and religion, something analogous happened with Driesch's vitalism. The German physiologist and pacifist activist Georg Friedrich Nicolai had been one of the four signers in October 1914 of the anti-militarist Manifesto to Europeans, also signed by Einstein. After losing his teaching position in Berlin and undertaking an adventurous escape from Europe, Nicolai ended up as professor in the University of Córdoba, which in 1918 had been shaken by a democratic and student-led secularist Reform Movement which eventually radicalised.[40] Here we should make a very short excursus on the University Reform, pushed forward by Yrigoyen, in

---

[39] Mac Donagh's articles on Hudson and Darwin are listed in Guillermo Furlong, "Emiliano J. Mac Donagh († 1º de Agosto de 1961)," *Anales de la Academia Argentina de Geografía* 5 (1961): 16–23; see also Emiliano Mac Donagh, *150 Años de Evolución Científica Argentino-Británica* (La Plata: CICBA, 1960).

[40] Richard J. Walter, *Student Politics in Argentina. The University Reform and Its Effects, 1918–1964* (New York: Basic Books, 1968), 39–62.

which students were given a place in the government of the houses of higher learning and appointments were done on the basis of public competitions.[41] Henry Ferns has remarked that the reform was "directed to opening up the universities, not to improving them as agents of higher education and research."[42] As regards secularisation, the reform movement has been seens as "the culminating point in the face-off between anti-liberal Catholicism and positive liberalism."[43]

In Córdoba, public discourse in support of the reform was full of grandiose denuntiations of clerical obscurantism stifling the students who longed for scientific enlightenment; the reform in Buenos Aires took a less anticlerical character.[44] On 15 June 1918, a group of students stormed a hall of the University of Córdoba and vandalised the oil paintings of priests who had been former presidents of the university at the cry of "No friars! No dogmas!".[45] In the manifesto of the reform, the leaders talked of a school subjected to "clerical dominion" and proclaimed that "the teaching was vitiated by a narrow dogmatism, which kept the university away from science."[46] In Parliament, the socialist representative Juan B. Justo reviled the former state of the University of Córdoba, "deeply infiltrated by orthodox and die-hard Catholicism and partisan in the worst sense of the word;" he called to "disengage the University of Córdoba from the Church of the Jesuits" to which in his view it was still intimately attached.[47] (Córdoba was a city fond of its Catholic traditions, but the Jesuits had long ceased to play any significant political role in its university.) The ubiquitous Ingenieros declared that with the advent of democracy, "modern scientific methods would predominate over mediaeval dialectic dogmatism."[48]

---

41 For Catholic reactions to the Reform, see Diego Mauro and José A. Zanca, eds., *La reforma universitaria cuestionada* (Rosario: HyA ediciones, 2018).
42 Henry S. Ferns, *Argentina* (New York: Praeger, 1969), 147.
43 Morello, "'It Takes Two to Tango'," 62.
44 Joseph A. Tulchin, "La Reforma Universitaria. Córdoba 1918," *Criterio* no. 1599–1600 (9 and 23 July 1970), https://www.researchgate.net/publication/324213193_La_Reforma_Universitaria_Cordoba_1918.
45 Zenón Bustos y Ferreyra, "Pastoral con motivo de la agitación antirreligiosa en Córdoba," *Revista Eclesiástica del Arzobispado de Buenos Aires* 18 (1918): 489–94; María I. De Ruschi Crespo "La Reforma Universitaria de 1918 y la Iglesia: algunas precisiones. Actuación de los Centros Católicos de Estudiantes," *Archivum* 25 (2006): 13–40.
46 "La juventud argentina de Córdoba a los hombres libres de Sud América," in *La Reforma Universitaria. Tomo I. El movimiento argentino*, ed. Gabriel del Mazo (Lima: Universidad Nacional Mayor de San Marcos, 1967), 4.
47 Congreso Nacional. Cámara de Diputados, *Diario de sesiones... Año 1918*, vol. 2 (Buenos Aires: Rosso, 1918), 458, 470.
48 José Ingenieros, "La Universidad del Porvenir," *Ateneo* no. 3 (1920): 68–69.

The arrival of Nicolai was perceived as a legitimation of the reform movement by a consecrated European savant. During his years in Córdoba, he wrote on relativity and the definition of life.[49] His 1922 opening lecture was deliberately provocative. He attacked the "childish belief of primitive peoples in the creative force of an omnipotent God" and affirmed that life was "a very complicated machinery" which arose from an increasingly complex array of molecules.[50] In a later article, Nicolai singled out Driesch as the perpetrator of the resurrection of the Aristotelian vital entelechy, while harping on the idea of the gods as personification of the natural forces, along the lines of Max Müller's speculations on the origin of religion.[51] Nicolai claimed that science and religion were one and the same, insofar as both expressed the mysteries of nature—but once it became embroiled in dogmas, religion lost its original source.[52]

The year of Nicolai's arrival in Córdoba, the young zoologist Emiliano Mac Donagh published an analysis of Driesch's vitalism in a nationalist journal in Buenos Aires; his article, written from an embryological point of view, was emphatically nonphilosophical.[53] In 1924, the Thomist philosopher and fascist ideologue Nimio de Anquin published in the obscure journal he himself edited in Córdoba a Spanish translation of a text by Driesch; that issue also brought a paper on Darwin's *Voyage* by Mac Donagh.[54] From the other side of the political and religious divide, Narciso C. Laclau, a chemist with philosophical inclinations and progressive ideas, contributed to the strongly secularist *Revista de Filosofía* an article critical of Driesch's vitalism. In his defence of the "mechanical theory of life," Laclau identified as the foe those Aristotelian philosophers "among us" who had been seduced by Driesch's notion of entelechy.[55]

---

49 Georg F. Nicolai, "La base biológica del relativismo científico y sus complementos absolutos," *Revista de la Universidad Nacional de Córdoba* 12, no. 1–3 (January–March 1925): 1–182. For a detailed account of Nicolai's years in Córdoba, see Wolf Zuelzer, *The Case Nicolai. A Biography* (Detroit: Wayne State University Press, 1982), 311–54, 375–91.

50 Georg F. Nicolai, "La vida," *Revista de Filosofía. Cultura, ciencias, educación* 16, no. 4 (July 1922): 65–76.

51 Georg F. Nicolai, "Las definiciones de la vida," *Revista de Filosofía. Cultura, ciencias, educación* 26, no. 4 (July 1927): 1–21.

52 Georg F. Nicolai, "La ciencia y la moral," *Revista de Filosofía. Cultura, ciencias, educación* 26, no. 3 (May 1927): 301–17.

53 Emiliano J. Mac Donagh, "Ensayo sobre la regulación orgánica en Hans Driesch," *Signo* no. 8 (1922): 246–59.

54 Hans Driesch, "Idealismo y vitalismo," *Arx* (Córdoba) 1, no. 1 (1924): 47–69; Emiliano J. Mac Donagh, "Sobre un estudio de Darwin por su *Voyage*," *Arx* 1, no. 1 (1924): 1–14.

55 Narciso C. Laclau, "El vitalismo de Hans Driesch," *Revista de Filosofía. Cultura, ciencias, educación* 23, no. 2 (March 1926): 299.

When the conservative forces finally succeeded in expelling Nicolai from the University of Córdoba, a socialist journal compared his forced resignation to the sufferings of Bruno and the process of Galileo "persecuted by the Inquisition."[56] Aníbal Ponce, a major figure in the University Reform movement who eventually became a fellow traveller of the Communist Party, also claimed that Nicolai had been seen as "a sacrilegious enemy of tradition."[57] In a previous article against Driesch, Ponce equated Driesch's ideas to "the survival of the mystical conscience;" the note ended prophesying that "the vogue of Driesch's vitalism shall leave us as soon as the conservative reaction that brought it comes to an end."[58] César Pico did not let the challenge go by. In a series of articles in *Criterio*, he pointed out Ponce's contradiction between a positivistic attitude and the simultaneous assertion of materialism which implied a metaphysical stance.[59]

Nicolai's departure from Córdoba coincided with Driesch's visit to Argentina. The German savant arrived in Buenos Aires in August 1928; he gave a course at the University of Buenos Aires and lectured in other cities.[60] Driesch enjoyed his stay in Argentina: he was paid handsomely and he was able to socialise with other international guests.[61] Soon after his departure, the liberal press published an article by the young Jewish philosopher León Dujovne (at that time working on his dissertation on Mach and Meyerson), who much like Laclau (whom he mentions approvingly) was sceptical of the notion of entelechy.[62] As regards this issue, Catholics were on the other side of the fence: it was Driesch's entelechy which they found particularly congenial.

In his second article on Driesch, Mac Donagh took the opportunity to defend the autonomy of biology with respect to both chemistry and philosophy and la-

---

[56] José Barcón Olesa, "A propósito de una conferencia del Dr. Nicolai," *Nosotros* 64, no. 240 (May 1929): 289–91.
[57] Aníbal Ponce, review of *Homenaje de despedida a la tradición de Córdoba docta y santa*, by G. F. Nicolai, *El Hogar* 24, no. 995 (9 November 1928): 8.
[58] Aníbal Ponce, "Hans Driesch y los fantasmas del vitalismo," *El Hogar* 24, no. 994 (2 November 1928): 16.
[59] César Pico, "Los fantasmas del vitalismo," *Criterio* no. 37 (15 November 1928): 201–203 and no. 38 (22 November 1928): 233–40; see idem, "¡Otra vez Aníbal Ponce!," *Criterio* no. 38 (22 November 1928): 240.
[60] "Ha llegado el filósofo alemán Dr. Hans Driesch, quien dará conferencias," *La Nación*, 16 August 1928; "Herr Professor Hans Driesch," *Deutsche La Plata Zeitung*, 16 August 1928; "Die Vorträge des Professors Dr. Hans Driesch," *Argentinisches Wochenblatt*, 25 August 1928.
[61] Hans Driesch, *Lebenserinnerungen: Aufzeichnungen eines Forschers und Denkers in entscheidender Zeit* (Munich: Ernst Reinhardt 1951), 251.
[62] León Dujovne, "Hans Driesch," *La Nación*, 2 December 1928, Literary section.

mented that Driesch's analysis of embryological facts was quasi-philosophical.[63] The Thomist lawyer Tomás Casares, one of the founders of the CCC and of *Criterio* and a very forceful figure in the Catholic renaissance of the thirties, claimed that Driesch's scientific work was "a model of intelligent objectivity" notwithstanding the "unacceptable" philosophical position of the author (Driesch was something of a neo-Kantian). Unlike Mac Donagh, Casares set strong limits to the empirical sciences while defending "the rights of philosophical speculation."[64] Although presenting a common front against the liberals, Catholics differed among themselves about the autonomy positive sciences should have with respect to philosophy.

Evolution and Belief

In the lecture on Darwin pronounced at the time he joined the Argentine Academy of Geography (1959), a sixty-three-year-old Mac Donagh claimed that evolutionism could be admitted provided there was "a possibility of conciliation with finalism."[65] In his articles for the general public on evolution published in *Criterio* from 1928 until 1952 (two years after the promulgation of Pius XII's *Humani generis*), he hinted at a version of evolutionism in harmony with Catholic doctrine, but he never spelled out the details of how such a synthesis might be reached. Many of his articles imply a defence of the rights of science as well as of the demands of traditional religion. In a 1936 conference at the CCC, Mac Donagh tried to strike a middle road between those Catholic "apologists" who championed Linnaeus's "static zoology" and an "excessively wide-awakened evolutionism."[66] He defended the legitimacy for a Catholic naturalist of carrying out research on the physical evolution of human beings on the basis of Aristotle's notion of the human being as an animal, but he also contended that evolution was not applicable to "man [sic] as a compound being." In his favouring of Aristotle over the Lutheran Linnaeus, he moved within the limits of the Thomist philosophy he had imbibed in the CCC. Mac Donagh preserved Dar-

---

[63] Emiliano J. Mac Donagh, "La difícil doctrina de Hans Driesch," *Criterio* no. 29 (20 September 1928): 363–65.

[64] Tomás D. Casares, "La naturaleza de la vida según Hans Driesch," *Criterio* no. 28 (13 September 1928): 337–40.

[65] Emiliano J. Mac Donagh, "Carlos Darwin y el Origen de las especies," *Anales de la Academia Argentina de Geografía* 3 (1959): 30.

[66] Emiliano J. Mac Donagh, "La libertad intelectual del investigador," *Criterio* no. 428 (14 May 1936): 36–39; no. 429 (21 May 1936): 63–65.

win from excessive criticism and, if possible, gave vent to his admiration for him. He blamed Darwin's "disciples" for what he saw as the evolutionist "underlying philosophy" in the first chapter of *The Descent of Man* (1871), which he distinguished from Darwin's "science." He affirmed that the author of the *Origin* had been "theist first, agnostic later, and always anti-philosophic;" and although Darwin had been pushed by his followers into claiming that the human being descended from an inferior being, "he never explicitly said that [this ancestor] was a monkey." In an earlier article, Mac Donagh regarded the famous final paragraph of the first edition of the *Origin* as "Darwin's unexpected final synthesis...with its admission of the greatness of the belief in a Creator as an explanation of the harmony he [Darwin] had found in nature;" he contrasted this to the view of "the pure selectionists, with their living world left to chance."[67] In later years, Mac Donagh remained faithful to this idea: in his review of Vera Barclay's anti-evolutionistic *Darwin is not for Children* (1950), he distinguished between the "well-known change" undergone by Darwin between his first conception, "respectful of the notion of creation," and his later "crude mechanistic asseveration (although he was not an atheist)."[68]

Summing up, Mac Donagh supported a limited view of the transformation of species, in which natural selection seems to have played a restricted role; evolution theory, he affirmed in 1929, was not "rational unless a finalist interpretation is added to the current mechanistic view."[69] In the case of human beings, it is evident that he considered evolution as applicable only to the human body.

These reflections upon evolution should be seen in the light of Mac Donagh's conception of the epistemological status of biology. He understood his was a balanced, middle-of-the-road solution: biology was an independent science, neither subordinated to metaphysics nor reducible to chemistry. For Mac Donagh, biologists should avoid indulging in an ill-advised religious apologetics as much as they should refrain from a reductionistic approach.[70] In particular, he warned against those apologists who thought they were defending religion from the attacks of science by highlighting the apparently contradictory results which were no more than the unavoidable comings and goings of scientific pro-

---

[67] Emiliano J. Mac Donagh, "A favor de los símiles," *Criterio* no. 56 (28 May 1929): 403–5.
[68] Emiliano J. Mac Donagh, "Darwin no es para los niños," *Criterio* no. 1169 (14 August 1952): 561–65.
[69] Emiliano J. Mac Donagh, "El peligro de un buen ejemplo," *Criterio* no. 44 (3 January 1929): 9–12.
[70] Mac Donagh, "La difícil doctrina."

cedure.⁷¹ Mac Donagh contributed articles to almost every single rightist Catholic magazine published in Argentina in the thirties, but in his writings he always remained within the boundaries of the natural sciences.

Maritain at the Crossroads

Before his visit to Argentina, in August–October 1936, Jacques Maritain was the brightest intellectual beacon for all stripes of Catholic intellectuals in Argentina; after his departure, he left a cleft that time only deepened. Pico and his likes thought they would receive the former follower of Charles Maurras and the author of "À propos de la question juive," which *Criterio* had belatedly translated.⁷² Unaware of Maritain's shift of political position around those years, the far-right integral Catholics would be bitterly disappointed with their former hero's fraternisation with Jews and liberals in Buenos Aires and his refusal to see in Franco the Catholic crusader they thought he was.⁷³ This story has already been told.⁷⁴ Here I shall discuss those aspects of Maritain's visit related to the question of the connections between science, philosophy, and religion.

Maritain's first lecture in Buenos Aires before the audience of the CCC was about science and philosophy. In his presentation of the illustrious guest, Pico perhaps smuggled his own obsessions when he underlined that the latter "had set epistemological limits" to the positive and physico-mathematical sciences.⁷⁵ Maritain's talk was a succinct exposition of the basics of *Les degrés du savoir* (1932) with its analogical conception of the notion of "science," preceded by a criticism of the "neo-positivist" philosophy of the Vienna Circle. In Maritain's view, theology, metaphysics, and philosophy of nature are "sciences" beyond the phenomena and akin to "wisdoms" i.e., modes of intellection which despite being different from the phenomenal sciences are nonetheless valid. Modern

---

71 Emiliano J. Mac Donagh, "Apología de un plural inusitado," *Criterio* no. 33 (18 October 1928): 83–84; idem, review of *Los orígenes de la vida*, by D. L. Saint-Ellier, *Criterio* no. 38 (22 November 1928): 251.
72 Jacques Maritain, "À propos de la question juive," *La Vie Spirituelle* 2 (1921): 305–10. Cf. idem, "Nota sobre la cuestión judía," *Criterio* no. 33 (9 August 1934): 356–57.
73 See Richard F. Crane, *Passion of Israel. Jacques Maritain, Catholic Conscience, and the Holocaust* (Eugene, OR: Wipf & Stock, 2014), 7–34.
74 The most exhaustive and balanced treatments of Maritain's visit are Olivier Compagnon, *Jacques Maritain et l'Amérique du Sud* (Villeneuve-d'Ascq: Presses Universitaires du Septentrion, 2003), 109–36 and José Zanca, *Cristianos antifascistas. Conflictos en la cultura católica argentina* (Buenos Aires: Siglo XXI, 2013).
75 "Jacques Maritain ha iniciado ayer sus disertaciones," *La Nación*, 20 August 1936.

age has separated science from wisdom and it is the task of Christian thought to reconcile them.[76]

Maritain developed this line of thought in another lecture, which was a rehearsal of the first chapter of his book *Science et sagesse* (1935). He traced the relationship between science and wisdom in three stages: the ancient world, the Christian world of the Middle Ages, and the modern world, which witnessed the "victory of science over wisdom."[77] While developing his argument on the hierarchy of knowledge, Maritain rallies against Descartes for having "deposed wisdom" and denying theology its scientific character, but at the same time laments wisdom's "imperialism" with respect to science during the Christian stage of his three-step economy.[78] The contents of the lecture (a synthesis of his thought on science in its relationship with Modernity) paled before the striking circumstances of the conference: Maritain was talking in the auditorium of the Argentine Hebraic Society. His defiant political stance was met with a number of crude anti-Semitic articles written in far-right nationalistic journals.[79]

At the time of presenting Maritain to Argentinian audiences, Pico called him "one of the greatest philosophical geniuses of the present age."[80] But the Argentine bacteriologist would soon be writing his *Letter to Jacques Maritain on the Collaboration of Catholics with Fascist Movements* in which he lamented that the anti-fascist attitude of his admired mentor had not been "prudent enough."[81] In this pamphlet, Pico reminded Maritain that in *Antimoderne* (1922) and *Trois réformateurs* (1925) he himself had claimed that the modern spirit led to "the subversion of hierarchy and the end of Christian culture."[82] Pico opposes Pius IX's *Syllabus* to the democratic society Maritain had pictured in *Humanisme intégral* (1936), chides the latter for the articles he had published in the liberal Argentine magazine *Sur*, and waives the flag of the traditionalist thinkers René Guénon and Berdiáyev.[83] The series of long-winded Thomist musings on philosophy of science and epistemology based upon Maritain's writings and published

---

[76] Ibid.
[77] Jacques Maritain, *Science and Wisdom*, trans. Bernard Wall (London: Geoffrey Bles, 1940), 27.
[78] Ibid., 27–29.
[79] For the anti-Semitic reactions against Maritain's conference, see Compagnon, *Jacques Maritain*, 121–26.
[80] "Dio su primera conferencia el profesor francés Jacques Maritain," *El Pueblo*, 20 August 1936.
[81] César Pico, *Carta a Jacques Maritain sobre la colaboración de los católicos con los movimientos de tipo fascista* (Buenos Aires: Adsum, 1937), 43.
[82] Ibid., 10.
[83] Ibid., 10 and 23.

by Pico between June 1938 and April 1939 in the magazine of the CCP he himself directed (*Iatría*) show no trace of political opinions.[84] While Maritain was writing about pluralism and the autonomy of the temporal order, Pico retained the strong emphasis on hierarchy and subordination, congruent with the strict limits he set to empirical knowledge, and with an integral view of Catholicism understood as the spiritual grounding of authoritarian regimes such as those of Oliveira Salazar in Portugal or Franco in Spain.

Against Draper

In his capacity as director of *Criterio*, Msgr Gustavo Franceschi was perhaps the most outstanding Catholic public intellectual of those times. His long career went through significant changes, but in the period under consideration he can best be seen as a representative of integral Catholicism.[85] A versatile author of quick intellect and an able communicator, early in his priestly career Franceschi became interested in social issues. His visit to Spain during the Civil War reinforced his support of the Franco regime and he saluted with enthusiasm the 1943 takeover of power by the military in Argentina. Since 1945, Franceschi gradually evolved into an active supporter of Christian Democracy. Born in Paris of a Corse father and a mother of Dutch descent, he arrived in Argentina as a child and attended the Seminary in Buenos Aires.[86] As a young cleric, Franceschi had been a serious amateur of marine biology. He got himself a yacht and organised expeditions for the collection of specimens around Mar del Plata, on the Atlantic coast. He was particularly interested in nudibranchia, of which he described a variety, now considered a species: *Polycera marplatensis* Franceschi (1928).[87] He spent much time in the Museum of Natural History of Buenos Aires, where he kept a terrarium with land snails (*Megalobulimus*).[88] His political opin-

---

[84] The first was César Pico, "Las ciencias. Introducción a su metodología y a su epistemología," *Iatría* 8 (June 1938): 3–11. The series was closed by idem, "Las ciencias. Las evidencias primordiales," *Iatría* 9 (April 1939): 3–18.

[85] Austen Ivereigh, "Franceschi y el movimiento católico integral, 1930–1943," *Criterio* no. 2081 (14 November 1991): 623–30; no. 2082 (28 November 18 1991): 660–68.

[86] "Gustavo Franceschi," in *Argentines of Today*, 2:873–74; Floria and Montserrat, "La política desde *Criterio*."

[87] Claudia Muniaín and Jesús Ortea, "The Taxonomic Status and Redescription of *Polycera marplatensis* Franceschi, 1928 (Nudibranchia: Polyceratidae) from Argentina," *The Veliger* 61 (1998): 142–47.

[88] Juan J. Parodiz and Enrique Balech, "El Museo Argentino de Ciencias Naturales 'B. Rivadavia'...en pantuflas" (unpublished manuscript, 1992), mimeo.

ions did not get in the way of his science: at one point he declared that he had learned a technique of histological staining for the nervous system from Pío del Río Hortega, a prestigious histologist and at that time a Republican exile from the Spanish civil war.[89]

In a 1944 paper on science and faith, Franceschi refers to Draper's famous "conflict view."[90] He postulates three alternative models of interaction between science and religion. In one of them, science and religion maintain an "external harmony," such as can be found in two tangent circles; in another model, they are "friends, but without any internal relationship." Franceschi opts for a relationship of "intrinsic harmony." From an "objective" point of view, this kind of connection results from the fact that the created world and its chain of secondary causes as well as the divine Revelation have God as their author. The "subjective" side of the intimate bond between science and religion is given by the unity of the act of knowing and the act of belief (faith) complementing each other. Without denying the autonomy of scientific disciplines, Christians should avoid the "secularisation of intelligence" and should try to integrate science into their lives. Franceschi's rendering of the Thomist account of faith and science tips the balance towards some kind of fusion or intimate connection between them.

Writing at the onset of the decade of the 1930s, Franceschi feels confident when he affirms that "scientism is dead: it has been killed by the joint action of true science and sane philosophy." In the articles published in the 1940s, he manifested once and again his reservations about and even rejection of technology and applied science, which society celebrated in detriment of "disinterested" research.[91] This tone would become harsher after Hiroshima. The atomic age, Franceschi claimed, was "the age of terror," for its symbol is the atomic bomb. The only solution before this impending threat was the regeneration of humanity in the Christian religion; the stark option was "Christ or death."[92]

---

[89] Gustavo J. Franceschi, "Los fenómenos físicos sobrenaturales ante la medicina," *Criterio* no. 972 (31 October 1946): 413.
[90] Gustavo J. Franceschi, "Ciencia y fe," *Criterio* no. 375 (9 May 1935): 37–39.
[91] Gustavo J. Franceschi, "El pensamiento 'científico' de la Edad Media," *Criterio* no. 831 (3 February 1944): 104.
[92] Gustavo J. Franceschi, "La era atómica," *Criterio* no. 941 (28 March 1946): 277–81.

Catholics and Eugenics

In the decades of 1920s and 1930s, enthusiasm for eugenics was on the rise in Argentina as well as elsewhere in the world. Unlike those Catholic doctors who tried to reach some accommodation with eugenics (not to be discussed here), the response of integral Catholics to this movement was definitely negative—their greater or lesser endorsement of fascist regimes was not connected to an acceptance of eugenics, which they rejected as unscientific and also on religious grounds.

In 1923, Father José M. Blanco compiled a series of articles on eugenics published in *Estudios*. Two groups promoting eugenics were active in Buenos Aires at that time: the Argentine Eugenics Society (1918) and the Argentine League of Social Prophylaxis founded in 1921 by Alfredo Fernández Verano, inspired in the ideas of Adolphe Pinard, a famous Parisian obstetrician of leftist ideas who made famous the notion of *puériculture* and favoured the natalist movement.[93] In the early 1920s, the pages of the medical journal *La semana médica* were clogged with contributions on eugenics, sexual education, and prophylaxis of venereal disease.[94] It was against them that Blanco wrote aiming at "the dark silhouettes of many doctors and eugenicists, who in the name of science...sow the germs of lust, tireless destroyers of the species."[95] The Jesuit rallied against the "'eugenics societies" which had "flourished in our capital city [Buenos Aires], whose goal is the improvement of the race through the suppression of the social scourges."[96] He decried the value of medical prophylaxis and recommended "the education of the will...based on the principles of the strictest morality."[97] The Jesuit argued extensively on the basis of traditional Catholic teaching against the secular agenda of regulating social sexual behaviour in terms of eugenics, a program connected to the then widely discussed issues of prostitu-

---

[93] Stepan, *"The Hour of Eugenics,"* 58–59, 82–83. For Pinard, see Marius Turda and Aaron Gillette, *Latin Eugenics in Comparative Perspective* (London: Bloomsbury, 2016), 33–35. See also Carolina Biermat, "La eugenesia argentina y el debate sobre el crecimiento de la población en los años de entreguerras," *Cuadernos del Sur* no. 34 (2005): 251–73.

[94] See, for example, Leopoldo Bard, "Educación especial y profilaxis antivenérea en los centros obreros," *La semana médica* 27, no. 17 (1920): 562–65; Alfredo Fernández Verano, "Ensayo de sistematización de las doctrinas eugénicas," *La semana médica* 27, no. 50 (1920): 800–12; Lázaro Sirlin, "Algo sobre eugenesia y educación," *La semana médica* 27, no. 52 (1920): 875–77; Jules Regnault, "La eugénica," *La semana médica* 29, no. 1 (5 January 1922): 22–25.

[95] José M. Blanco, *La eugenia naturalista y la campaña profiláctica contra la lúe* (Buenos Aires: Amorrortu, 1923).

[96] Ibid., 9.

[97] Ibid., 15.

tion, free love, pornography, neo-Malthusianism, and sexual education. Blanco declares himself against "negative" eugenics (measures such as coercive sterilisation and state control of human reproduction) and also against "positive" eugenics (the encouragement of fit individuals to procreate). He adduces the failure of the prohibition laws in the United States as a proof that compulsory legislation is in the end ineffective but his main argument is humanitarian, not pragmatic: the supporters of negative eugenics "in the end favour violent measures, which humanity will never condone."[98] In an unconscious anticipation of things to come, Blanco compares these proposals to the gas chambers used by the societies for animal protection to give the pets a "more human" treatment.[99] In his view, religious faith was based upon the "very nature of things" and eugenics went "against common sense and the most elementary doctrines of natural law and humanitarian sentiment."[100] His book can be seen as an allegation to place the socially disrupting questions posed by eugenics well under the control of Catholic moral theology and away from medicine. But Catholic doctors had also something to say on the question.

In 1928, César Pico had published a brief anti-eugenics article in which he stated that eugenics was a result of scientism, itself a manifestation of the modern "confusion of spirits" which failed to distinguish between *scientia* (which deals with "the phenomenal order of secondary causation") and *philosophia* (which is concerned with essences and remote causes).[101] Pico commented on a recent incident in Madrid. Gregorio Marañón, a famous liberal Spanish endocrinologist and leader of the sexual reform movement of the 1920s, was one of the lecturers in the First Spanish Course of Eugenics at the School of Medicine of the Central University of Madrid in February 1928. Due to the strong opposition of the Church, the dictatorial government of Primo de Rivera suspended the course.[102] Pico sided with the Spanish authorities and piled abuse on Marañón.

The Consortium of Catholic Physicians (henceforward CMC after its name in Spanish) was founded in September 1929 with the approval of the Archbishop of Buenos Aires Fr José M. Bottaro and the support of the Jesuits (Guillermo Furlong was the moving force behind it and Father Ubach was the first censor of its jour-

---

98 Ibid., 29.
99 Ibid., 48.
100 Ibid., 56–57.
101 César E. Pico, "Marañón y los excesos eugénicos," *Criterio* no. 4 (29 March 1928): 114.
102 Alejandra Ferrándiz and Enrique Lafuente, "El pensamiento eugénico de Marañón," *Asclepio* 51, no. 2 (1999): 133–48. See also Thomas F. Glick, "Marañón, Intersexuality and the Biological Construction of Gender in Spain," *Cronos* 8 (2005): 121–38.

nal *Iatría*).¹⁰³ Created in the wake of the "Catholic revival," the CMC elected Michael [Miguel] Petty as its president; in 1938, Pico became the editor of the journal.¹⁰⁴ The main objective of the CMC was the furtherance of the "true and objective deontology," resting upon a philosophical and a natural theological foundation. The declaration of principles of the new medical association shows its affinity with the ideals of integral Catholicism: it made explicit mention of "dogmatic [theology] and the institution of the perfect society, binding for all persons, which is the holy [Roman Catholic] Church."¹⁰⁵

The articles on eugenics published in *Iatría* make a striking contrast with those published in *La semana médica* and show the extent to which the doctors of the CMC disowned eugenics. Since Petty had made all his career in Britain, it comes as no surprise that the first issue of the journal included a note translated from English in which the author argued that the only "moral" way of obtaining a "vigorous and healthy race" consisted in fighting the causes of mental deficiency: sexually-transmitted diseases, alcoholism, venery, working pregnant mothers, and malnutrition.¹⁰⁶ A two-part article written by the Jesuit professor of Moral Theology at the Seminary of Buenos Aires protested against eugenic male sterilisation imposed by the state to "criminals and degenerates."¹⁰⁷ This was followed by the translation of a note about the medical debate carried on in *The Times* (London) between a group of doctors in favour of the sterilisation of "mentally abnormal" people and those who were against that measure.¹⁰⁸ In a paper on inheritance in mental disease, two psychiatrists of the Women's Asylum in Buenos Aires denied any role to heredity in psychopathology and argued against practicing abortions or sterilising patients with mental illness.¹⁰⁹ The Uruguayan physician Salvador García Pintos states that eugenics based on Mendel's laws "lacks scientific rigour" because pathological inheritance is mostly polygenic in nature.¹¹⁰ García Pintos affirms that eugenics "benefits from the prestige granted to it by genetics, but has in itself no demonstrable scientific ob-

---

103 "Aprobación del Consorcio," *Iatría* 1, no. 1 (April 1930): 1–2.
104 There is some factual information about the CMC in Ana María T. Rodríguez, "Los médicos católicos argentinos en los años treinta," *Sociedad y Religión* 20, nos. 30–31 (2008): 137–70.
105 "El Consorcio de Médicos Católicos. Su programa cultural," *Iatría* 1, no. 1 (April 1930): 5.
106 F. J. Colvin, "Los católicos y la eugenesia. Males de la civilización," *Iatría* 1, no. 1 (April 1930): 5–10.
107 Secundino Lezaun, "Esterilización masculina," *Iatría* 5, no. 1 (1934): 3–15.
108 "La esterilización de los anormales mentales," *Iatría* 5, no. 1 (1934): 16–19.
109 Luis Esteves Balado and Juan Carlos Arizabalo, "La herencia en las enfermedades mentales," *Iatría* 9 (November 1939): 3–14.
110 Salvador García Pintos, "¿Qué hay de positivo en materia de eugenesia?," *Iatría* 11 (April 1941): 3–27.

jectivity."¹¹¹ It is the virtues taught by Christian morals which are "the best safeguard of desirable biological inheritance;" through moral education, the Church has made "authentic eugenics." García Pintos declares that science rejects "materialist eugenics" and sterilisation as anti-scientific, while the Church condemns them as contrary to natural and divine law. The voices of Church and science are in perfect accordance because they express the harmony of God's truth.

Pius XI's encyclical *Casti connubii*, promulgated on 31 December 1930, was a defence of the traditional teaching of the Catholic Church on marriage and reproduction in the face of the declining birth rates in Europe (particularly in France). It condemns birth control, neo-Malthusianism, eugenic sterilisation, abortion, and marriage restriction by the state; it has been observed that the encyclical represents "one element of the struggle for authority that marked the church's relationship with modern nation-states in the nineteenth and twentieth centuries."¹¹² *Criterio* published a series of articles on the encyclical.¹¹³ The overall editorial stance was succinctly expressed by Franceschi: "It is the divine law and not the [pharmaceutical] concoctions which will contribute to a healthy race."¹¹⁴ One of the authors, Javier Ochoa, argued that Christian purity culture had made eugenics unnecessary for centuries.¹¹⁵ He argued further that the "moral disorder of peoples" such as those of England, the United States, Germany, Belgium, and France was the cause of an "excess of idiots," and indirectly of the rise of eugenics.¹¹⁶ (This geography of eugenics corresponds well to the magazine's admiration for Iberian Catholicism.) Ochoa also attacked Bertrand Russell's eugenic dream (or nightmare) in *The Scientific Outlook* (1931) as wholly "unscientific."¹¹⁷ *Criterio* also commented approvingly on the decree of the Holy Office of 21 March 1931 condemning sexual education and positive and neg-

---

111 Ibid., 25.
112 Sharon M. Leon, *An Image of God. The Catholic Struggle with Eugenics* (Chicago: The University of Chicago Press, 2013), 96; cf. John T. Noonan, Jr., *Contraconception. A History of Its Treatment by Catholic Theologians and Canonists* (Cambridge, MA: Harvard University Press, 1965).
113 The first was "La encíclica *Casti connubii*," *Criterio* no. 150 (8 January 1931): 85–86.
114 Gustavo J. Franceschi, "La salud de la raza," *Criterio* no. 494 (19 August 1937): 373–75.
115 Javier Ochoa, "Al margen de la última encíclica. De eugenesia," *Criterio* no. 152 (29 January 1931): 147–48.
116 Javier Ochoa, "Al margen de la última encíclica. Las taras de la civilización," *Criterio* no. 157 (5 March 1931): 309–10.
117 Javier Ochoa, "Después de la *Casti connubii*. Pornografía y falsa ciencia," *Criterio* no. 200 (31 December 1931): 475–76.

ative eugenics.[118] In relation to the fragile scientific basis of eugenics, an anonymous author declared that "pseudo-scientific arrogance" had been responsible for more crimes than the Great War.[119] A Spanish Jesuit priest and doctor who contributed articles to *Criterio* under a pseudonym in 1936 criticised the German eugenic legislation while praising the former Jesuit and eugenicist Hermann Muckermann for having renounced to the Kaiser-Wilhelm-Institut für Anthropologie, menschliche Erblehre und Eugenik in 1933.[120] The author, who had participated in the Second International Congress of Catholic Doctors in Vienna (1936) and was awed by the German-speaking medical establishment, informed that the conference had declared that "science confirmed in the form of a biological postulate what he [Pius XI] had advised and ordered" (the Pope is mentioned as "the great friend" and "the loving father" of all doctors).[121]

In an article on the freedom of the research scientist, Mac Donagh contrasts the triumph of Mendel's genetics with the failure of Galton's theory of inheritance and declares that eugenics is a kind of "heresy." Mac Donagh strongly criticises eugenics in the United States and the project of elimination of "a pauper stock" considered as a race (a group of poor interrelated families in a given region). Commenting on the papers of the Second International Congress of Eugenics held in the American Museum of Natural History (New York) in 1921, Mac Donagh uses Mendelian heredity (and implicitly rejects neo-Lamarckism) to disqualify the arguments of the eugenicists: "They try to convince us that inheritance in the moral order is the same as inheritance in the physical order, a lie whose monstrous character can be appreciated...when it is shown that after almost eight decades of research the inheritance of acquired traits could not be proved."[122] Mac Donagh urges the reader to compare the "unfair moral condemnation of the paupers" with the care of the poor by the members of the Congregation of Saint Vincent de Paul (to which he belonged). Eugenicists, he goes on, have two kinds of arguments: one of them is biological; the other is social and economic i.e., the "undesirables" cost too much to society in terms of jails, hospitals, and asylums.[123]

---

118 Enrique Osés, "La Santa Sede, la educación sexual y la eugenesia," *Criterio* no. 174 (2 July 1931): 7.
119 "La eugenesia," *Criterio* no. 218 (5 May 1932): 130.
120 Petrus Canisius [pseud.], "Eugenica," *Criterio* no. 425 (26 April 1936): 396–97.
121 Petrus Canisius [pseud.], "Segundo Congreso Internacional de Médicos Católicos, Viena, 28 de mayo a 3 de junio. Su significación," *Criterio* no. 457 (3 December 1936): 327–29.
122 Mac Donagh, "La libertad intelectual del investigador," 63.
123 Ibid., 64.

Pico and Mac Donagh argued that eugenics was not scientifically valid. The former saw it as mere "scientism," while the later drew upon Mendelian genetics to counter the prevalent neo-Lamarckian ideas of Argentine eugenicists. These sympathisers of integral Catholicism favoured strong authoritarian political regimes, but their condemnation of eugenics was unambiguous. Although both were rightist nationalists, they did not feel attracted to the discourse on race and nation dear to the liberal reformists of the 1910s.[124] They also skewed the "Italian" type of eugenics that flourished in Argentina in the 1930s imported by the Genoese endocrinologist and eugenicist Nicola Pende and promoted by the Argentine Association of Biotypology, Eugenics, and Social Medicine, founded in the early 30s. Stepan has claimed that "the Argentine eugenicists much admired Mussolini's state rhetoric and program of 'family protection,' a program whose pro-natalist, anti-abortion, and anti-feminist ideology was highly congruent with their own."[125] This was not the case with Pico, Mac Donagh, the members of the CMC or the Jesuits who wrote on eugenics. At least, there is no mention in their writings of the Franciscan Agostino Gemelli, the doctor and psychologist who founded the University of the Sacred Heart in Milan, favoured a "Latin" type of positive eugenics, and was in good terms with Mussolini until 1938.[126] But some Catholics in the medical profession were sympathetic to Gemelli or to the type of approach he embodied. The president of the board of the Argentine Association of Biotypology was the Catholic Mariano Castex and there were other prominent Catholic physicians among its members.[127] The acting president of the association, Octavio López, had an interview with Gemelli in the course of his visit to Pende's institute in Italy. He later said that Gemelli had declared that the Catholic Church "far from denying science…only objects to its more irritating edges." The "truly scientific eugenics," goes on López, "neither

---

124 Eduardo Zimmerman, "Racial Ideas and Social Reform: Argentina, 1890–1916," *The Hispanic American Historical Review* 72, no. 1 (1992): 23–46.
125 Nancy Stepan, "Race, Gender and Nation in Argentina: The Influence of Italian Eugenics," *History of European Ideas* 15, no. 4–6 (1992): 749–56. See also idem, *"The Hour of Eugenics,"* 114–122. Cf. Turda and Gillette, *Latin Eugenics*, 129–136 and Andrés H. Reggiani, "Dépopulation, fascisme et eugénisme 'Latin' dans l'Argentine des années 1930," *Le Mouvement Social* no. 230 (2010): 7–26.
126 Turda and Gillette, *Latin Eugenics*, 117–121; Aaron Gillette, "Agostino Gemelli and the Latin Eugenics Movement," *Römische Quartalschrift für Christliche Altertumskunde und Kirchengeschichte* 109, no. 1–2 (2014): 92–102.
127 Gustavo Vallejo, "La hora cero de la eugenesia en la Argentina," *História, Ciências, Saúde. Manghinos* 25, suppl. 1 (August 2018): 15–32.

violates the laws of marriage nor resorts to sterilisation."[128] López could hardly be called a Catholic—he calls eugenics "an Emersonian song of force and creative energy"—but his reference to a Catholic authority suggests that many of his readers were sensitive to this kind of argument. Briefly, there was a wide range and many shades of Catholic opinion on eugenics. While some of them might have felt identified with the type of "Latin" eugenics represented by Gemelli, it seems that others—the CMC, the Jesuits, Mac Donagh, and even those who like Pico openly supported fascism—chose to steer clear from it.

Conclusions

In this period of triumphant Catholicism, Catholics imagined themselves as reverting the partly secularised, liberal nineteenth-century Argentina to what they construed as a "Catholic" society. Integral Catholicism in interwar Argentina was an experiment in the Catholicisation of society tied to a restricted democracy or to authoritarian political regimes. In this attempt at infusing all the dimensions of society with confessional values, science had a contentious place. Those Catholic scientists who enthusiastically supported this flight from secularism were in the quandary of granting a meaning to modern science while at the same time undermining the basis of Modernity, which in the first place had given birth to it. As a whole, this kind of criticism of secularism was not antiscientific per se: it sought to disentangle science from what they saw as its secularist adhesions, to grab the banner of science from liberals and socialists and integrate it into a Catholic whole, as every other aspect of a new "organic" polity.

Franceschi argued in favour of something like an intimate connection between science and religion. His fourfold categorisation of the relationships between both shows that his was an educated opinion. Among integral Catholics, the question of science and religion was not discussed as such. Pico and Mac Donagh, who belonged to the same political and religious circles but never referred to each other, might have felt that the issue was the preserve of the clergy and that as laypersons they were not entitled to pronounce themselves about it. Pico surely understood the relationships between science and religion in terms of Maritain's degrees of knowledge. Aquinas' epistemological hierarchy can be interpreted in two ways: either stressing the (always relative) autonomy of science or underlining its subordination to the upper levels of the hierarchy.

---

128 Octavio V. López, "Cristianismo y eugenesia," *Anales de biotipología, eugenesia y medicina social* 1, no. 9 (1 August 1933): 6–7.

Pico followed the second way, which was in tune with his political options. Notwithstanding his politically reactionary positions, Mac Donagh was comparatively open with respect to the theory of evolution. His zealous defence of the autonomy of biology (fed perhaps by worries related to the professionalisation of the natural sciences) might have contributed to this. But the freedom and independence of his discipline was conceived by him within the broad limits of Catholic doctrine.

A third point to be remarked is that reactionary political positions did not necessarily entail backward scientific options. Mac Donagh's finalist interpretation of biological evolution was of a kind that many Christian scientists held at that time.[129] Opinions among Catholics varied. For instance, a brief uninformed note criticising evolution which appeared in *Criterio* in 1939 was answered point by point by an article by two agronomists who saw themselves as treading a path between those "who think they can sink the Church in the name of transformism" and "the overzealous apologists."[130] In an article written when he was a medical student, Pico made fun of an essay by the Spanish Jesuit and anti-evolutionist Jaime Pujiula.[131] We have seen that Ángel Gallardo, perhaps the foremost Argentine naturalist at that time and a conservative Catholic, had no qualms in accepting evolution as a fact. An analogous disconnection between politics and scientific theories applies to eugenics: for all his nationalism, his writings do not show racist attitudes; the same can be said of Mac Donagh.

Science was not left out of the vast program of integral Catholicism to permeate all levels of society. Catholic scientists were willing to engage in contentious questions and make clear that their disciplines were not the preserve of a secularised world view, that science was not intrinsically secular. The answers they gave on the relationships between science and religion and on crucial issues such as the theory of evolution were not uniform. On the contrary, the wide array of options defies any attempt to trace a pattern correlating political

---

129 Harry W. Paul, "Religion and Darwinism: Varieties of Catholic Reaction," in *The Comparative Reception of Darwinism*, ed. Thomas F. Glick (Chicago: The University of Chicago Press, 1988), 407–36; R. S. Appleby, "Exposing Darwin's 'Hidden Agenda': Roman Catholic Responses to Evolution, 1875–1925," in *Disseminating Darwinism*, 173–207; Stefaan Blancke, "Catholic Responses to Evolution, 1859–2009: Local Influences and Mid-Scale Patterns," *Journal of Religious History* 37 (2013): 353–68.
130 Félix Albani and Juan L. Pedelaborde, "¿Católicos contra transformistas?," *Criterio* no. 584 (11 May 1939): 35–36. The article they criticise is César Soler, "La función crea el órgano," *Criterio* no. 582 (27 April 1939): 393–94.
131 César Pico, "Sobre la significación morfo-fisiológica del apéndice ileo-cecal," *Tribuna Universitaria* 3, no. 18 (15 June 1916): 14–23; Cf. Jaime Pujiula, "El apéndice vermiforme del ciego," *Estudios* 6 (June 1916): 397–404.

and scientific positions. Here as elsewhere, when it comes to the relationships between science, religion, and politics, things get unavoidably complex. (Brooke has already warned against overemphasizing the political factor at the time of analysing the complexity involved in the relationshps between science and religion.)[132] Integral Catholic scientists put into question the notion that science was intrinsically linked to secularism and sought to integrate scientific knowledge with the teachings of the Church. The following chapter will show that the most prestigious Catholic scientists in the country held different views of the relationships between science, religion, and secularism.

---

[132] Brooke, "Afterword," 240.

Chapter 8
Argentine Catholic Democratic Scientists and their Projects of a Research University (1932–1959)

Two groups of Catholic scientists with high public profile can be identified in Argentina since the late 1920s and early 1930s. On the one hand, those who adhered to integral Catholicism and far-right nationalism: they supported Franco in the Spanish Civil War and defended Argentine neutrality or showed sympathy for the Axis during the Second World War (they have been treated in the previous chapter). On the other hand, those committed to Christian democracy, who did not consider that the Spanish Civil War was a Catholic crusade and from the start sided with the Allies.[1] Juan Traherne Lewis, Eduardo Braun Menéndez, and Augusto J. Durelli belonged to the latter group. Lewis and Braun were physiologists and associates of Nobel Prize winner Bernardo Houssay; Durelli was an engineer who spent the largest part of his professional career in the United States. The three of them were politically active during the years of the ascendancy of Juan D. Perón (1943–46), his first and second presidential terms (1946–55), and the period following his demise (1955–59).[2] The ambiguous and in the end conflictive relationships between the Catholic Church and Perón has been the subject of many studies and lays beyond the scope of this work.[3] While adopting many themes of Catholic social doctrine in his first programmatic years, Perón eventually tried to subject the Church to state control. The violent confrontation between the government and the Church in the last

---

[1] For democratic Catholics in Argentina, see Zanca, *Cristianos antifascistas* and idem, *Los intelectuales católicos y el fin de la cristiandad, 1955–1966* (Buenos Aires: FCE, 2006).
[2] Braun Menéndez, Durelli, and Lewis were among those who signed the February 1946 "Manifesto of Christian Democrats" supporting the candidates of the Democratic Union in the presidential elections. See "Manifiesto de los demócratas cristianos en auspicio de la fórmula presidencial Tamborini-Mosca," *Orden Cristiano* no. 104 (February 1946): 412–15.
[3] See Kennedy, *Catholicism, Nationalism, and Democracy in Argentina*; Robert McGeagh, *Relaciones entre el poder politico y eclesiástico en la Argentina* (Buenos Aires: Itinerarium, 1987), 23–115; Lila Caimari, *Perón y la iglesia católica* (Buenos Aires: Sudamericana, 1994); Ivereigh, *Catholicism and Politics in Argentina*, 143–182; Burdick, *For God and the Fatherland*, 45–109; Roberto Bosca, *La iglesia nacional peronista. Factor religioso y poder político* (Buenos Aires: Sudamericana, 1997); Loris Zanatta, *Perón y el mito de la nación católica* (Buenos Aires: Sudamericana, 1999); Susana Bianchi, *Catolicismo y peronismo. Religión y política en la Argentina, 1943–1955* (Tandil: IEHS, 2001).

two years of his second presidency (1954–55) is generally recognised as one of the motives that contributed to his fall. As members of the intelligentsia that opposed Peronism, prestigious Catholic scientists marginalised from state-controlled universities participated in the creation of privately funded centers of scientific research while refloating proposals for the creation of a Catholic university.[4] These plans would materialise in the founding of the Catholic Institute of Sciences (1953–54). Braun would play a leading role during the protracted conflict that developed in the post-Peronist era over the proposals for the creation of private confessional institutions of higher learning. With their plans of creating Catholic centres of advance research and teaching, Lewis, Braun, and Durelli challenged the entrenched view that saw science as the secular turf of liberals and socialists.

Lewis: An Early Proposal for "Free" Universities

In the 1930s, Bernardo A. Houssay was at the head of the indisputably leading scientific research group in Argentina.[5] As a result of his work on the role of the pituitary in sugar metabolism, he would be granted the Nobel Prize in medicine in 1947. Juan T. Lewis (1898–1976) was the second author of Houssay's standard textbook of physiology, which he and his wife Olive Mary Lewis née Thomas, a graduate of the London School of Medicine for Women, had translated into English.[6] Lewis's early research focused on the physiology of adrenal glands, the action of insulin, and the regulation of blood sugar, but his original articles covered a wide range of physiological topics.[7] Born in Argentina to a family of Welsh descent, he had studied medicine in the School of Medicine of the University of Buenos Aires. While a student, he worked as research assistant in the Bacteriological Institute of the National Department of Hygiene, where he would later be the head of the Section of Pharmacology (1924–28). After graduating

---

[4] This question (particularly as regards Braun and Durelli) has been explored from a different viewpoint in Diego Hurtado de Mendoza and Analía Busala, *Los ideales de universidad "científica" (1931–1959)* (Buenos Aires: Libros del Rojas, 2002).
[5] For Houssay, see Frank Young and Virgilio G. Foglia, "Bernardo Alberto Houssay, 1887–1971," *Biographical Memoirs of the Royal Society* 20 (1974): 246–70; Marcos Cueto, "Laboratory Styles in Argentine Physiology," *Isis* 85 (1994): 228–46.
[6] A. J. Carlson, review of *Human Physiology*, by Bernardo A. Houssay et al., trans. Juan T. Lewis and Olive M. Lewis (New York, 1951), *Science* 113, no. 2946 (15 June 1951): 700.
[7] Juan T. Lewis, *Antecedentes, títulos y trabajos presentados a la Universidad Nacional del Litoral al hacerse cargo de la Cátedra de Fisiología de la Escuela de Medicina de Rosario* (Buenos Aires: Frascoli & Bindi, 1929).

in 1921, Lewis joined Houssay's recently created Institute of Physiology; he was named substitute professor of physiology in 1930.[8] Supported by a grant from the Rockefeller Foundation, Lewis spent the months from April through November 1925 as a research fellow in Harvard working under Walter B. Cannon, who had trained several young promising physiologists from Latin America.[9] Cannon told Houssay that he "had never had any advanced student come to the laboratory who was better prepared for research, because of knowledge, insight, and technical skills than Dr. Lewis."[10] Diagnosed with tuberculosis in late 1925, Lewis spent some time in the Trudeau Sanatorium at Saranac Lake, New York, and later in Chamonix, Switzerland. Once discharged, he took the opportunity to visit research institutes in that country, France, and the United Kingdom.[11] When in 1928 he traveled to England from Argentina to marry his cousin Mary, he took opportunity to meet his father's cousin, the physiologist Sir Thomas Lewis at University College, London and the pharmacologist Sir Henry H. Dale, then at the National Institute for Medical Research, with whom he had planned to work supported by his Rockefeller fellowship before falling ill.[12] Upon returning to his country in 1929, Lewis founded the Institute of Physiology in the medical school of Rosario, the third largest city in Argentina, which harboured a large British colony.

The steady growth and consolidation of integral Catholicism in Argentina in the 1930s and 1940s has been discussed in the previous chapter. Politically liberal Catholics, never more than an influential minority, entered the limelight with the controversies triggered by Maritain's visit to the country in 1936 and managed to maintain visibility through publications such as the bimonthly *Orden Cristiano* (1941–48).[13] Lewis met Maritain presumably when the latter lectured in Rosario.[14] When interviewed by *Orden Cristiano* in 1945, Lewis declared

---

8 Juan T. Lewis, *Antecedentes*; "Lewis, Juan T.," in *Quién es quién en la Argentina*, 7th ed. (Buenos Aires: Kraft, 1958), 443–44; Jimmy Lewis, "El doctor Juan T. Lewis," *Rosario. Su historia y región* no. 83 (March 2010): 8–10.
9 Marcos Cueto, "The Rockefeller Foundation's Medical Policy and Scientific Research in Latin America: the Case of Physiology," *Social Studies of Science* 20 (1990): 229–54; idem, "Laboratory Styles," 236–37.
10 Walter Cannon to Bernardo Houssay, 4 February 1926, file 8044 "Lewis, Juan T.," Archive of the Faculty of Medicine, University of Buenos Aires (henceforward cited as AFM).
11 Lewis, "El doctor Juan T. Lewis."
12 Lewis, "El doctor Juan T. Lewis;" Clifford W. Wells to John T. Lewis, 1st October 1925, file 8044, AFM.
13 Zanca, *Cristianos antifascistas*, 55–128.
14 As a consequence, Lewis organised a reading group of *Humanisme intégral* (1936). Lewis, "El doctor Juan T. Lewis."

that the publication had contributed to counter the views of "those who have tried to identify the Church with reactionary politics" in a context in which "Catholic democrats will have to fulfil the duty of defending the Church."[15] In that year, Lewis joined the Social Democratic Action in Santa Fe, one of the several political groups inspired by the ideals of Christian Democracy, which arose in several large cities in Argentina in opposition to Peronism.[16] During February 1946, he campaigned actively in Rosario against the candidacy of Perón, whom he thought had slim chances of winning.[17] Two days after the presidential elections in which Perón had a sweeping victory, Houssay, Braun, and others sent a letter to Lewis declaring their sympathy with him, "at a time of injustice for you and shame and sorrow for the country."[18]

Since early in his professional life, Lewis wrote about medical research and his conception of what universities in Argentina should be in pages at times informed by his religious beliefs. In a lecture on the occasion of Houssay's reception of the Nobel Prize, he described science teachers and their "influence on the spiritual formation of human beings" with terms evocative of monastic life.[19] A few years later, he portrayed the intellectual vocation as based "upon an asceticism" which brings with it the cultivation of virtue.[20] Lewis pictured himself as a demanding teacher "who has tried to be fair, with that kind of justice informed by charity, which is a profound love of the neighbour resulting from the love of God."[21]

Lewis's religious commitments were far from those of integral Catholicism. On the contrary, he defended civil freedom and felt comfortable to use that freedom to proclaim Catholic values. This is evident in his defence of the creation of "free universities," by which he meant private institutions of higher learning. It has already been pointed out that an early attempt at the creation of a Catholic university in the second decade of the twentieth century (1909–1920) had failed because the UBA refused to allow a private institution the granting of professio-

---

15 Juan T. Lewis, [Answer to a survey] in *Orden Cristiano* 5, no. 97 (1 October 1945): 22.
16 Ricardo G. Parera, *Democracia Cristiana en la Argentina* (Buenos Aires: Nahuel, 1967), 62–63; Enrique Ghirardi, *La Democracia Cristiana* (Buenos Aires: CEAL, 1983), 75–77.
17 Juan Lewis to Bernardo Houssay, 20 and 25 February 1946, box 4, 08–6, Archive Bernardo Houssay, Buenos Aires (henceforward ABH).
18 Bernardo Houssay et al. to Juan Lewis, 26 February 1946, box 4, 08–6, ABH.
19 Juan T. Lewis, "El hombre de ciencia," *Revista de la Asociación Médica Argentina* 62, no. 631–632 (15 and 30 June 1948): 279–80.
20 Juan T. Lewis, "La vocación intelectual," *Criterio* no. 1201–1202 (24 December 1953): 990–93.
21 Juan T. Lewis, "La Universidad no debe callar," *Mundo Médico* 10, no. 108 (1945): 21–23.

nal degrees.[22] In May 1932, under the presidency of General Agustín P. Justo (a liberal conservative who came to power through rigged elections), the government launched a proposal for a new law regulating the functioning of state universities.[23] Lewis criticised the project and made a proposal of his own inspired by the kind of schools he had known in the anglophone world. A major failure of the government's plan, according to Lewis, was that it did not provide for the creation of free universities, which "cannot admit of any control other than that necessary for the attainment of their ends."[24] They should be allowed to exist "in order to open the way to private initiative" and because measures that "grant freedom and space to the intellectual activities of the persons" are never in excess.[25] Houssay and those around him were of a mind that the country needed private schools created along the lines of those in the United States and the United Kingdom.[26]

While Lewis's defence of "free," private schools was in line with the principle of subsidiarity of Catholic social teaching, his idea of a university rested upon the Aristotelian notion that the distinctive characteristic of the human being is an "imperious necessity of knowing."[27] The university is the sphere where that quest for knowledge takes place; its goal is "the cult of truth."[28] Lewis's ideals were obviously influenced by his personal experiences in Harvard and England and also by his readings of Abraham Flexner, John H. Newman, Alfred Whitehead, and Bertrand Russell—a reading list quite alien to the French model of higher education prevalent in Argentina. In his comments on the 1932 government's bill for a university reform law, Lewis also criticised the insistence upon discipline and nationalism proclaimed by the then recently nominated rector of the University of Buenos Aires, Ángel Gallardo, for that approach implied that

---

22 Zuretti, "La fundación de la primera Universidad Católica;" Pelosi and Mackintosh, "El ensayo de Universidad Católica."
23 Ministerio de Justicia e Instrucción Pública, *Ley Universitaria. Mensaje y proyecto de ley* (Buenos Aires: Talleres Gráficos de la Penitenciaría Nacional, 1932); Walter, *Student Politics in Argentina*, 87–98.
24 Juan T. Lewis, *Comentarios sobre el proyecto ministerial de ley universitaria. Un anteproyecto de ley universitaria* (Rosario, 1932), 10.
25 Ibid., 31.
26 Bernardo Houssay, "Universidades, alma y cuerpo," *La Nación*, 23 October 1939, reproduced as "Concepto de la Universidad," in idem, *Escritos y Discursos* (Buenos Aires: El Ateneo, 1942), 45.
27 Juan T. Lewis, "Objeto y fin de la Universidad," *Universidad* (Universidad del Litoral) 1 (October 1935): 47. Cf. Juan T. Lewis, *Influencia de la investigación científica sobre la sociedad* (Buenos Aires: Las Ciencias, 1934).
28 Lewis, "La Universidad no debe callar," 21.

universities would be just "departments of the State and subjected to its policy."[29] While Lewis distrusted any kind of state intervention in the universities, he also disliked that students participated in the government of those institutions, a state of affairs sanctioned by the 1918 University Reform and which he found "not only unnecessary in theory but also uncalled for."[30] Students' unions, according to him, "should have nothing to do with political action, be it national or international."[31]

Braun Menéndez: Research Institutes as the Basis of the University

The Army coup that took power in Argentina in June 1943 inaugurated a three-year authoritarian and nationalistic regime which strove to maintain the country's neutrality in the world conflict. Democracy was restored when Colonel Juan D. Perón, the strong man in Farrell's government, triumphed in the general elections of 24 February 1946.[32] He would lead the country in two successive presidencies from June 1946 to September 1955. On 14 October 1943, during the de facto rule of General Ramírez, Gustavo Martínez Zuviría (a far-right Catholic writer) was put in charge of the Ministry of Justice and Public Instruction. That same day a group of academics and politicians published a declaration demanding the return to democracy and "Latin American solidarity" (which meant the breaking of diplomatic relations with the Axis powers).[33] Houssay and Lewis were among those who signed the declaration.[34] As a result, both were dismissed from their positions in the School of Medicine of the UBA.[35] Houssay gave a last

---

29 Lewis, *Comentarios sobre el proyecto*, 11.
30 Ibid., 15.
31 Ibid., 29.
32 Rock, *Argentina, 1516 – 1987*, 214 – 61.
33 For the international context of the period, see for example Harold F. Peterson, *Argentina and the United States, 1810 – 1960* (New York: SUNY Press, 1964), 428 – 45; Joseph A. Tulchin, *La Argentina y los Estados Unidos. Historia de una desconfianza* (Buenos Aires: Planeta, 1990), 185 – 205.
34 "Una Declaración sobre democracia efectiva y solidaridad Americana," *La Prensa*, 15 October 1943; "Una manifestación de anhelos respecto a la situación política," *La Nación*, 15 October 1943.
35 [Pedro P. Ramírez, Gustavo Martínez Zuviría], Decree 127 – 43 (28 October 1943), *Boletín Oficial*, 5 November 1943, 3 – 4; A. Labougle to Juan Bacigalupo, 20 October 1943, in file 8044, AFM (letter from the deputy rector of the university to the dean of the School of Medicine ordering the discharge of Lewis and other four teachers); "Comunicóse la cesantía a cuatro profesores

lecture (it the event, it turn out not to be so) before a multitudinous auditory of local medical celebrities.[36] The Catholic daily *El Pueblo* limited itself to reproducing the official decree and published the list of professors expelled from the national universities; the liberal newspaper *La Nación* run an editorial on "The apostolate of science."[37] Lewis was also discharged from his position as professor in the School of Medicine in Rosario.[38] The affair had repercussions in the broader scientific community, to the point of reaching the pages of *Science*.[39] It is perhaps ironic that two months before, Lewis had written to Houssay telling him that he was "convinced of the honesty and good intentions" of the government.[40] Houssay was alarmed by the shift towards extreme nationalism of the military and feared that the support of the clergy could eventually result in an anticlerical backlash.[41]

Braun Menéndez, at that time associate professor of physiology in Buenos Aires, was not among those dismissed in 1943, but he resigned in solidarity alleging "personal reasons."[42] Eduardo Braun Menéndez (1903–1959), born in Punta Arenas (Chile), was raised in Buenos Aires; his family was among the wealthiest in the country at that time, owning immense estancias in southern Patagonia.[43] After he graduated as a physician from the University of Buenos Aires in 1929, he joined the cardiology department of one of the city hospitals and began his research work in the Institute of Physiology under Houssay on the in-

---

de la Facultad de Medicina," *La Nación*, 21 October 1943; "En la Universidad comunicóse la cesantía a varios profesores," *La Nación*, 22 October 1943; "Comunicóse la cesantía de otros profesores de la universidad local," *La Prensa*, 21 October 1943.

36 "Una memorable hora científica vivió el viejo anfiteatro de fisiología," *La Nación*, 20 October 1943.

37 "Fueron dispuestas cesantías de funcionarios y profesores," *El Pueblo*, 21 October 1943; "Diéronse a conocer las cesantías de profesores de la Universidad," *El Pueblo*, 22 October 1943; "El apostolado de la ciencia," *La Nación*, 25 October 1943.

38 Juan Lewis to Bernardo Houssay, 18 October 1943; Houssay to Lewis, 18 October 1943, box 4, 08–6, ABH.

39 E. S. H., "Declaration of Citizens of Argentina," *Science* 98, no. 2552 (26 November 1943): 467–68; the note refers to Houssay, Lewis, and Orías; Philip Bard, C. H. Best et al., "The Argentine Citizens Declaration," *Science* 99, no. 2565 (25 February 1944): 166 (this note expressed the concern of the American Physiological Society); Kirtley F. Mather and Harry Grundfest, "The Argentine Citizens Declaration," *Science* 99, no. 2566 (3 March 1944): 176.

40 Juan Lewis to Bernardo Houssay, 5 August 1943, box 4, 08–6, ABH.

41 Bernardo Houssay to Juan Lewis, 24 August 1943, box 4, 08–6, ABH.

42 Eduardo Braun to Juan Bacigalupo, 20 October 1943, file 18173 "Braun Menéndez, Eduardo," AFM.

43 For his life and work, see Ignacio Peña and Guillermo Jaim Etcheverry, eds., *Eduardo Braun Menéndez. Ciencia y conciencia* (Buenos Aires: MINCYT, 2015).

fluence of the hypophysis in hypertension, which resulted in his dissertation presented in 1934. By the end of 1937, he joined the Department of Physiology of University College, London, in order to work with Charles Lovatt Evans on heart metabolism. He stayed there until June 1938 and during the next three months visited several centers of physiological research in the United Kingdom.[44] His work was well received and his book on heart sounds was soon translated into English.[45] Upon his return to Buenos Aires, Braun joined one of Houssay's teams (Juan Fasciolo and Alberto Taquini) who were looking for a humoral factor in hypertension. Together with Juan M. Muñoz and future Nobel Prize Luis F. Leloir, they were able to discover the mechanism of nephrogenic hypertension simultaneously with the American Irvine H. Page. Using the ligature of the renal artery, they isolated the substance "hypertensina" afterwards called angiotensin by an agreement with Page, reached "after two Martinis."[46] Braun and his group identified an enzyme (renin) produced by the kidney, which transforms a protein in the blood in angiotensin, a substance which causes hypertension. The book summarising all this work was also translated into English.[47] Braun was invited to give the May 1942 Hersztein lectures sponsored by the University of California and Stanford; he embarked on a lecture tour delivering 15 lectures in the course of a month and a half in the West Coast, the Midwest, and the East.[48]

As a result of continuous unrest in Argentine universities, in February 1945 President Farrell sought to bring some calm and quietly reinstated the professors dismissed by his predecessor.[49] On the occasion of Lewis's reincorporation to his chair in Rosario, he proclaimed his conviction that universities should not be "an organ of the state."[50] Against the backdrop of the statist policies of the na-

---

[44] Letter from C. Lovatt Evans (no addressee), 29 September 1938; Eduardo Braun Menéndez to the Dean of the School of Medicine, 8 October 1938, file 18173, AFM.
[45] Oscar Orías and Eduardo Braun Menéndez, *The Heart-Sounds in Normal and Pathological Conditions* (Oxford: Oxford University Press, 1939).
[46] Eduardo Braun Menéndez and Irvine Page, "Suggested Revision of Nomenclature-Angiotensin," *Science* 127, no. 3292 (31 January 1958): 242. Nidia Basso and Norberto Terragno, "History about the Discovery of the Renin-Angiotensin System," *Hypertension* 38 (2001): 1246–49.
[47] Eduardo Braun-Menéndez, Juan C. Fasciolo, Luis F. Leloir et al., *Renal Hypertension*, trans. Lewis Dexter (Springfield, IL: Charles C. Thomas Pub., 1946).
[48] Besides several conferences in California, he lectured in Michigan, Western Reserve University, Chicago, Columbia, the Rockefeller Institute, Harvard, and Yale. Eduardo Braun Menéndez, Report to the School of his trip to the United States, 17 April 1942, file 18173, AFM.
[49] [Edelmiro Farrell], Decree 3157–45 (10 February 1945), *Boletín Oficial*, 20 February 1945, 4.
[50] Lewis, "La Universidad no debe callar." Cf. Robert A. Lambert, "The Argentine Declaration," *Science* 101, no. 2629 (18 May 1945): 502. Lewis was professor in Rosario and substitute professor of physiology in Buenos Aires, which put him into a difficult position for the School of Medicine in Rosario would not grant him leave to substitute Houssay when the latter travelled abroad.

tionalistic military regime, Houssay and his associates, notably Lewis and Braun, fostered the creation of a number of research institutions supported by private local and international funding.[51] The first and largest of them was the Institute of Biology and Experimental Medicine (IBYME) in Buenos Aires, founded in 1944, into which Houssay's group moved *en masse*.[52] It was at that time that Braun Menéndez, Lewis, and Durelli (see below) launched a campaign in favour of the establishment of private universities on the basis of one or more research institutes.

From April 1945 through December 1946, most of the editorials of the journal of the Argentine Association for the Progress of Science (AAPC)—created in 1933 under Houssay's leadership—discussed the nature of universities, scientific research, and the importance of private institutes of scientific research.[53] The notes were not signed, but their style suggests that they were written by Lewis; Braun could also have had a hand in them. In any case, they express the common concerns of the group to secure institutions of higher learning and research free from the encroachment of the government. The title of Houssay's March 1946 editorial "Science needs a climate of freedom" aptly summarises the opinions of the group.[54]

In a conference for the general public given on 5 September 1945, Braun made explicit his vision of the future of university research in Argentina; his model was Daniel Coit Gilman's Johns Hopkins.[55] He admitted that private universities faced the risk of turning into mere professional schools for lack of financial support. The best way out of this was to establish small scientific research centres analogous to those of the Kaiser Wilhelm Gesellschaft, "institutes with complete freedom of research and associated to teaching."[56] The recent creation of the IBYME surely contributed to this idea. Universities should be "free," which

---

"Informe de la Pro-Secretaría sobre la actuación docente del Prof. Lewis en esta Facultad," 27 September 1946, file 18173, AFM.
51 Asúa, *Una gloria silenciosa*, 199–201.
52 "The Institute of Biology and Experimental Medicine at Buenos Aires," *Science* 99, no. 2575 (5 May 1944): 360–61; Virgilio Foglia, "The History of Bernardo A. Houssay's Research Laboratory, Instituto de Biología y Medicina Experimental: The First Twenty Years, 1944–1963," *Journal of History of Medicine and Allied Sciences* 35, no. 4 (1980): 380–96.
53 See Diego Hurtado de Mendoza, "Los primeros años de la AAPC (1933–1945)," *Ciencia e Investigación* 56, no. 2 (2004): 35–40.
54 Bernardo Houssay, "La ciencia necesita un ambiente de libertad," *Ciencia e Investigación* 2, no. 3 (March 1946): 97–98.
55 Eduardo Braun Menéndez, *Universidades no oficiales e institutos privados de investigación* (Buenos Aires, 1945).
56 Ibid., 24.

meant "freedom from the intromission of the governments, the onslaught of politics, ideological violence, and sectarian narrowness."[57] Braun distinguished the granting of academic diplomas from the granting of professional degrees; he considered the latter an undeniable responsibility of the state.[58]

In September 1946, the inaugural year of Perón's presidency (he had accessed to power in June), Houssay was again relieved from his post with the specious excuse of having reached retirement age.[59] His disciples, among them Braun Menéndez and Lewis, resigned in solidarity from their positions in the University of Buenos Aires.[60] By the end of 1946, about 1200 teachers had quit their university positions (around one third was dismissed and the rest resigned).[61] Just as the first dismissal of Houssay had resulted in the creation of the IBYME, this time an analogous institution was created in Rosario by Lewis in 1948: the Institute for Medical Research.[62]

In November 1947, Congress approved a new legislation for the universities, which in the eyes of the opposition transformed them in something very much like state dependencies.[63] The journal of the AAPC answered with an editorial piece in all likelihood written by Lewis, which rehearsed his ideas about the creation of "free universities" as "associations devoted to the search and propagation of knowledge, supported and regulated by the state, but not subjected to the government;" they should be supervised by the national academies.[64] A second

---

57 Ibid., 16.
58 Ibid., 19.
59 "Plan progresivo de cesantías," *La Nación*, 3 October 1946; "Respecto del límite de edad para ejercer la cátedra universitaria," *La Nación*, 4 September 1946; "Comunicóse al Dr. Houssay que ha cesado en su cátedra," *La Prensa*, 7 September 1946; "Universities of the Argentine: Retirement of Prof. Bernardo A. Houssay, For. Mem R.S.," *Nature* 158, no. 4021 (23 November 1946): 739.
60 For Braun, see letter of resignation, Eduardo Braun Menéndez to Carlos Bergara, 7 September 1946, file 18173, AFM. Cf. "Los alumnos del Dr. Houssay abstuviéronse de asistir ayer a clases," *La Prensa*, 10 September 1946; "Motivó protestas la cesantía del profesor Bernardo A. Houssay," *La Nación*, 10 September 1946. For Lewis, see Juan Lewis to Carlos Bergara, 10 September 1946, file 8044, AFM; "Renunció otro profesor solidarizándose con el Dr. Bernardo Houssay," *La Prensa*, 13 September 1946.
61 Pablo Buchbinder, *Historia de las universidades argentinas* (Buenos Aires: Sudamericana, 2005), 149–50. See the list in Federación de agrupaciones para la defensa y progreso de la universidad democrática y autónoma, *Avasallamiento de la universidad argentina* (Buenos Aires, 1947).
62 Juan T. Lewis, "El Instituto de Investigaciones Médicas de Rosario," *Ciencia e Investigación* 5, no. 1 (1949): 1–2, 37–39.
63 "Ley 13.031. Régimen universitario," *Boletín Oficial* 55, no. 15905 (4 November 1947): 2–6.
64 "Las universidades libres," *Ciencia e Investigación* 3, no. 1 (January 1947): 1–2.

editorial note, probably authored by Braun Menéndez, came back with the idea of creating free universities on the basis of small institutes of scientific research which should also provide post-graduate teaching and whose sole purpose should be "the disinterested pursuit of truth."[65]

Durelli

Augusto J. Durelli (1910–2000) was born in Buenos Aires to a family of comfortable means—his father owned a building company. He received his Degree of Engineer from the University of Buenos Aires where he had been strongly involved in student politics. During the last months of General Uriburu's de facto rule and the first months of General Agustín P. Justo's accession to the presidency (1931–32), Durelli was the elected leader of the engineering students union, whose course he steered, he says, under a "dictatorial government and university authorities servile to that government."[66] In the police files he was identified as "a communist and a professional agitator" although it is quite clear that he was a Catholic student without party allegiances of any sort.[67] After graduation he went to Paris, where by 1936 he obtained a doctor's degree in engineering from the University of Paris and a second doctor's degree in social and political sciences from the Institut Catholique.[68] (While in Europe, in 1935 he made a short visit to the U.S.S.R.).[69] His return to Buenos Aires coincided with Jacques Maritain's visit to Argentina (August–September 1936), which opened a cleft between integral Catholics and those who sided with the French philosopher's ideas of a democratic and pluralistic society inspired by Christian values. The polemic, fueled by Maritain's open defence of the Jewish people and his negative to ascribe to Franco's side in the Spanish Civil War, resulted in the publication of Durelli's articles in *Sur*, the liberal literary review directed by the woman of let-

---

65 "La iniciativa privada en los estudios superiores," *Ciencia e Investigación* 3, no. 2 (February 1947): 45–46.
66 Augusto J. Durelli, *Discursos al iniciar y dar término al período presidencial 1931–32 en el Centro de Estudiantes de Ingeniería* (Buenos Aires, 1932), 5.
67 Augusto J. Durelli, *Carta abierta al Sr. Decano de la Facultad de Ciencias Exactas, Físicas y Naturales. Sobre el concepto de Patria* (Buenos Aires, 1931), 3.
68 Both dissertations were published: Augusto J. Durelli, *Contribution à l'étude du béton traité. Essai photoelastique* (Saint-Amand [Cher]: R. Bussière, 1936); *Essai sur les "mentalités" contemporaines: bourgeoisisme, capitalisme, nationalisme, christianisme, communisme*, 2nd ed. (Paris: Librairie Montsouris, 1939).
69 Zanca, *Cristianos antifascistas*, 96–98.

ters and socialite Victoria Ocampo.[70] Following the steps of his master, between 1936 and 1938 Durelli published four articles in that review, in the first of which he protested against "the horrible 'nationalist' conception of Catholicism."[71] These articles brought upon him the accusation of publishing Catholic material in a liberal magazine, levelled by one of the contributors to *Criterio*, whose editorial line, laid down by Msgr Gustavo Franceschi, was openly on the side of Franco.[72] This period in Durelli's life was closed by the publication of a critique of Fascism and rightist nationalism.[73]

With the support of a Guggenheim fellowship, Durelli spent a year at MIT (1941–1942), upon which he obtained a position as visiting professor at Polytéchnique Montréal, where he remained until 1944. In that year, he published his last book on Christian social philosophy, much influenced by Emmanuel Mounier's *personalisme*.[74] He returned to Buenos Aires with his wife and child during General Farrell's government. Reliving his years of student activism, Durelli authored a pamphlet denouncing the brutal police repression of a students' strike in October 1945 at his former school, which had resulted in some casualties.[75] Durelli's remaining year in Argentina was signed by his political anti-Peronist activism.[76] By the beginning of 1946 he was exonerated from his position at the Laboratory for the Testing of Materials of the Municipality of Buenos Aires, on account of his opposition to the government.[77] By then he felt identified with the Catholic Pop-

---

[70] For the magazine, see John King, *Sur. A Study of the Argentine Literary Journal* (Cambridge: Cambridge University Press, 1986).

[71] Augusto J. Durelli, "La unidad entre los católicos," *Sur* 8, no. 47 (August 1938): 72–80.

[72] Héctor A. Llambías, "Límites de la libertad," *Criterio* no. 554 (13 October 1938): 171–72; Augusto J. Durelli, "Toma de posición," *Criterio* no. 556 (27 October 1938): 217–20. The editor, Gustavo Franceschi added two and a half pages of notes to Durelli's article, backing Llambías's position, here 220–23.

[73] Augusto J. Durelli, *El nacionalismo frente al cristianismo* (Buenos Aires: Losada, 1940).

[74] Augusto J. Durelli, *La libération de la liberté* (Montreal: Editions de l'Arbre, 1944). On Durelli's personalist ideas, see *Enciclopedia della persona nel XX secolo*, ed. Antonio Pavan (Naples: Edizioni Scientifiche Italiane, 2009), s. v. "Durelli, José Augusto" (article by José Zanca).

[75] Augusto J. Durelli, *Forma y sentido de la resistencia universitaria de octubre de 1945* (Buenos Aires, 1945). See Robert A. Potash, *El ejército y la política en la Argentina*, vol. 1 (Buenos Aires: Hyspamérica, 1985), 341–401.

[76] One of his books became in its time a small classic of anti-Peronist literature: Augusto J. Durelli, *La mochila del coronel* (Buenos Aires: Acción Democrática de Ingenieros, Técnicos y Agrimensores, 1946).

[77] Augusto J. Durelli, "La relación entre el estado y los empleados," *Orden Cristiano* 5, no. 105 (March 1946): 532–34.

ular Party, the Argentine version of Luigi Sturzo's Italian movement.[78] His adieu to Houssay, with whom he had kept strained relations, was of a formal kind.[79] Leaving his country for good, he obtained a position at the Illinois Institute of Technology supported by the Armour Foundation. He then moved to the Catholic University of America where he developed an outstanding career in the area of stress and strain analysis, a field in which he became a leading world expert.[80]

Around the mid-forties, Durelli published a number of articles in Argentina, which he collected in two books, on technological and scientific research and on the nature and goals of universities.[81] Against was is the case in "Hegelian" states, he envisages an ideal situation in which the state favours research but room is left for the action of individuals.[82] Durelli seconds Braun's ideas for the creation of very small institutes with highly qualified scientists financed by private philanthropy.[83] His proposal is to send 15 to 20 graduates to the United States to be trained so that they could come back and establish a research institute in technology.[84] Durelli denounces Perón's universities, in which the teachers are employees of the state bureaucracy but he also argues against the traditional state monopoly of higher education in Argentina. He propounds a plan for the creation of private universities, one for each "spiritual family."[85] Granting of professional degrees should be carried out by each professional association and the state should award a subsidy to each student, who could decide where to enroll.[86] When compared with Lewis and Braun's proposal, Durelli's project, couched in the language of French personalist philosophy, strikes as more polit-

---

[78] Augusto J. Durelli, "Democracia y cristianismo," *Orden Cristiano* 5, no. 103 (January 1946): 352–62. See Diego Mauro, "*I popolari* en la Argentina. Luigi Sturzo y el antifascismo católico de entreguerras," *Anuario IEHS* no. 29–30 (2011): 267–88.
[79] Augusto Durelli to Bernardo Houssay, Chicago, 10 July 1946; Houssay to Durelli, Buenos Aires, 10 August 1946, box 4 08–6, ABH.
[80] Vincent J. Parks, "Professor August J. Durelli: A Short Biography," in *Progress in Experimental Mechanics. Durelli Anniversary Volume. Contributions in Honor of the Sixty-Fifth Birthday of Professor A. J. Durelli*, ed. V. J. Parks (Washington, D.C.: The Catholic University of America, 1975), ix–xi; see list of papers and books, xiii–xxvii; César A. Sciammarella, "Augusto J. Durelli," *Experimental Techniques* 25, no. 4 (July–August 2001), https://sem.org/Files/about/sem_history25_4.pdf; Patricio Laura, "Augusto J. Durelli," *Anales de la Academia Nacional de Ciencias Exactas, Físicas y Naturales* 39 (1987): 247–55.
[81] Augusto J. Durelli, *La investigación científico-técnica* (Buenos Aires: El Ateneo, 1945); *Del universo de la universidad al universo del hombre* (Buenos Aires: Palumbo, 1947).
[82] Durelli, *La investigación*, 30–35.
[83] Ibid., 36–39.
[84] Ibid., 87–95.
[85] Durelli, *Universo del hombre*, 50.
[86] Ibid., 99–102.

ically embittered and less in touch with local circumstances—perhaps an expression of the author's ambivalent sense of belonging.

The Catholic Institute of Sciences

Braun's Catholic Institute of Sciences (from now on ICC after its name in Spanish) was inaugurated on 9 June 1953 in the presence of the Archbishop of Buenos Aires Msgr Santiago L. Copello.[87] It was a very modest affair: a few classrooms in the facilities of the Catholic Institute of Culture of Buenos Aires (ICCBA), then in charge of Father Luis Maria Etcheverry Boneo.[88] The ICC was conceived as one of the schools of the ICCBA; there were also a Professorship of Religion and Morals, a School of Social Sciences, a School of Philosophy, and a Higher School of Organ Music. The whole constituted a blueprint of a future Catholic University. Braun was accompanied in the board of directors by Venancio Deulofeu (who had been dismissed in 1952 from his position as professor of organic chemistry in the Faculty of Exact and Natural Sciences of the UBA) and Emiliano Mac Donagh.[89] Most of the members and students in the ICC were opponents to Perón, to the point that the Institute was surrounded by a "subversive climate."[90] Braun was persuaded that the Institute should count with the best the country could offer in terms of science and disregarding political allegiances had invited to lecture the party-liner director of the National Institute of Endocrinology, Dr Rodolfo Pasqualini.[91] The objectives of the ICC were the establishment of laboratories, the organisation of lecture series by recognised experts in each field, and the teaching of courses for students and young graduates. This structure of labs, research seminaries, and general courses was consistent with Braun's project of founding a research university.[92] It was hoped that the institute would be finan-

---

[87] "Instituto Católico de Ciencias," *Ciencia e Investigación* 9, no. 6 (June 1953): 278; see also *Criterio* no. 1190 (June 1953): 486.

[88] That year, the ICCBA under the direction of Father Luis Etcheverry Boneo had replaced the Courses of Catholic Culture. Etcheverry Boneo would play a role in the creation of the Catholic University of Argentina. Cayetano Bruno, *Vida y acción del Padre Luis María Etcheverry Boneo* (Buenos Aires: Asociación Cristo Sacerdote, 1996), 43–47, 65–69.

[89] Rosa M. de Lederkremer and Eduardo G. Gros, "Venancio Deulofeu, 1902–1984," *Advances in Carbohydrate Chemistry and Biochemistry* 46 (1988): 11–15.

[90] Marcelino Cereijido, *La nuca de Houssay* (Buenos Aires: FCE, 1990), 41.

[91] Other member of the ICC held less conciliatory positions. See Guillermo Jaim Etcheverry, ed., *Retratos. Eduardo Braun Menéndez (1903–1959)* (Buenos Aires, 1989), 38–39.

[92] Guillermo Jaim Etcheverry, "La concepción universitaria de Eduardo Braun Menéndez," *Medicina* 60 (2000): 149–54.

cially supported by course fees and donations, either institutional or from individuals.

Braun believed that it was absurd to affirm "a Catholic science as opposed to non-confessional science… Science cultivated in the Institute should be like that which is cultivated in any authentic scientific institute." He explains that the institute is called "Catholic" for it is supported by the hierarchy and at least initially, financed by Catholics. Braun claimed that "any deviation from the objective search for scientific truth" would lead to "the denial of the scientific spirit, to the negation of the meaning of the term 'catholic' which means universal, and to the negation of the spirit of charity, which is the very essence of Christianity."[93] Many of those who taught in the ICC were non-Catholics who had been forced to leave their positions in the UBA due to political reasons.[94] During its ephemeral existence, the ICC offered a large number of courses in all the fields of science and even some of technology; the teachers were certainly well qualified and distinguished in their fields. There was a concentration on biomedicine, biochemistry, and physiology, but mathematics, physics, engineering, inorganic and organic chemistry, agronomy, biology, and zoology, were also covered.[95] The last courses were announced in September 1954; in 1955, with the worsening of the relationships between Church and state, the ICC did not offered courses.

The First Catholic Universities

Very soon after the fall of Perón, Braun was reinstated to his position as associate professor of physiology.[96] By then, he authored a series of editorial notes criticising the massive number of students in the universities, particularly in the School of Medicine of the UBA.[97] In early December 1955, Braun participated in a meeting in the city of Tucumán convened to discuss the future of universities in post-Peronist Argentina. He defended that "free" universities could grant di-

---

[93] "El Instituto Católico de Ciencias," *Ciencia e Investigación* 9, no. 9 (August 1953): 337–38.
[94] Caimari, *Perón y la Iglesia* Católica, 298n. 21.
[95] Miguel de Asúa and Analía Busala, "Instituto Católico de Ciencias (1953–1954). Más en la leyenda que en la historia," *Criterio* no. 2368 (March 2011): 40–44.
[96] Nerio Rojas to Eduardo Braun Menéndez, 24 October 1955 and Braun Menéndez to Nerio Rojas, 26 October 1955, file 18173, AFM.
[97] Eduardo Braun Menéndez, "¿Se cumplen los fines de la universidad?," *Ciencia e Investigación* 11, no. 9 (September 1955): 385–87; idem, "Una facultad monstruosa," *Ciencia e Investigación* 11, no. 12 (December 1955): 529–31. The deposed government had introduced free admittance to higher education, which had increased enormously the number of university students.

plomas but should neither habilitate (license) the exercise of the liberal professions nor receive funds from the state.[98] On 23 December of that year, president General Pedro E. Aramburu and his minister of education, the Catholic intellectual Atilio Dell'Oro Maini, signed a decree that granted autonomy to the state universities, so far under the control of the government; the resolution included a clause (the much discussed "article 28") that allowed the creation of private universities with the power of granting degrees habilitating the practicing of professions such as medicine, law, and engineering.[99] The initiative was the beginning of a long battle over the issue between those who defended the state's exclusive control of higher education and the principles of secularism (the student movement, the various strands of the left, many liberal politicians) and Catholics, who aspired to wrestle from the government the right to establishing confessional schools.[100] Braun sought a middle ground, along the lines he had first expounded a decade ago: "I am an advocate of the creation of private universities but I am talking of universities, not of professional schools."[101]

In February 1956, the government named an advisory commission to regulate article 28. The commission was headed by Houssay and included Braun Menéndez and Lewis among a small number of prominent scientists.[102] Lewis was established in Rosario, but he was eager to participate in the commission; Houssay obliged keeping him abreast of the events.[103] For all practical purposes, the commission remained dormant, for Houssay was not willing to sanction the creation of universities which could grant professional degrees. Braun took every opportunity to defend his vision of confessional research universities. He set himself against those projects who planned to build up from a cluster of profes-

---

98 Comisión Organizadora, *Jornadas Universitarias de Tucumán* (San Miguel de Tucumán: Universidad Nacional de Tucumán, 1957), 94, 100.
99 "Organización de las universidades nacionales. Decreto Ley 6403," *Boletín Oficial*, 3 January 1956, 1–2. For "article 28," see Virginia W. Leonard, *Politicians, Pupils, and Priests. Argentine Education since 1943* (New York: Peter Lang, 1989), 177–86.
100 Leonard, *Politicians, Pupils, and Priests*, 189–218; Horacio Sanguinetti, "Laica o libre. Los alborotos estudiantiles de 1958," *Todo es Historia* no. 80 (January 1974): 8–23; Zanca, *Los intelectuales católicos*, 85–135. For the role of the Communist Party in street agitation, see Isidoro Gilbert, *La Fede. Alistándose para la revolución. La Federación Juvenil Comunista 1921–2005* (Buenos Aires: Sudamericana, 2009), 368–70.
101 Eduardo Braun Menéndez, "Los títulos habilitantes y la universidad," *Ciencia e Investigación* 12, no. 1 (January 1956): 1–4.
102 Ministerio de Educación y Justicia, *Antecedentes sobre reglamentación legal de las universidades privadas* (Buenos Aires, 1958), 5–6; Leonard, *Politicians, Pupils, and Priests*, 202–204.
103 Juan Lewis to Bernardo Houssay, 5 March 1956; Houssay to Lewis, 23 May 1956 and 26 May 1956, box 4 08–6, ABH.

sional schools and revitalised his old idea of setting up advance centres of research in which "the teachers and intellectual creation should occupy the first place."[104] A year later, he expounded again his compromise solution: neither the state monopoly of higher education nor the creation of massive professional schools that would pass as universities. At this time, he suggested the creation of technical state boards for the granting of professional titles, with representatives from professional associations, the ministry of education, and the scientific academies.[105]

In September 1956, Lewis resigned from his position as associate professor of physiology in Buenos Aires.[106] One month later Braun was designed full professor of physiology.[107] In June 1957 he lectured at the University of Michigan in a symposium on renal hypertension.[108] By then, Durelli paid a visit to Buenos Aires to teach a course on experimental elasticity in the School of Engineering of UBA.[109] On 18 November of that year Houssay renounced to the chair of the advisory commission, for he felt his attributions were constrained but on 1st April 1958 the minister confirmed him in his post.[110] Two weeks later, the commission finally issued its pronouncement, which recommended that graduates of private universities in the professional careers should take an exam administered by the state.[111] The Catholic journal *Criterio* criticised this view as "Pharisaic," for the state lacked the competence to judge on university issues; the note claimed that the licensing should be in the hands of each professional association.[112]

Arturo Frondizi took charge of the presidency of the country on 1 May 1958. For political reasons, he was determined to allow the creation of private, confessional universities and he succeeded, but the road was harsh.[113] In September of that year, the streets of Buenos Aires were the scene of demonstrations, riots, and at times violent confrontations between the supporters of private (i.e. Catholic) universities and those who wished to continue with the usual way of things

---

104 Eduardo Braun Menéndez, "Las etapas para la creación de una universidad privada," *Ciencia e Investigación* 13, no. 3 (March 1957): 97–99.
105 Eduardo Braun Menéndez, "La ley universitaria," *Ciencia e Investigación* 14, no. 7 (July 1958): 289–90.
106 Juan Lewis to Nerio Rojas, 15 September 1956, file 8044, AFM.
107 Eduardo Braun Menéndez to Nerio Rojas, 18 October 1956, file 18173, AFM.
108 John M. Sheldon to Braun Menéndez, 26 April 1957, file 18173, AFM.
109 "Ingeniero Augusto J. Durelli. Su visita a nuestro país," *Ciencia y Técnica* 124, no. 620 (July 1957): 234.
110 Ministerio de Educación y Justicia, *Antecedentes*, 15–18.
111 Ibid., 25–26.
112 "En torno al artículo 28," *Criterio* no. 1307 (8 May 1958): 332.
113 Leonard, *Politicians, Pupils, and Priests*, 202–18.

(i.e. the state holding the monopoly of university education). In a massive demonstration on 19 September, José L. Romero, former president of UBA in the months that followed the fall of Perón and one of the leaders of the cause against the creation of confessional universities, gave a speech. In three passages he referred with indignation to "clerical" forces allied to the "oligarchy," to "anti-Argentine" interests, and to "reactionary" and "anti-popular" sectors of society.[114] Risieri Frondizi, by then rector of the University of Buenos Aires and brother of the president of the nation, talked of "dark forces," of private universities "ascribed to certain sects," and of "the empire of dogmatism."[115] It is telling that in his speech he appropriated Braun's rhetoric and arguments. But it is quite clear that Braun's position was not that of Risieri Frondizi's. In an article published in a liberal newspaper, Braun decried the aspiration of many Catholic groups "to obtain the privilege of licensing professional practice" but he also complained about "the anti-Catholic secularist campaign" and accused the authorities of the University of Buenos Aires of participating in public demonstrations "which involved attacks on freedom of teaching and on religion."[116] The day after this note was published, Braun departed for a tour that took him to Rome, Brussels, and London.[117] Parliament discussed the bill and in the teeth of bitter opposition passed the corresponding legislation in the last week of September, which involved a modification of article 28 broadly along the lines of Braun and Houssay's ideas.[118] An editorial article in *Criterio* by its director Father (later Cardinal) Jorge M. Mejía exposed the deepest layers of the conflict. The state monopoly of education, he claimed, turned it into a "church" which imposed its own dogmas upon the citizens. Those who were claiming for secular education did not mean a neutral attitude of the state with respect to religion; they meant "a contempt of the religious phenomenon as such and its overcom-

---

114 "El acto de la F.U.A. en Plaza del Congreso," *La Nación*, 20 September 1958.
115 Risieri Frondizi, "La Enseñanza libre," in Federación Universitaria de Buenos Aires, *La Reforma Universitaria, 1918–1958* (Buenos Aires: FUBA 1959), 257–65.
116 "Del Dr. Braun Menéndez," *La Nación*, 7 September 1958.
117 Eduardo Braun Menéndez to Florencio Escardó, 20 October 1958, file 18173, AFM.
118 Congreso Nacional, Cámara de Diputados, *Diario de sesiones… Año 1958*, vol. 7 (Buenos Aires: Imprenta del Congreso, 1959), 5128–39, 5178–79, 5342–45; cf. Horacio C. Domingorena, *Artículo 28. Universidades privadas en la Argentina. Sus antecedentes* (Buenos Aires: Editorial Americana, 1959), 75–83. "Universidades. Ley 14.557," *Boletín Oficial*, 24 October 1958, 1. The law was approved by Congress on 30 September and promulgated by the president on 17 October 1958.

ing by the power of 'science'."[119] The creation of Catholic universities—which lies beyond the scope of this inquiry—should be seen in the context of an increase vitality of the Catholic Church in the late 1950s: "the Catholic movement of the 1930s–1940s had by the 1960s succeeded in reversing the laicist trend."[120] Twenty-five new dioceses were created between 1957 and 1961 and Perón's 1954 secularising push, by which he "dismantled the plausibility structures that Catholicism had laboriously reconstructed over the past 30 years," was controlled.[121]

On 12 January 1959, Braun was received by the nuncio Msgr Umberto Mozzoni who wished to hear his opinions about the creation of the Catholic university in Buenos Aires, which was already in operation. An account of that interview has survived.[122] Msgr Mozzoni recognised that the school had begun on the wrong foot and asked Braun whether he saw any possibility of correcting its course. The intention of the Vatican diplomat was to ask Braun to organise an Institute for Basic Scientific Research and it would seem he understood and even approved of the latter's idea of university. Braun abandoned the residence of the nuncio in a hopeful mood, thinking that it was "possible to obtain the 'Great Change' necessary" for things to improve. His projects came to nil: a few days later, on 16 January, Eduardo Braun Menéndez died in a plane crash.

At that time, there had been a more or less open confrontation between Dell'Oro Maini, the former minister responsible for the fateful wording of article 28, and the commission presided by Houssay, who felt that a clean new legislation was needed.[123] But it is quite clear that Dell'Oro was not averse to implementing a system controlling the licensing of professional titles by the state.[124] He as well as Braun Menéndez wished that the Archdiocesan Catholic University could begin as a graduate school with a focus on research.[125] Things turned out other-

---

119 Jorge M. Mejía, "Educación y libertad," *Criterio* no. 1316 (25 September 1958): 683–87. Mejía's take on this issue is better understood in the light of the role of scientific discourse in the secularisation of education in late nineteenth-century.
120 Ivereigh, *Catholicism and Politics in Argentina*, 200.
121 Morello, "'It Takes Two to Tango'," 65, 69; Zuretti, *Nueva historia eclesiástica*, 424–31.
122 Eduardo Braun Menéndez to Rafael Braun, 16 January 1959. A transcription of Eduardo Braun's letter to his son was generously given to me by the late Father Rafael Braun in July 2011.
123 See Dell'Oro's note of 23 April 1958 in Ministerio de Educación y Justicia, *Antecedentes*, 27–31.
124 See his radial speech on 2 March 1956 in Ministerio de Educación y Justicia, *La Revolución Libertadora y la Universidad, 1955–1957* (Buenos Aires: Ministerio de Educación y Justicia, 1957), 166–73.
125 Emilio F. Mignone, *Política y universidad. El estado legislador* (Buenos Aires: Lugar Editorial, 1998), 43–44.

wise. The Catholic University of Argentina "Santa María de los Buenos Aires" was formally established by the episcopate on 7 March 1958 (henceforward UCA, after its name in Spanish). Its Higher Council (*Consejo Superior*) had begun its meetings in that same month. Braun Menéndez had been appointed to it but he refused the nomination.[126] In the first preliminary meeting, Dell'Oro pointed out that "the episcopal documents are quite clear that we should do science and that should be the starting point of the schools;" implicitly referring to Braun, he also manifested the importance of creating institutes for "there is a movement of persons devoted to the sciences who have volunteered their collaboration."[127] On 11 April, the Higher Council decided to create two institutes, one of Mathematics, Physics, and Engineering (more on this below) and another of Biology; the direction of the latter would be offered to Braun Menéndez by Msgr Derisi, the rector of the university.[128] There is not further reference to Braun in the proceedings.

Lewis's house, situated in the neighborhood of Fisherton, Rosario, had been the cradle of the Christian Democratic Party, which took form in meetings held from 8 July through 10 in 1954, 14 months before the fall of Perón. Lewis was member of the National Committee of the party for the period 1956–1957 and the delegate from the province of Santa Fe to the 1957 Constitutional Convention convened by the revolutionary government which had deposed Perón.[129] He was involved in politics to the point that by July 1959 he lamented he was not able to finish on time the translation of Houssay's speech for the International Congress of Physiology to be held in Buenos Aires that year, for he had been campaigning and "visiting those little towns, to which you [Houssay] at one time referred contemptuously."[130] Lewis died in 1976. In the homage the Senate of the Nation paid him, one of the orators quoted him as affirming the "necessity of political commitment of those devoted to science."[131]

---

126 Octavio M. Derisi, *La Universidad Católica Argentina en el recuerdo: a los 25 años de su fundación* (Buenos Aires: UCA, 1983), 34.
127 *Libro de Actas del Consejo Superior de UCA*, fol. 3 and 4 (meeting of 14 March 1958). Archive Universidad Católica Argentina, Buenos Aires (henceforward AUCA).
128 *Libro de Actas*, fol. 26 (meeting of 11 April 1958), AUCA.
129 Parera, *Democracia Cristiana en la Argentina*, 62–63, 82–83; Ghirardi, *La Democracia Cristiana*, 75–77, 85.
130 Juan Lewis to Bernardo Houssay, 15 July 1959, box 4 08–6, ABH.
131 José A. Allende, "Homenaje a Juan T. Lewis," Congreso Nacional, Cámara de Senadores. *Diario de sesiones… Año 1975*, vol. 6 (Buenos Aires: Imprenta del Congreso de la Nación, 1976), 3878–79.

The Other Side of the Coin

Not all Catholic scientists were opposed to Perón, as shown by the case of the Jesuit astronomer Juan A. Bussolini, who succeeded Father Puig as director of the Observatory of San Miguel. Bussolini entered the Society of Jesus in Argentina, took his philosophy courses in the Seminary in Buenos Aires, taught in a secondary school in Santiago de Chile for three years, and was then sent to Valkenburg (a large Jesuit house with an important observatory) to study theology. After being ordained in August 1933, he visited observatories in Germany, Belgium, and the United Kingdom. Because of the war, Bussolini was not able to study astronomy in Berlin as he had planned. When back in his country, he attended courses at the School of Astronomical Sciences of the Observatory of La Plata, which had been created by Félix Aguilar, a close associate of Msgr Devoto.[132] Bussolini was a congenial teacher with a talent for institution building—an asteroid has been named in his honour.[133] In 1948, he contributed to the erection of an astronomical observatory in the province of San Juan.[134] He was also the force behind the growth of the Jesuit observatory in San Miguel, which he directed between 1944 and his death in 1966. In the 1950s, he organised the department of Solar Physics, which initiated the systematic observation of solar sunspots and the patrolling of solar activity. Solar Physics would become the observatory's main claim to fame while the institution remained in Jesuits hands (it was transferred to the Air Force in the mid 1970s). In the International Years of the Quiet Sun (1964–1965) new equipment was incorporated (a Lyot filter attached to a Zeiss refractor and a radiotelescope). Bussolini planned the construction of a new building with a vacuum spectroheliograph (built by the American Optical Co. and modelled after the one installed in the Jesuit Observatory in Manila by the Jesuit Richard Miller, who was briefly engaged in San Miguel); the tower with the instrument was installed soon after Bussolini's death in 1966.[135]

---

[132] Hugo Storni, "Bussolini, Juan Antonio," in *Diccionario Histórico de la Compañía de Jesús*, 1:579; "Reverendo padre Juan Antonio Bussolini. Su fallecimiento," *La Nación*, 11 September 1966; Otto Schneider, *Geofísica y geodesia*, vol. 8, *Evolución de las ciencias en la República Argentina, 1923–1972* (Buenos Aires: Sociedad Científica Argentina, 1980), 272–73.
[133] Schmadel, *Dictionary of Minor Planets*, 2:193.
[134] José A. López, "Observatorio Astronómico Félix Aguilar," in *Astronomía*, ed. Simón Gershanik and Luis A. Milone, vol. 7, *Evolución de las ciencias en la República Argentina, 1923–1972* (Buenos Aires: Sociedad Científica Argentina, 1979), 204.
[135] Schneider, *Geofísica y geodesia*, 85–88; Udías, *Searching the Heavens*, 142–45; Edmund Benedetti, "Radioastronomía solar y actividades del Observatorio de San Miguel," *Boletín de la Asociación Argentina de Astronomía* 13 (1968): 6–9; M. A. Greco, H. Grassi Gallegos and J. R. Seibold, "Solar Physics at the Observatorio Nacional de Física Cósmica, San Miguel, Argen-

By the end of the 1960s a large group of Jesuit scientists and enginners worked there: Carlos Esponda (geophysics), Arturo Yriberry (geoelectricity), Ricardo Cocito (theoretical physics), Nilo Arriaga (astronomy), Edmund Benedetti (radioheliography), Richard Miller, Jorge Seibold, and Thomas Paneth (solar optics).[136] The Observatory eventually grew into a large Jesuit scientific research facility, with buildings for Father Williner's entomology institute and laboratories of biology and human physiology.

Three years after Perón was elected president of the country, Bussolini was nominated member of the National Board of Intellectuals (1948) together with Catholic nationalists who had held public charges in the regime that followed the June 1943 military coup.[137] He endorsed the scientific policy of the government, geared to foster applied science and technology with a view to creating a national industry and developing military projects in the areas of aeronautics and atomic physics.[138] In a 1949 interview, Bussolini declared that there was at that time in Argentina "a lively interest for everything related to nuclear technology," while adverting that the destructive use of atomic energy was "a crime against humanity."[139] The next year, an official publication destined to popularise atomic energy and science published a short note by Bussolini in which he attempted to harmonise the scientific corpuscular view of matter with scholastic natural philosophy.[140] Later, he declared that since 1948 he had "encouraged any

---

tina," *Solar Physics* 14 (1970): 503–7; Richard A. Miller, "New Spectroheliograph at Manila Observatory," *Applied Optics* 4, no. 9 (1965): 1085–89; Jorge Seibold and Tomás Paneth, "El espectroheliógrafo del Observatorio de Física Cósmica de San Miguel," *Boletín de la Asociación Argentina de Astronomía* 15 (1970): 37–38; Tomás Paneth, "Ajuste del celóstato del espectroheliógrafo de San Miguel," *Boletín de la Asociación Argentina de Astronomía* 15 (1970): 38.
136 *Observatorio Nacional de Física Cósmica* (San Miguel, 1970) (pamphlet).
137 Ministerio de Educación, Secretaría de Cultura, *Junta Nacional de Intelectuales. Antecedentes de su creación, decretos y reglamento interno* (Buenos Aires, 1949).
138 Diego Hurtado de Mendoza and Analía Busala, "De la 'movilización industrial' a la 'Argentina científica': la organización de la ciencia durante el peronismo (1943–1955)," *Revista da Sociedade Brasileira de História da Ciência* 4, no. 1 (2006): 17–33; Jonathan Hagood, "Arming and Industrializing Perón's 'New Argentina': The Transfer of German Scientists and Technology after World War II," *Icon* 11 (2005): 63–78; Diego Hurtado de Mendoza, "Autonomy, even Regional Hegemony: Argentina and the 'Hard Way' toward Its First Research Reactor (1945–1958)," *Science in Context* 18, no. 2 (2005): 285–308.
139 "En busca del rayo supersónico," *Argentina. Revista mensual* 1, no. 3 (1949): 77–81.
140 Juan A. Bussolini, "Filosofía y física de los corpúsculos," *Mundo atómico* 1, no. 1 (September–October 1950): 78. Other Jesuits from San Miguel also contributed to this magazine, see for example, Arturo J. Yriberry, "Cámaras de niebla en la investigación atómica," *Mundo atómico* 2, no. 5 (1951): 11–13, 83; Nilo Arriaga, "El sol y sus relaciones con los fenómenos terrestres," *Mundo atómico* 4, no. 13 (1953): 25–28.

project that could imply Argentina's participation in the international trade of atomic energy."¹⁴¹ Bussolini's allegiance to Perón was confirmed when the latter decided to back the project of the Austrian physicist Ronald Richter to obtain nuclear energy from controlled nuclear fusion under laboratory conditions. When after setting up a secret laboratory in a lake island in Bariloche (western Patagonia) scientists aligned with the government began to suspect that Richter's project was a fiasco, a long process of inquiry began: investigative commissions of experts were named and Bussolini formed part of them.¹⁴² While prestigious physicists and engineers cogently showed that Richter's experiments were an imposture, Bussolini persisted in defending them as successful, qualified his colleagues of "unworthy of scientifically [sic] tying Richter's shoelaces," and tried to the last to convince Perón of supporting the Austrian physicist, "a scientist in the full sense of the word."¹⁴³ Bussolini's blunder did not seriously hampered his career (in fact, it was during the post-Peronist years when the San Miguel Observatory grew the most).

In an article published in 1963, Bussolini distinguishes between scientific theories, which "describe" the phenomena and attend to "proximate causes," and philosophy, which "explains" things and searches for "ultimate causes." Revelation—which "presides and precedes human reason" and is the object of theology—is above philosophy. Bussolini calls upon Augustine's illumination theory to affirm that science and philosophy, while retaining their epistemological independence, are "subsidiary" of theology. It is significant that these neoscholastic commentaries conclude with a nationalistic appeal by criticising the "expert scientists...brought from abroad to direct us with total ignorance of the national soul."¹⁴⁴ In a 1956 article critical of the then recently founded National

---

141 Vicepresidencia de la Nación. Comisión Nacional de Investigaciones, *Documentación, autores y cómplices de las irregularidades cometidas durante la segunda tiranía*, vol. 1 (Buenos Aires, 1958), 34.
142 Mario Mariscotti, *El secreto atómico de Huemul. Crónica del origen de la energía atómica en la Argentina* (Buenos Aires: Sudamericana, 1987), 203–238; Jacques E. C. Hymans, *Achieving Nuclear Ambitions. Scientists, Politicians, and Proliferation* (Cambridge: Cambridge University Press, 2012), 203–238.
143 Vicepresidencia, *Documentación*, 635–84. This is the report of the investigative commission set up by the regime that deposed Perón to investigate the alleged corruption and mismanagement of his government. Although for different motives, all the parties involved (the anti-Peronist members of the commission and the witnesses, including Bussolini) were interested in providing factual reports of the events.
144 Juan A. Bussolini, "Las ciencias del espíritu, de la materia y de la vida," *Estudios* no. 543 (May 1963): 201–10. For nationalism, Peronism, and Catholicism, see, for example, Cristián Buchrucker, *Nacionalismo y peronismo. La Argentina en la crisis ideológica mundial (1927–1955)*

Research Council of Science and Technology (CONICET) by Houssay and the authorities that forcibly deposed Perón, Bussolini recalls the latter's creation of what in his view was a similar governmental agency for promoting science. His defence of the "social-humanist solution" as over against "liberalism, individualism, and capitalism" on the one hand and "atheism, collectivism, and communism" on the other is entirely in line with Perón's discourse.[145]

Bussolini's view about the relationships between science, philosophy, and theology do not necessarily imply (and are not necessarily implied by) his concrete political options. But in his hierarchical conception of the articulation between religion and science it is possible to hear distant echoes of the kind of integral Catholicism that vast sectors of the Church in Argentina favoured since the thirties—some of which would eventually turn into supporters of Peronism.

Science and Religion

As was perhaps the case with Bussolini, the political choices of the other Catholic scientists discussed in this chapter were in tune with their views about the relationships between science and religion. Durelli affirmed that "science and technology are good in themselves" and cultivating them is just fulfilling the divine mandate of subduing the Earth. He criticised the pseudo-spiritual attitude of Catholic nationalist intellectuals who while extolling "Hispanic spiritualism" and disparaging "Anglo-Saxon materialism" set themselves against the cultivation of science and technology.[146]

Lewis claimed that the object of scientific research was the determination of "the relationships between the phenomena" and the elucidation of "how they are produced." This implied that "speculations on the first cause [of natural phenomena]" and their "essential or transcendent signification" did not belong to science.[147] He allowed that there was a sphere of knowledge different from science, for he admitted that "the basis upon which a statement can be made,

---

(Buenos Aires: Sudamericana, 1987); Richard J. Walter, "The Right and the Peronists," in *The Argentine Right. Its History and Intellectual Origins, 1910 to the Present*, ed. Sandra McGee Deutsch and Ronald H. Dolkart (Wilmington, DE: SR Books, 1993), 99–118; Zanatta, *Perón y el mito de la nación católica*.

145 Juan A. Bussolini, "A propósito de un Consejo Nacional de Investigaciones Científicas y Técnicas," *Estudios* no. 477 (August–September 1956): 27–31.
146 Durelli, *La investigación*, 16–18.
147 Lewis, *Para el progreso de las ciencias*, 3.

the proof of its truth can be physical or metaphysical certainty."[148] In an article on the magisterium of the Church, which testifies to his familiarity with the Bible and the Pontifical documents, he urged that Catholic intellectuals should know their theology.[149]

For Braun, "science is a method and a faith. It is a method that employs observations and experiments objectively registered with absolute honesty. It is also a faith...that truth exists and can be discovered and deserves to be discovered."[150] Science and faith, he affirms, belong to different domains of human knowledge. Rehearsing the traditional apologetic argument in Duilhé de Saint-Projet's *Apologie scientifique de la foi chrétienne* (1885), Braun distinguishes three realms of knowledge: science, metaphysics, and theology.[151] Taking his cue from Louis Pasteur, he claimed that "science is positivist and objective, rational and experimental."[152] He attributes a large part of the alleged conflict between science and theology to taking as definitive truth in the scientific field what is no more than a provisory hypothesis. In a paragraph that conflates his politico-ecclesiastical and scientific positions, Braun admits that some members of the hierarchy had been "obscurantist and intolerant" and concedes that the Church has had "evil and heretic priests, just as it shelters today Nazis and anti-Semites," but on the whole it has "cultivated and fostered science."[153]

Conclusions

Braun Menéndez became a high-profile personality in the post-Peronist polemics surrounding the creation of private confessional universities before his tragic demise in 1959. By that time, Durelli was Professor in the Department of Civil Engineering at the Illinois Institute of Technology; in 1961 he would become Professor of Civil Engineering and Head of Stress Analysis Laboratories in the Catholic University of America.[154] Lewis, the senior of them all, lived in Rosario and was

---

148 Lewis, "Objeto y fin de la universidad," 47.
149 Juan T. Lewis, "El magisterio de la Iglesia," *Criterio* no. 1230 (24 February 1955): 127–29.
150 Rafael Braun, "Compilación de las ideas de Eduardo Braun Menéndez sobre ciencia y universidad" (unpublished manuscript, 1961), Microsoft Word file. Father Rafael Braun graciously gave me a copy of this document.
151 Eduardo Braun Menéndez, *Bases para el progreso de las ciencias en la Argentina* (Buenos Aires: Ateneo del Club Universitario de Buenos Aires, 1946), 15–17.
152 Ibid., 17.
153 Ibid., 18.
154 Parks, "Professor August J. Durelli," ix.

much engaged in the organisation of the Christian Democratic Party. Despite the striking resemblance of their views and their belonging to the same social, academic, and political circles, these figures never attempted at the creation of an advocacy group; they were individualist personalities who shared a Catholic identity, a scientific vocation, and an anti-Peronist position associated to a Christian liberal-democratic view of society. Did their ideas impinge on the creation of the Catholic universities allowed by the 1958 legislation?

Engineering was the only area concerned with science and technology in the first years of the Catholic University of Argentina. The distinguished mathematician Agustín Durañona y Vedia, recently retired from the University of Buenos Aires and member of the Higher Council of UCA, accepted to head the Institute of Mathematics, Physics, and Engineering. This was the core of the School of Engineering which began teaching courses in 1960; the plan foresaw sections of physics and mathematics, applied mechanics, electricity, and engineering. The new school soon incorporated some bright young faculty, among them engineers Hilario Fernández Long (future president of the University of Buenos Aires, 1965–1966) and Horacio C. Reggini.[155] Trained in Columbia University's Watson Scientific Computer Laboratory (1959), the latter organized UCA's Computing Department, which in 1962 acquired an IBM 1620, the first large transistor computer in the country—as a result of which the university stood briefly at the head of computer research in Argentina.[156] Durelli was not connected with these developments, which can be seen as part of the wave of modernisation that swept over the Argentine system of science and technology between 1956 and 1966.[157] In 1958, Emiliano Mac Donagh, also a member of UCA's Higher Council, volunteered to organise an Institute of Natural Sciences as a substitute of the failed Institute of Biology (see above); the undertaking did not prosper for he died shortly after (1961).[158]

Briefly, *there were* institutional scientific and technological initiatives in UCA, but the (in view of the circumstances perhaps utopic) projects of Lewis, Durelli, and Braun Menéndez of founding an institution of higher learning upon the basis of a number of advanced research centres did not determine the design of

---

155 The Institute of Mathematics, Physics and Engineering absorbed the Institute of Applied Mechanics, of the Corporation of Catholic Engineers. *Libro de Actas*, fols. 26, 46, and 68–69 (meetings of 11 April, 2 May, and 23 May 1958), AUCA; Luis A. Santaló, "Agustín Durañona y Vedia," *Anales de la Academia Nacional de Ciencias Exactas, Físicas y Naturales* 33 (1981): 315–17.
156 Nicolás Babini, *La informática en la Argentina* (Buenos Aires: Letra Buena, 1991), 96–99.
157 Asúa, *Una gloria silenciosa*, 243–51.
158 *Libro de Actas*, fol. 33 (meeting of April 18, 1958), AUCA.

the newly created Catholic universities, which for reasons of expediency chose to set up soon a small number of professional careers.[159] The issue of scientific research versus teaching was hotly debated in the first crucial meetings of UCA's Higher Council. As discussed above, ideas akin to those of Lewis, Durelli, and Braun were forcefully defended by Dell'Oro Maini and Durañona y Vedia, against the opposition of Faustino J. Legón (the first dean of the Law School) and those who urged the immediate opening of the courses over the plan of first consolidating institutes of scientific or scholarly research. In his memories, Msgr Derisi recalled that "There were those who wished to make of the university an advanced institute of pure research, but the majority of the council saw, in accord with the declaration and intentions of the bishops, that the university should be first a teaching institution without necessarily abandoning research."[160] Derisi's "teaching first" position triumphed but the creation of the Institute of Engineering (which in any case also became a professional school) suggests that the "research first" position did not go entirely unheeded.[161]

Paradoxically, the message of Lewis, Durelli, and Braun Menéndez might have had broader repercussions *ad extra*, outside the Catholic world, in the mil-

---

[159] The other large Catholic institution created in that period in Buenos Aires was the Jesuit Universidad del Salvador (1959). See Julio R. Lascano, *Los estudios superiores en la historia de Buenos Aires* (Buenos Aires: Municipalidad de la Ciudad de Buenos Aires, 1981), 273–90. The Catholic University of Córdoba was also created in that year (the institute upon which it was based had been founded in 1956). After 1958, there was a proliferation of Catholic universities in Argentina (by 1962, seven of them had been created) with the result of "watering down standards...they chose the path of least resistance and set up institutions that offered courses in the humanities and social sciences, programs less costly than those in physical sciences." Leonard, *Politicians, Pupils, and Priests*, 259–60. See also Juan C. Del Bello, Osvaldo Barsky and Graciela Giménez, *La Universidad privada argentina* (Buenos Aires: Zorzal, 2007), 91–92, 121–132.
[160] Derisi, *La Universidad Católica Argentina en el recuerdo*, 35. Father Guillermo Blanco (first dean of the School of Philosophy) and Canon Luis M. Etcheverry Boneo (in charge of the Institute of Culture and Extension Courses) were aligned with Derisi's position.
[161] *Libro de Actas*, fols. 4, 8–10, and 12–13 (meetings of March 14 and 19, 1958), AUCA. Besides the institutes of engineering and natural sciences, there were the Institute of Culture and Extension Courses and the Institute of Linguistic and Literature, but these were essentially geared to teaching. Other authors have already called attention to the polemics between teaching and research in the origins of UCA. Emilio Mignone, a leading and experienced Argentine Catholic authority on universities, affirmed: "Dell'Oro Maini as well as other Catholic intellectuals and scientists of those times (among them the distinguished physiologist Eduardo Braun Menéndez...) promoted the idea that the Catholic University should begin as a graduate school, concentrating on research... But the Catholic hierarchy opted for Msgr Derisi's project oriented toward a massive, conventional, and tidy undergraduate education." Mignone, *Política y universidad*, 43–44. For an institutional history of UCA, see Florencio Hubeñak, *Historia de la Universidad Católica Argentina* (Buenos Aires: Universidad Católica Argentina, 2016).

itantly secular circles that since the fall of Perón were becoming strong in the universities and institutions of scientific research supported by the state. Scientists with staunch anti-Catholic positions still respected and admired Braun Menéndez for his ideas and his moral integrity.[162]

Since the battles of the late nineteenth-century over secularisation, science in Argentina had become a symbol and an instrument of the secularist ideology of progress, with anticlerical resonances. The tide began to turn with the "Catholic Renaissance" of the 1920s and the slightly earlier coming onto the public scene of the first Catholic scientists: the naturalist Ángel Gallardo and the astronomer Father (later Msgr) Fortunato Devoto. With their outstanding scientific credentials and their articulated projects of higher education, Lewis, Braun, and Durelli further confirmed that it was possible for Catholics to carry out sound science while contributing to call into question the myth of science as a secular pursuit, which was getting a new lease of life with the post-1955 secularist drift which opposed the creation of confessional universities.[163]

---

[162] See the testimonies in Jaim Etcheverry, *Retratos*. and Cereijido, *La nuca de Houssay*.
[163] The opposition to the creation of private confessional universities alienated Catholics from liberals, socialists, and communists contributing to further dislocate in the cultural sphere the unstable political alliance that had toppled Perón.

Conclusions

In Catholic majority countries (France, some parts of German-speaking Europe, Italy, and the Iberian world), the historical relationships between science and religion are hardly intelligible if secularism is left out of the equation. A large share of modern science was born in this cultural geography. Besides their intrinsic value, the study of cases such as that of Argentina might be worth pursuing because they function as a distant mirror or a scale model of this historical pattern. Natural theology is a core organising notion in the panoramic narratives of science and religion in anglophones societies. While it is true that natural theology was present in Río de la Plata during the period of Catholic Enlightenment, the key to the interpretation of the relationships between science and religion in "paleo-Durkheimian" societies is secularisation. Our long-term study allows to discern three crucial historical episodes in this regard (it should be remarked that these three moments concern the arena of science and religion and should not be taken as a periodisation of the general process of secularisation in Argentina).[1] In the first, as a result of the ecclesiastical reform of the 1820s secularisation took the form of expropriation of church buildings by the state, some of which were recycled as science teaching and research facilities. In the second, in the 1880s, secularisation pushed forward by the conservative liberal elite meant the differentiation of the ecclesiastical sphere from the other spheres of society. The dominant ideology of progress, the introduction of the first large scientific institutions in the country, and the impulse of modernisation enthroned the image of science as a secular enterprise opposed to tradition and religion. In the third episode, broadly between the 1900s and 1920s, freethinkers and socialist groups made attempts at infusing secular science with a religious valence.

The complementary side of the vector of secularisation driving the dynamics of science and religion is the sequence of the fall and rise of Catholics in science. They were obviously the main actors in the period of Jesuit science and, in a lesser degree, of Catholic Enlightenment. In the 1820s, science became for the first time identified with secular values and this was in part due to the arrival in Ar-

---

[1] Three stages in the confrontation between Catholicism and liberalism in Latin America have been described: (1) 1880–1930 (liberal authoritarian secularism); (2) 1930–1980 (integral Catholicism); (3) from 1980 onwards (globalisation with increasing pluralisation and privatisation of the religious). Fortunato Mallimaci, "Catolicismo y liberalismo. Las etapas del enfrentamiento por la definición de la modernidad religiosa en América Latina," in *La modernidad religiosa. Europa latina y América Latina en perspectiva comparada*, ed. Jean-Pierre Bastian, trans. Dulce M. López Vega (Mexico: FCE, 2004), 19–44.

gentina of science teachers disaffected with the post-Napoleonic reactionary regimes. During the high tide of secularisation and the sharpening of the conflict between science and religion in the 1880s, there were no Catholic scientists in Argentina. In the first decades of the twentieth century, a small group of Jesuits from Spain and members of religious congregations expelled from France strengthened science teaching in Catholic schools and engaged in resonant polemics on questions of human evolution and neuroscience. The appearance on stage of Ángel Gallardo and bishop Fortunato Devoto, the first Catholic professional scientists in the history of the country, marked a turning point. From that moment on, being a scientist and a Catholic ceased to be a contradiction. The Catholic revival of the 1930s brought with it a small numer of scientists who conceived the relationships between science and religion in neo-scholastic terms, were committed to the positions of integral Catholicism, and politically sided with nationalist and rightist governments—but their reactionary politics did not affect their scientific opinions. A group of liberal Catholics, active in the middle decades of the twentieth century, were among the most prestigious scientists in Argentina, helped to build the scientific system of the country, and advanced proposals for the creation of a research Catholic university.

In the following three sections, I shall recapitulate the interplay between science and religion instanced with the three episodes of secularisation mentioned above; afterwards, I shall briefly rehearse the comings and goings of Catholic science in Argentina.

Religious and Secular Space

Insofar as science in the Jesuit missions of historical Paraguay and Río de la Plata was at the service of Christianisation, the relationships between science and religion during this period can be characterised as harmonious and even amicable (chapter 1). In the years surrounding the Revolution of independence, Bourbon military engineers and crown's bureaucrats, patriots, and those revolutionary priests engaged in natural history shared a positive view of natural theology and an enlightened belief in the value of useful science. In this period of Catholic Enlightenment it is already possible to discern tensions arising from the uneasy balance between a predominant Christian worldview and the emergence of deist philosophies (chapter 2).

The first signs of conflict between religion and science emerged in the 1820s during Rivadavia's rule, with the introduction of the new currents of thought of the English Enlightenment and the French Revolution and the arrival of science teachers escaping from the reactionary regimes of the Holy Alliance (chapter 3).

Rivadavia's initiative of ecclesiastical reform of 1822 was conceived along Gallican lines; it was not anti-Catholic. Its objective was the lessening of the power of the religious orders and the strengthening of the secular clergy controlled by the state. Science as such did not play a causal role in Rivadavia's policy of secularisation (in its original meaning of seizure of ecclesiastical property by a secular power).[2] But it seems symbolically significant that his scientific foundations were installed in confiscated church property. Kim Knott has extensively discussed the "shifting fortunes" of religious and secular spaces.[3] Kong and Woods distinguished between "sacred space," a result of the insider's standpoint, and "religious space," a social construction of the outsider's standpoint; it is with religious space we are concerned here.[4] The full significance of this shifting of ecclesiastical space to secular uses is made evident if we recall that Jesuit science had been carried out primarily in the mission towns. Guillermo Wilde has characterised "mission space" as "a symbolic and social process, involving several levels or scales of organization, representation, and performance," where "levels or scales are connected in a network of links and activity-places, where particular agents operate."[5] Jesuit cartography and astronomy contributed to the definition of a space in which it was not always possible to distinguish the religious from the secular, a "mission space" in which science was at the service of a religious project and in harmony with it. The 1820s transformation of the convent of Santo Domingo in the city's "science centre" (the reconversion of space from religious into secular) was an emblematic gesture of Rivadavia's ecclesiastical reform: science dislodging religion as a locus (physical as well as symbolic) of authority. It was also a fit scenario for an ideology that associated scientific culture with the lessening of church social and political influence, a development so far unheard of in the country. This ideological construction would remain as a reference point for those who would build a narrative naturalising the link between science and secularism.

---

2 David Herlihy, "Secularization of Church Property," in *New Catholic Encyclopedia*, 2nd ed. (Detroit: Thomson/Gale; Washington, D.C.: The Catholic University of America, 2003), 12:869–74.
3 Kim Knott, *The Location of Religion. A Spatial Analysis* (London: Equinox, 2001), 61.
4 Lily Kong and Orlando Woods, *Religion and Space. Competition, Conflict, and Violence in the Contemporary World* (London: Bloomsbury, 2016), 4–7.
5 Guillermo Wilde, "The Political Dimensions of Space-Time Categories in the Jesuit Missions," 167.

Science and Secularisation as Social Differentiation

The first open clash between science and religion took place in the 1880s, at the beginning of the presidency of Julio A. Roca, whose secularising policy culminated in the breaking of diplomatic relations with the Vatican (chapter 4). The spark of the controversy was the project of educational reform which made elementary schooling mandatory, co-educational, and free. The law, passed in 1884, also contemplated the suppression of the teaching of religion (Catholic catechism). The other secularising laws were the creation of civil registers (1884), the law of civil marriage (1888), and the establishment of civil cemeteries (at different times in the various provinces). It was a large process of differentiation of functions transferred from the Catholic Church to the secular state. From then on, it was the public civic authorities and not the priests who marked the rhythms of the vital cycle: birth, formal enculturation, marriage, and death.

As important as the legislation passed in the 1880s was the heated political and ideological debate between liberals and Catholics carried out in Congress and the newspapers. Hugh McLeod has remarked that "secularisation did not just happen: it was welcomed or actively promoted by some social groups."[6] The laws approved by Congress in Argentina in the 1880s were driven by a political and intellectual elite. In some ways, the scenario was not entirely unlike that described by Christian Smith for the United States in the 1920s: a secularist enlightened group which through political action and the shaping of public opinion crystallised the social perception of science as an enterprise "which considered religion to be irrelevant and often an obscuring impediment to knowledge."[7] Draper's work, which was enthusiastically received by liberal audiences, confirmed the already held conviction that what was at stake in the configuration of the new nation was a battle between the forces of progress, urban civilisation, enlightenment, and science, and on the other side, backwardness, tradition, ignorance, superstition, and the Catholic church—the world of the interior country ruled by local caudillos, one of whom, Facundo, led his gaucho troops to the rallying cry of "Religion or Death." This opposition, which Shumway has called one of the "guiding fictions" of the country, was immortalised in the subtitle of Domingo F. Sarmiento's classic: *Civilisation or Barbarism* (1845).[8] In late Victorian Britain the conflict of science and religion was in great measure related

---

[6] McLeod, *Secularization in Western Europe*, 85.
[7] Christian Smith, "Introduction. Rethinking the Secularization of American Public Life," in *The Secular Revolution. Power, Interests, and Conflict in the Secularization of American Public Life*, ed. Christian Smith (Berkeley: University of California Press, 2003), 2.
[8] Nicholas Shumway, *The Invention of Argentina* (Berkeley: University of California Press, 1993).

to the efforts of natural scientists to carve for themselves a secular professional space out of one dominated by natural theologians.⁹ In Argentina, the situation was more like that in France, where "the leading politicians [of the Third Republic]...were believers in the liberating potential of an anti-religious science."¹⁰

Between 1850 and 1880, Argentina became a modern nation-state. The rise of observatories, museums, and academies; campaigns of geographical reconnaissance and the "conquest of the Desert;" and the cataloguing of the natural resources of the country are instances of the role played by scientific experts in laying the ideal and material blueprint of the new nation. Sarmiento, the driving force behind many of these developments, saw science as the generator of progress and modernity. In his address inaugurating the 1871 International Exposition in Córdoba, he eulogised progress with Biblical references: "the divine breath, God's spirit marching over the waters."¹¹ Sarmiento hired prestigious German and North-American scientists to sow the seed of science in Córdoba, but the Protestant and liberal background of the newcomers, representative of the atmosphere of German universities at that time, endeared them neither to the population nor to the conservative dignataries of the local University. The creation of the School of Medicine in that city was a controversial issue. Its organiser and first professor was the Dutch zoologist Hendrik Weyenbergh; the German botanist Georg Hieronymus, another member of the newly created Academy of Sciences, claimed defiantly that opposition to establishing the school (which began functioning in 1878) was due to "the rotten brains of the friends of obscurantism, who wish to poison and dim the clear light emanating from the study of nature."¹²

The ground was ready for the germination of a Darwinian garden in the Pampas. The reception of Darwinism in Río de la Plata took place in the intellectual atmosphere fostered by the liberal anticlerical elite that rose to prominence with Roca. It is no surprise that Darwin's theory was decoded in terms of the ongoing battles for secularisation. In the 1880s, there were very few naturalists in Argentina who had first-hand technical knowledge of Darwin's work (Hendrik

---

9 Frank M. Turner, *Contesting Natural Authority. Essays in Victorian Cultural Life* (Cambridge: Cambridge University Press, 1993); Bernard Lightman and Michael S. Reidy, *The Age of Scientific Naturalism* (London: Pickering and Chatto, 2014).
10 McLeod, *Secularization in Western Europe*, 5.
11 Domingo F. Sarmiento, "Discurso inaugural de la Exposición de Córdoba," in idem, *Obras completas*, ed. Augusto Belin Sarmiento, vol. 21 (Buenos Aires: Imprenta Mariano Moreno, 1899), 319.
12 Abraham Willink, "Vida y obra de Hendrik Weyenbergh," *Boletín de la Academia Nacional de Ciencias* 49 (1972): 56.

Weyenbergh was one of them). From the start, as it is evident in Holmberg's satirical novelettes, evolutionism was used as a rhetorical weapon in the culture wars over secularism. Although Darwin's name was duly mentioned in public discourse, what really worked was evolution as conceived by Spencer and Haeckel insofar as these authors offered the great intellectual vistas which fed the local enthusiasm for essay writing.[13] Two native Argentinians from patrician families and an Italian immigrant of humble origins fostered a national, evolutionistic, and public-friendly style of natural sciences. Their chosen enemy was Burmeister, whose anti-Darwinism had methodological, not religious origins. The Young Turk naturalists excelled in field work and exploration and opted for different forms of secularism: Doctor Eduardo L. Holmberg was proud of being a materialist; the explorer and anthropologist Francisco P. Moreno followed something approaching pantheism; and Florentino Ameghino added to his perdurable work on paleontology his volatile theoretical speculations in the spirit of Ernst Haeckel's *Die Welträthsel* (1899). The fields of inquiry of these natural scientists merged imperceptible with many politically charged issues and helped to build up an atmosphere of secularism. Catholic answers to evolution were undistinguished and derivative. There were actually no Catholic scientists; those Catholics who entered the fray—all of them lawyers and politicians—tackled the questions of the nature of humanity, social Darwinism, and like issues, eschewing technicalities. Based on the traditional Catholic doctrine of the unity of truth, they defended progress and science in abstract terms and wielded a positivist notion of scientific evidence against the hypothetical nature of evolutionary theory.

Positivism was ubiquitous among the intellectual elite of the 1880s and 1890s but Comte's Religion of Humanity never obtained more than a few followers and no positivist chapels were erected in Argentina. Positivism constituted the core of what has been called a "scientific culture"—the expression is apposite if "scientific" is construed as ideological scientism.[14] The reading list of any would-be intellectual would include Comte, Cousin, Taine, Renan, Sainte-Beuve, Michelet, Victor Hugo, and other "Parisian prophets" with a generous dose of Spencer and a sprinkle of the representatives of mid-nineteenth-century German scientific materialism.[15] The strongholds of positivism were the teachers' schools but this philosophy also permeated discourse related to the nascent so-

---

13 Levine and Novoa, *¡Darwinistas!*, 26–28.
14 Terán, *Vida intelectual en Buenos Aires fin-de-siglo*.
15 Frank E. Manuel, *The Prophets of Paris* (Cambridge, MA: Harvard University Press, 1962); Frederick Gregory, *Scientific Materialism in Nineteenth Century Germany* (Dordrecht: Reidel, 1977); Novoa, "The Rise and Fall of Spencer's Evolutionary Ideas in Argentina."

cial sciences in the areas of race and social control, the kind of issue that attracted the attention of the liberal elite trying to come to grips with national identity and a massive immigration. Eventually, a more "spiritualised" positivism emerged, either in the form of a washout version of Comte's Religion of Humanity or of an elaboration of Spencer's Unknowable.

In the 1880s, Argentine society underwent a process of secularisation resulting in a drastic transformation and reduction of the public role and functions of the Catholic Church. The liberal elite which led these changes was dazzled by visions of unending progress, the result of positive science sweeping the last remnants of religious superstition and prejudice. Evolutionism, positivism, and scientific materialism were the ingredients of English, French, and German ideologies of science combined in a local heady mixture. The transformations stopped short of the separation of Church and estate. In any case, they were accompanied by an effort to impress upon the social imaginary the glorious narrative of a secular modernity powered by scientific rationality superseding a benighted religious past. The atmosphere which nurtured the growing of the first scientific institutions in Argentina was suffused with the hallowed myth of the natural opposition between science and religion. The humbler truth was that the kind of liberal state integrated into the open world economy and welcoming immigrants from Protestant Europe required a new settlement with the Catholic Church, which as a political power was forced to redefine its future place in the novel scheme of things. Science was not a cause of the secularising policy fostered by the intelligentsia and implemented by the government, but it was a key element in the ideology of progress at the core of their project of nation-building. Novoa and Levine have affirmed that "in the specific case of Argentina, scientific thought always served political ends."[16] In the 1880s, those ends were the secularisation of society.

Secularisation as Appropriation of Religious Rituals and Symbols

Science in fin-de siècle Argentina was not only part of the cult of progress, it was also a component of secularist discourse (chapter 5). In a 1910 article, the Catholic social thinker Emilio Lamarca made fun of his adversaries imagining that the *pithecanthropi*, once shaved and tailless, had met to sign the Rousseaunian social contract that allowed them to join into tribes, which in due time evolved

---

16 Levine and Novoa, ¡*Darwinistas!*, 77.

into the state, which was nothing else than "humanised matter."[17] His critique of the historical-materialist notion of the state was heavily tinted by a nostalgia for a society informed by Catholic principles and lost forever.

By the first decade of the twentieth century the secularising momentum began slowly to wane: the divorce bill discussed in the Chamber of Deputies in 1902 did not pass. Positivism also fell out of fashion; new currents of thought expressing vague spiritual yearnings began to blow over Río de la Plata. As a result of the massive immigration a large urban proletariat gathered in the cities of Buenos Aires and Rosario. Much of the efforts of Catholic lay social leaders and the clergy in the years preceding and following the beginning of the twentieth century were concentrated in reaching out to this mass of workers and channelling the forces of protest into confessional labour unions, such as those promoted by the German priest Friedrich Grote, also a critic of spiritualism. Anarchism, which was initially strong, eventually declined while socialism grew into an organised political party. Two styles of secularism have been described for this period: the combative and strongly anticlerical secularism of the left, and the conciliatory secularism of a conservative liberal type, which had reached some kind of compromise with the Catholic hierarchy in order to counteract the rising tide of social unrest.[18] Many of the men and women who led the Socialist Party, founded in 1896, were doctors active in public health issues.[19] The socialists developed an active campaign of popularisation of science, which involved the creation of a cultural and educative centre in Buenos Aires and the publication of cheap editions of evolutionist literature, mostly Haeckel and the classics of mid-nineteenth century German scientific materialism.[20] But the most celebrated author in the pages of the socialist daily (*La Vanguardia*), which began publication in 1894, was Ameghino.

Ameghino's hometown, Luján, became the subject of a symbolic operation geared to the appropriation for secular purposes of the sacred meanings of rites and sanctuaries, in order to generate what Kim Knott called a "secular sa-

---

17 Lamarca, "Necesidad de la acción social," 19.
18 Mallimaci, *El mito de la Argentina laica*, 54.
19 Richard J. Walter, *The Socialist Party of Argentina, 1890–1930* (Texas: Institute of Latin American Studies, 1977), 43.
20 Alfred Kelly, *Descent of Darwin. The Popularization of Darwinism in Germany, 1860–1914* (Chapel Hill: The University of North Carolina Press, 1981); Andreas Daum, "Naturwissenschaften und Öffentlichkeit in der deutschen Gesellschaft. Zu den Anfängen einer Populärwissenschaft nach der Revolution von 1848," *Historische Zeitschrift* 267, no. 1 (1998): 57–90; Gregory, *Scientific Materialism*; Claus Spenninger, "A Movement That Never Materialized: The Perception of Scientific Materialism as a Secular Movement in Nineteenth-Century Germany," in *Freethinkers in Europe*, 273–96.

cred."²¹ Our Lady of Luján, whose image is venerated in the city's neo-gothic basilica, had become an icon in which religious and nationalistic overtones merged. But Luján was also the town of the national hero of evolutionism and its countryside hid paleontological treasures which it was expected would prove the theory of evolution. This simmering ambivalence between the secular and the religious came to the surface in 1916, when in the Chamber of Deputies of the province of Buenos Aires the nationality of Ameghino became a disputed issue, on the occasion of the proposal by a socialist deputy to assign a certain amount of the budget to transform Ameghino's home into a centre of secular worship toward which the school children should peregrinate. The attempt was thwarted by the Catholic deputies, who exhibited documentary evidence which showed that the savant had been born in Genoa. Soon the question became enmeshed with the accusations (eventually proved true) of scientific fraud on the part of some followers of Ameghino concerning the theories of their teacher about the Tertiary origin of humanity in the pampas. These accusations were raised by the strongly anti-evolutionist Jesuit amateur anthropologist José María Blanco. The construction of the image of a secular scientific figure invested with sacred meanings and representing the moral virtues of the nation had begun early upon Ameghino's death in 1911. In the "civic funeral" organised to honour his memory by the University of La Plata—a citadel of secularism and late positivism—the Italian-Argentine doctor and at that time socialist intellectual José Ingenieros (Giuseppe Ingegnieri) eulogised the deceased paleontologist as a model of the "modern sanctity," which he re-interpreted by enumerating the secular equivalents of the Christian virtues. The pomp of Ameghino's civic funeral, which featured a coffin set against a solemn stage reminiscent of Republican Rome, was enhanced by the presence of the famous French socialist leader Jean Jaurès, a symbol of the confrontation of the secular state with the church.

Thomas Lessl has applied rhetorical analysis to discuss evolutionism as discourse that wraps up secular matters into mythical narrative form with religious overtones.²² In his classic study of nationalism as a religion, Carlton Hayes affirmed that "a nationalist theology of intellectuals" easily becomes "a nationalist mythology for the masses."²³ Jeffrey Cox sees secularisation as "the only master

---

21 For a discussion about the decoupling of "sacred" and "religious" and the "secular sacred," see Kim Knott, "The Secular Sacred: In between or both/and?," in *Social Identities between the Sacred and the Secular*, ed. Abby Day, Giselle Vincett and Christopher R. Cotter (Farnham: Ashgate, 2013), 145–160.
22 Lessl, *Rhetorical Darwinism*, 38.
23 Carlton Hayes, *Nationalism: A Religion* (New York: Macmillan, 1960), 168.

narrative of religion in modern history": it is partly hidden; it explains a lot to a lot of people; and it is the only grand narrative of religion in modern history.[24] During the years preceding the Great War, socialists and a sector of liberals in Argentina strove to unite these three threads (evolutionism, nationalism, and secularism) into a single and powerful narrative which implied that science and religion were intrinsically at odds. As we will see in the next section, history told otherwise.

In the first decades of the twentieth century, socialists and anarchists raised the banner of radical secularism in Argentina. While large sectors of society began slowly to shift toward various combinations of spiritualism, nationalism, and traditionalism, in some culturally influential quarters an alliance of socialism, freethought, late positivism, and evolutionism evolved into a matrix for anti-religious positions. Just as had been the case in the French Third Republic, aggressive secularism of this type was popular among school teachers and pedagogues. There were inorganic attempts to turn this ideology into some kind of secular religion, although this should not be understood in the strong, church-like sense in which Saint-Simonism or Comte's positivism evolved in their homeland. In the Argentine case, freethinkers and socialists sought to created secular cults around scientific figures while investing them with sacred auras. Insofar as these symbolic operations were not supported by the state, their influence was relatively limited.

Catholic Scientists

In the first half of the twentieth century the schools in the city of Buenos Aires run by French religious congregations offered quality scientific education modelled upon the tradition of the French *lycée* with imported personnel and equipment.[25] It was these congregations and the Jesuits (those born in Río de la Plata to Italian and German immigrant families, and a small group of Spanish Jesuits trained in the laboratories of Tortosa and the Ebro Observatory) who carried the

---

24 Jeffrey Cox, "Master Narratives of Long-term Religious Change," in *The Decline of Christendom in Western Europe, 1750–2000*, ed. Hugh McLeod and Werner Ustorf (Cambridge: Cambridge University Press, 2003), 201–17.

25 The literature that pictures Catholic scientific teaching as scientifically backwards has been written from the point of view of the warfare thesis and is in need of substantial revising. See, for example, Silvina Gvirtz, M. Aisenstein, J. Cornejo and A. Valerani, "The Natural Sciences in the Schools: Tension in the Modernization Process of Argentine Society (1870–1960)," *Science and Education* 10, no. 6 (2001): 545–58.

faint torch of Catholic representation in the sciences during this period. Catholic teachers cultivated astronomy and natural history but for the most they were anti-evolutionistic and rejected relativity theory. The first two professional Catholic scientists in the country were a layperson (Ángel Gallardo) and a secular priest (Fortunato Devoto), eventually auxiliary bishop of Buenos Aires (chapter 6). In tune with the liberal type of Catholicism he represented, Gallardo kept his science and his religion apart and held evolutionary ideas (he spoke highly of Ameghino). Devoto was a professional astronomer educated in Paris who in the wake of nationalistic politics promoted the creation of the Jesuit Observatory of San Miguel as a daughter institution to the Ebro Observatory.

The 1918 student reform movement which erupted in the University of Córdoba and soon radicalised and extended to other Latin American countries had an anticlerical side to it. But if by the end of the second decade of the twentieth century secularism felt at home in the universities, the society as a whole looked in the opposite direction. In the 1920s and 1930s, the Catholic Church experienced a cultural Renaissance and a resurgence of massive expressions of devotion inspired by the ideal of integral Catholicism. The association of the Church with authoritarian governments was at a maximum following the military coup of 1943, when right-wing integral Catholics took control of many areas of social life and a renewed Catholic society seemed to be at hand. By the end of the 1930s, the notion of "Catholic scientist" had ceased being an oxymoron. By the 1940s, two groups of Catholic scientists could be distinguished. One of them was formed by those who adhered to integral Catholicism, were sympathetic to the Axis powers, and supported the regimes of Franco in Spain and Salazar in Portugal. But the adherence to right-wing politics did not necessarily imply scientific obscurantism (chapter 7). The ichthyologist Emiliano Mac Donagh, who was Director of Schools in Buenos Aires during the military regime which followed the coup of 1943, was an admirer of Darwin who endorsed a version of theistic evolutionism and strongly opposed eugenics.

The other group was composed by scientists with solid international careers; two of them (Juan T. Lewis and Eduardo Braun Menéndez) were associates of Nobel Prize Bernardo Houssay (chapter 8). They favoured the Allies, rejected Franco's regime and were aligned with the ideas of liberal Catholicism; Lewis was one of the founders of Christian Democracy in Argentina. In the last years of President Perón's government (1954), Braun set up a Catholic Institute of Sciences backed by the Church hierarchy. His idea was to scale this up into a small number of research institutes which in time could form the basis of a university— his model was Johns Hopkins. Braun made explicit that his aim was to foster sound science and not "Catholic science" (the lecturers were chosen on the grounds of their qualifications and irrespective of confessional convictions). In the turbulent months that preceded the fall of Perón, the institute was dissolved. With their flawless scientific reputations and unblemished democratic credentials, Lewis, Braun Menéndez, and engineer Augusto Durelli (who eventually left the country and settled in the United States) not only confuted the belief that science was a secular pursuit, they also demonstrated that some of the best science in the country was done by Catholics. But not all Catholic scientists were opponents to Perón, as shown by the Jesuit Juan A. Bussolini, director of the Observatory of San Miguel. Despite his unfortunate intervention in the Richter imposture of the 1940s, he was an indefatigable institution builder who fostered Solar Physics in his own observatory and helped with the erection of other astronomical facilities.

In late 1955, during the de facto government that suceeded Perón, Catholics campaigned for the creation of "free" (private and confessional) universities while secularists (some liberals and those on the left) claimed that only the state should grant university diplomas. The struggle was protracted and fought in the media and on the streets, even with some casualties. Braun Menéndez sought to cut a middle path: he advocated for the creation of a research Catholic university together with a state system of exams regulating the granting of diplomas for professional careers. In 1958, the Chamber of Deputies approved the bill that allowed the creation of private universities, with the support of president Arturo Frondizi (who also initiated the negotiations for the Concordat between Argentina and the Holy See signed in 1966). Despite some efforts to the contrary, the new Catholic universities began functioning for the most as professional schools, a far cry from the ideal of research universities cherished by Lewis and Braun. In the late 1950s and early 1960s, Catholics managed to organise a strong student movement in the state universities and were even able to have a Catholic professor elected president of the UBA (a development which is beyond the scope of the present inquiry).[26] The battle over the creation of confessional universities was an important confrontation over the questions of secularism but the particular issue of science and religion was never seriously discussed; the episode was for the most a political affair.

---

[26] Luisa A. Brignardello, *El movimiento estudiantil argentino* (Buenos Aires: Macchi, 1972); José Zanca, *Los Humanistas universitarios. Historia y memoria (1950–1966)* (Buenos Aires: Eudeba, 2019).

A Last Word

Writing in 1907, the Argentine intellectual Ernesto Quesada attributed the "deep disturbance of the spirits" and the scandal provoked by the publication of Robert Chambers's *Vestiges of the Natural History of Creation* (1844, published anonymously) to the speculative character of a work which contrasted starkly with the descriptive and factual science characteristic of the "English spirit." The latter, he claimed, aimed at "a conciliation with the Biblical text...through a *most particular* kind of literature, prototype of which are the Bridgewater Treatises" (my emphasis).[27] A few years later, engineer Nicolás Besio Moreno, in a lecture before the members of the most prestigious scientific association of the country (which he presided), claimed that "the defeat of dogmatism has been a result of the overwhelming outbreak of philosophy and science...in this stormy combat the [Catholic] Church has ended by wounding itself with its own poisonous dagger; and languishes, forgotten by philosophy and ignored by science."[28] Besio goes on to sing the glories of the *philosophes*, the French Revolution, and the "Rights of Man and the Citizen." The educated public in Buenos Aires was aware of two historical models of relationships between science and religion: what has been called the British "holy alliance" and what could be characterised as the French "unholy alienation."[29] Quesada—for all his cosmopolitanism and his familiarity with Spencer's work—could barely conceive of the former as little more than as a British eccentricity.

In the late 1980s, John H. Brooke explored the alternative patterns (British and French) of the relationships between science and religion, in connection with the development of natural theology. In the aftermath of the French Revolution, he argues, English scientists were compelled to pre-empt accusations of conspiring against the order of society and manoeuvred to dissociate their work

---

27 Ernesto Quesada, *Herbert Spencer y sus doctrinas sociológicas* (Buenos Aires: Talleres Gráficos de la Penitenciaría Nacional, 1907), 6–7; for the reception of *Vestiges*, see James A. Secord, *Victorian Sensation: The Extraordinary Publication, Reception, and Secret Authorship of Vestiges of the Natural History of Creation* (Chicago: University of Chicago Press, 2000).
28 Nicolás Besio Moreno, "El sistema filosófico de Agustín Álvarez," in Agustín Álvarez, *¿Adónde vamos?* (Buenos Aires: La Cultura Argentina, 1915), 14.
29 Basil Willey, *The Eighteenth-Century Background* (Harmondsworth: Penguin, 1962), 133, 162; John H. Brooke, "Scientific Thought and Its Meaning for Religion: the Impact of French Science on British Natural Theology, 1827–1859," *Revue de Synthèse* 110, no. 1 (1989): 34–35; John Gascoigne, *Cambridge in the Age of Enlightenment. Science, Religion, and Politics from the Restoration to the French Revolution* (Cambridge: Cambridge University Press, 1988).

from subversive meanings.³⁰ British scientists who popularised French theories in England, neutralised them blunting their secularist edge.³¹ Certainly French Enlightenment science was densely charged with anti-religious meanings, although, as Brooke pointed out, "much of the venom that went into attacks on established Christianity" was not a result of the contents of theories but "produced by a sense of political injustice."³² With this background, he claimed that the anti-religious aspects of certain forms of science "may reflect a set of secular attitudes that have already come about through quite distinct social and economic forces;" in other words, "a secular mentality determines how science should be interpreted."³³ The argument developed in this book can be seen as an elaboration of this latter statement from the point of view of the pattern of secularisation of Catholic countries, the model of which was France—more precisely, the city of Paris.

In Latin America, the relationships between science and secularism seem to involve the following aspects: (a) contrary to the many-valued logic of Protestant Pluralism, the two-valued logic of the Latin pattern (Catholic or non-Catholic) leads to a neater division of the scientific community into a secular majority and a small confessional minority; (b) secularism is at first a top-bottom affair promoted by the same elites who also advocate science as an ingredient of the ideology of progress; (c) not infrequently, the particular kind of relationship between state and Church (which varies in different countries and evolves with time) modulates the interactions between science and religion. All these factors are manifest in the case under study.

Although the historical contexts of the episodes of secularisation analysed in the previous chapters were separated in time, there is an element of continuity in that science always had a part in them. In the 1820s, Church property was secularised (expropriated by the state) in the wake of a Gallican reform of the clergy. Convents were put at the service of a political project that saw science as an important part in the construction of an "Enlightened Republic" and converted the religious spaces into scientific spaces (laboratories, an observatory, a museum and an agricultural garden). In the 1880s, science became a condensing metaphor of progress and urban civilisation against the barbarian desert, a beacon of light dispelling the shadows of religious superstition. Science was popularly

---

30 John H. Brooke, "Why Did the English Mix Their Science and Their Religion?," in *Science and Imagination in Eighteenth-Century British Culture*, ed. Sergio Rossi (Milan: Unicopli, 1987), 57–78.
31 Brooke, "Scientific Thought and Its Meaning for Religion."
32 Brooke, *Science and Religion*, 166.
33 John H. Brooke, "Science and the Secularization of Knowledge: Perspectives on Some Eighteenth-Century Transformations," *Nuncius* 4, no. 1 (1989): 50, 55.

identified as a secular pursuit: it had arrived in the country together with the gospel of progress, was cultivated by liberals and anticlericals, and could be readily used as a rhetorical instrument in secularist polemics. At the beginning of the twentieth century, the torch of secularism passed from the old liberal conservative elite to the emerging socialist movement, which invested secular savants with an aura of saintliness and science with sacred meanings. A battle was joined about the symbolic custody of a centre of Catholic devotion: could the pious caravans of pilgrims visiting the basilica be replaced by flocks of schoolchildren honoring the humble natal abode of the secular savant par excellence, prototype of the virtues of a *morale laïque?*

This long history shows that in a society with a French pattern of secularisation the confrontation of science and religion was an important ingredient in the secularising agenda of the liberal elite and later of the left. In Argentina, science as such (the contents of scientific theories, the actual practice of science in laboratories and the field) did not function as an engine of secularisation. Secularisation was motivated by a complex of political, social, and ideological causes related to the construction of the state, the history of the local Church, the social transformations of the country, and the beliefs (or more properly, unbelief) of the ruling elites.[34]

The historical actors who in the nineteenth century pushed a secularist agenda also promoted scientific research with the conviction that both things were necessary foundations for progress. The legitimazing discourse that accompanied these actions drew upon the prestige of science—to which even Catholic leaders bowed—to underpin political and no-religious or antireligious programmes.[35] The decades under study could be seen as the Argentine version of a never ending *banquet Berthelot*.[36] Ironically, it was the Catholic Ángel Gallardo who in his capacity as Argentine Minister of Foreign Affairs pronounced on 25 October 1927 the *éloge* of the fiercely anticlerical chemist Marcellin Berthelot

---

[34] Although the discussion of the general process of secularisation in Argentina is not one of the objectives of this paper, it is worth quoting the following remark: "[In Argentina] the religious has never been absent ... there is a long history of utilisation, negotiation, and instrumentalisation of religion on the part of politicians and the state. Also, of politics and the state by religious groups." Mallimaci, *El mito de la Argentina laica*, 245.

[35] Colin Campbell, *Toward a Sociology of Irreligion* (London: Macmillan, 1971), 21; Lee, *Recognizing the Non-Religious*, 31.

[36] Lalouette, "La querelle de la foi de de la science."

in the act celebrating the centenary of his birth, presided in the Panthéon by the president of the French Republic, Raymond Poincaré.[37]

The history of "Catholic science" (or, more precisely, Catholic scientists) in Argentina is an inverted mirror image of the narrative of secularist science. Practically inexistent during the secularising period of fin-de-siècle, Catholic scientists began to appear in the twilight of the belle époque. The inorganic group who emerged in the heat of the 1930s Catholic revival moved within the boundaries of the kind of integral Catholicism to which they adhered. It was with the democratic and liberal Catholic scientists of the 1940s and 1950s that the spell of the conflict view of science and religion was definitively broken. Not only did they form part of mainstream science: they themselves helped to lay the foundations of the modern system of scientific research in Argentina.

The prevailing systems of thought, the political situation, and the configuration of the relationships between the Catholic Church and the state contributed to shaping the relationships between science and religion in Argentina during the nineteenth and the extended first half of the twentieth century. The "conflict view" rose and declined just as secularist policies surged and ebbed. It was the dynamics of a particular pattern of secularisation which modulated the interplay between religion and science, not the other way around.

---

[37] P. Roche, "Le Centenaire du Berthelot," *Le Galois*, 26 October 1927; Robert Fox, "Science, Celebrity, Diplomacy: The Marcellin Bethelot Centenary, 1927," *Revue d'histoire des sciences* 69, no. 1 (2016): 77–115.

Abbreviations used in this book

AAPC:	Asociación Argentina para el Progreso de las Ciencias
ABH:	Archive Bernardo Houssay, Buenos Aires.
AFM:	Archive of the Faculty of Medicine, University of Buenos Aires.
IHSI:	Institutum Historicum Societatis Iesu, Rome.
AUCA:	Archive Universidad Católica Argentina, Buenos Aires.
CCC:	Cursos de Cultura Católica (Courses of Catholic Culture).
CER:	Centro de Estudios Religiosos (Centre of Religious Studies).
CMA:	Círculo Médico Argentino (Argentine Medical Circle).
CMC:	Consorcio de Médicos Católicos (Consortium of Catholic Physicians).
CNE:	Consejo Nacional de Educación (National Council of Education).
CNO:	Consejo Nacional de Observatorios (National Council of Observatories).
IBYME:	Instituto de Biología y Medicina Experimental
ICC:	Instituto Católico de Ciencias (Catholic Institute of Sciences).
ICCBA:	Instituto Católico de Cultura de Buenos Aires (Catholic Institute of Culture of Buenos Aires).
MACN:	Museo Argentino de Ciencias Naturales, Buenos Aires.
SCA:	Sociedad Científica Argentina (Argentina Scientific Society).
UBA:	Universidad de Buenos Aires.
UCA:	Universidad Católica Argentina, Buenos Aires.

Bibliography

Manuscripts

"Braun Menéndez, Eduardo." File 18173. Archive of the Faculty of Medicine, University of Buenos Aires.
Braun, Rafael. "Compilación de las ideas de Eduardo Braun Menéndez sobre ciencia y universidad." Unpublished manuscript, 1961, Microsoft Word file.
Correspondence Juan T. Lewis. Box 4, 08–6. Archive Bernardo Houssay, Buenos Aires.
Larrañaga, Dámaso. *Diario de Historia Natural, 1808–1814*. Instituto de Historia Argentina y Americana Emilio Ravignani, University of Buenos Aires, Buenos Aires.
"Lewis, Juan T." File 8044. Archive of the Faculty of Medicine, University of Buenos Aires.
Libro de Actas del Consejo Superior de UCA. Archive Universidad Católica Argentina, Buenos Aires
Riva, Benito. [*Cursus physicae*]. *De mundo et caelo*. Fondo Antiguo de la Compañía de Jesús en Argentina (FACJA), Buenos Aires.
Rodríguez, Cayetano. *Secunda phisicae pars seu Phisica Particularis*. FACJA, Buenos Aires.
Sánchez Labrador, José. *El Paraguay Natural*, 4 vols. Paraquaria 16–19. Institutum Historicum Societatis Iesu (IHSI), Rome.
Parodiz, Juan J. and Enrique Balech. "El Museo Argentino de Ciencias Naturales 'B. Rivadavia'...en pantuflas." Unpublished manuscript, 1992, mimeo. Buenos Aires: Museo Argentino de Ciencias Naturales.
Unsigned documents related to the dealings between Bartolomé D. Muñoz's executor and the University of Buenos Aires, 1831–32. In "Universidad 1828–1835." Box X-6–3–1. Archivo General de la Nación (AGN), Buenos Aires.

Newspapers

Argentinisches Wochenblatt.
Buenos Aires Herald.
Correo de Comercio.
Crónica Política y Literaria de Buenos Aires.
Deutsche La Plata Zeitung.
El Americano.
El Diario.
El Granizo. Diario político, literario y comercial.
El Librepensador.
El Lucero.
El Mosquito.
El Nacional.
El Pueblo.
Illinois State Register.
La América del Sud.
La Gaceta/Gazeta Ministerial.

La Gaceta Mercantil.
La Libertad.
La Nación.
La Prensa.
La Prensa Argentina.
La Tribuna.
La Unión.
La Vanguardia.
Le Galois.
The Standard (Buenos Aires).

Government Series

Boletín Oficial de la República Argentina.
Registro Nacional. Provincias Unidas del Río de la Plata.
Registro Oficial de la Provincia de Buenos Aires.

Printed Works

Ablard, Jonathan P. *Madness in Buenos Aires. Patients, Psychiatrists, and the Argentine State, 1880–1983.* Calgary: University of Calgary Press, 2008.
Adelman, Jeremy, ed. *Essays in Argentine Labour History, 1870–1930.* Basingstoke: Macmillan, 1992.
Adelman, Jeremy. *Republic of Capital. Buenos Aires and the Legal Transformation of the Atlantic World.* Stanford: Stanford University Press, 1999.
Aguilar, Félix. "La medición de un arco de meridiano en la República Argentina." *Anales del Instituto Popular de Conferencias* 22 (1937): 199–214.
Aguilar, Félix. "Monseñor Fortunato Devoto." *Revista Astronómica* 13 (1941): 259–63.
Aguirre Elorriaga, Manuel. *El abate de Pradt en la emancipación hispanoamericana (1800–1830)*, 2nd ed. Buenos Aires: Huarpes, 1946.
Albani, Félix and Juan L. Pedelaborde. "¿Católicos contra transformistas?" *Criterio* no. 584 (11 May 1939): 35–36.
Alberdi, Juan B. *Organización de la Confederación Argentina*, 2 vols. Besanzon: J. Jacquin, 1858.
Alberdi, Juan B. "Mi vida privada." In idem, *Escritos póstumos*, ed. Francisco Cruz, vol. 15, 262–312. Buenos Aires: Imprenta J. B. Alberdi, 1900.
Alberini, Coriolano. "Contemporary Philosophic Tendencies in South America, with Special Reference to Argentina." *The Monist* 37, no. 3 (1927): 328–34.
Alberini, Coriolano. *Die deutsche Philosophie in Argentinien.* Berlin: Hendriock, 1930.
Album biográfico de los Libre-Pensadores de la República Argentina en el Primer Centenario de su Independencia. Buenos Aires: "El Progreso." Periódico racionalista, 1910.
Alcacer, Pedro S. *La vida y el transformismo moderno.* Buenos Aires: Imprenta de la Penitenciaría, 1882.
Alcacer, Pedro S. *Locura y crimen.* Buenos Aires: Imprenta de la Penitenciaría, 1883.

Algorta Camusso, Rafael. *El padre Dámaso Antonio Larrañaga*. Montevideo: Barreiro y Ramos, 1922.
Allen, Phyllis. "The Royal Society and Latin America as Reflected in the *Philosophical Transactions*, 1665–1730." *Isis* 37, no. 3–4 (1947): 132–38.
Almeida, Teodoro de. *Recreação filosófica*, 10 vols. Lisbon: Officina de Miguel Rodrigues, Regia Officina Typographica, 1780–1791.
Álvarez Díaz de Vivar, Julio César. *Rivadavia: piedra angular de la enseñanza agrícola en las Provincias Unidas del Río de la Plata*. Buenos Aires: Suelo Argentino, 1945.
Amaral, Samuel. *The Rise of Capitalism in the Pampas. The Estancias of Buenos Aires, 1785–1870*. Cambridge: Cambridge University Press, 1998.
Ambrosetti, Juan B. "Prof. Pedro Scalabrini (1849–1916). Fundador y director de los museos provinciales de Entre Ríos y Corrientes." *Anales del Museo Nacional de Historia Natural de Buenos Aires* 28 (1916): 227–40.
Ameghino, Florentino. *La antigüedad del hombre en el Plata*, 2 vols. Paris: Masson; Buenos Aires: Igon, 1880–1881.
Ameghino, Florentino. "Un recuerdo a la memoria de Darwin. El transformismo como ciencia exacta." *Boletín del Instituto Geográfico Argentino* 3 (1882): 213–25.
Ameghino, Florentino. "Mi Credo." *Anales de la Sociedad Científica Argentina* 52, no. 2 (1906): 64–95.
Ameghino, Florentino. *Filogenia. Principios de clasificación transformista basados sobre leyes naturales y proporciones matemáticas*. Buenos Aires: La Cultura Argentina, 1915.
Ameghino, Florentino. "Origine et persistence de la vie (la matière, la vie, la mort et l'immortalité)." In idem, *Obras completas*, edited by Alfredo J. Torcelli, vol. 19, 184–581. La Plata: Taller de Impresiones Oficiales, 1935.
Ameghino, Florentino. "Una virgen falsificada." In idem, *Obras completas*, edited by Alfredo J. Torcelli, vol. 23, 505–9. La Plata: Taller de Impresiones Oficiales, 1936.
Anagnostou, Sabine. *Jesuiten in Spanisch-Amerika als Übermittler von heilkundlichem Wissen*, Quellen und Studien zur Geschichte der Pharmazie, 78. Stuttgart: Wissenschaftliche Verlagsgesellschaft, 2000.
Andrade, Olegario V. "Prometeo." In idem, *Obras Poéticas*, 85–108. Buenos Aires and La Plata: Peuser, 1887.
Andrews, George R. *The Afro-Argentines of Buenos Aires, 1800–1900*. Madison, WI: University of Wisconsin Press, 1980.
Appleby, R. S. "Exposing Darwin's 'Hidden Agenda': Roman Catholic Responses to Evolution, 1875–1925." In *Disseminating Darwinism: The Role of Place, Race, Religion, and Gender*, edited by Ronald L. Numbers and John Stenhouse, 173–207. Cambridge: Cambridge University Press, 2001.
"Aprobación del Consorcio." *Iatría* 1, no. 1 (April 1930): 1–2.
Arata, Pedro. "Banquete de la Universidad de Buenos Aires." *Revista de la Universidad de Buenos Aires* 13 (1910): 135–38.
Arboleda, Luis C. and Diana Soto Arango. "The Theories of Copernicus and Newton in the Viceroyship of Nueva Granada and the Audiencia de Charcas during the 18th Century." In *Universities and Science in the Early Modern Period*, edited by Mordechai Feingold and Víctor Navarro-Brotons, 289–309. Dordrecht: Springer, 2006.
"Argentine Prelate Honoured by France." *National Catholic Welfare Conference News Service* no. 33–2640 (18 March 1933).

Argentines of Today, edited by William Belmont Parker, 2 vols. Buenos Aires and New York: The Hispanic Society of America, 1920.
Argerich, Cosme M. "Un comunicado del Dr. Cosme Argerich." *El Americano* no. 27, 1st October 1819, 10–15.
Ariès, Philippe. *The Hour of our Death*, trans. Helen Weaver. New York: Knopf, 1981.
Arriaga, Nilo. "El sol y sus relaciones con los fenómenos terrestres." *Mundo atómico* 4, no. 13 (1953): 25–28.
Artigas, Mariano, Thomas F. Glick and Rafael A. Martínez. *Negotiating Darwin: The Vatican Confronts Evolution, 1877–1902*. Baltimore: The Johns Hopkins University Press, 2006.
Asad, Talal. *Genealogies of Religion: Discipline and Reasons of Power in Christianity and Islam*. Baltimore: The Johns Hopkins University Press, 1993.
Ashworth, William B. "Catholicism and Early Modern Science." In *God and Nature. Historical Essays on the Encounter between Christianity and Science*, edited by David C. Lindberg and Ronald L. Numbers, 136–66. Berkeley: University of California Press, 1986.
Asúa, Miguel de. "El apoyo oficial a la *Description Physique de la République Argentine* de H. Burmeister." *Quipu* 6, no. 3 (1989): 339–53.
Asúa, Miguel de. "Acerca de la biografía, obra y actividad médica de Thomas Falkner S.I. (1707–1784)." *Stromata* 62, no. 3–4 (2006): 227–54.
Asúa, Miguel de. "Linneo entre nosotros." *Ciencia Hoy* 18, no. 104 (2008): 19–27.
Asúa, Miguel de. "Historia de la Astronomía en la Argentina." In *Historia de la astronomía argentina*, edited by Gustavo E. Romero, Sergio A. Cellone and Sofía Cora, 1–19, Asociación Argentina de Astronomía. Book Series, 2. La Plata: Asociación Argentina de Astronomía, 2009.
Asúa, Miguel de. "Los artículos del P. José María Blanco S.I. en la revista *Estudios* sobre la evolución y las teorías antropológicas de Ameghino." *Stromata* 65, no. 3–4 (2009): 313–35.
Asúa, Miguel de. 2010. *La ciencia de Mayo. La cultura científica en el Río de la Plata, 1800–1820*. Buenos Aires: FCE, 2010.
Asúa, Miguel de. *Una gloria silenciosa. Dos siglos de ciencia en Argentina*. Buenos Aires: Zorzal, 2010.
Asúa, Miguel de. "El Congreso Científico Internacional Americano." *Ciencia Hoy* 21, no. 125 (2011): 18–24; 22, no. 126 (2012): 14–20.
Asúa, Miguel de. "Dos siglos y un museo." In *El Museo Argentino de Ciencias Naturales. 200 años*, edited by Pablo Penchaszadeh, 13–69. Buenos Aires: MACN-Conicet, 2012.
Asúa, Miguel de. *Science in the Vanished Arcadia. Knowledge of Nature in the Jesuit Missions of Paraguay and Río de la Plata*. Leiden: Brill, 2014.
Asúa, Miguel de. "'Los phisicos modernos quasi todos son copernicanos.' Copernicanism and its Discontents in Colonial Río de la Plata." *Journal for the History of Astronomy* 48 (2017): 160–79.
Asúa, Miguel de. "Los entomólogos de la Compañía de Jesús en Argentina." *Stromata* 73, no. 2 (2017): 231–43.
Asúa, Miguel de. "The 'Conflict Thesis' and Positivist History of Science. A View from the Periphery." *Zygon* 53, no. 4 (2018): 1131–48.
Asúa, Miguel de. "Natural History in the Jesuits Missions." In *The Oxford Handbook of the Jesuits*, edited by Ines G. Županov, 708–36. New York: Oxford University Press, 2019.

Asúa, Miguel de. "La 'ciencia' en las misiones jesuitas como encrucijada de saber global y culturas locales." In *A Experiência da Missão Jesuítica na Primeira Modernidade*, edited by Luiz F. Medeiros Rodrigues and Maria C. Bohn Martins, 180–200. São Leopoldo: Oikos, 2019.

Asúa, Miguel de. "Belgrano y la ciencia." In *Belgrano*, edited by Marcelo U. Salerno and Roberto L. Elissalde, 8–19. Buenos Aires: Academia Nacional de Ciencias de Buenos Aires, 2020, https://www.ciencias.org.ar/contenido.asp?id=790.

Asúa, Miguel de and Analía Busala. "Instituto Católico de Ciencias (1953–1954). Más en la leyenda que en la historia." *Criterio* no. 2368 (March 2011): 40–44.

Asúa Miguel de and Diego Hurtado de Mendoza, trans. "The Size of Space (An Essay on Mathematical Psychology). Leopoldo Lugones." *Science in Context* 18, nº 2 (2005): 317–36.

Asúa Miguel de and Diego Hurtado de Mendoza. *Imágenes de Einstein. Relatividad y cultura en el mundo y en la Argentina*. Buenos Aires: Eudeba, 2006.

Asúa, Miguel de and Roger K. French. *A New World of Animals. Early Modern Europeans and the Creatures of Iberian America*. Aldershot: Ashgate, 2006.

Atran, Scott. *Cognitive Foundations of Natural History*. Cambridge: Cambridge University Press; Paris: Éditions de la Maison des sciences de l'homme, 1990.

Auza, Néstor T. *Grote y la estrategia social*. Vol. 1, *Aciertos y fracasos sociales del catolicismo argentino*. Buenos Aires: Docencia, Don Bosco and Guadalupe, 1987.

Auza, Néstor T. *El proyecto episcopal y lo social*. Vol. 3, *Aciertos y fracasos sociales del catolicismo argentino*. Buenos Aires: Docencia, 1988.

Auza, Néstor T. "Racionalismo y tradicionalismo en el Río de la Plata. Gustavo Minelli-José Manuel Estrada." *Teología* 37, no. 73 (1999): 99–121.

Auza, Néstor T. *Católicos y liberales en la generación del ochenta*. Buenos Aires: Educa, 2007.

Avellaneda, Nicolás. "*El Gran Chaco*. Libro de don Luis J. Fontana." In idem, *Escritos y Discursos*, vol. 1, 107–126. Buenos Aires: Compañía Sudamericana de Billetes de Banco, 1910.

Avellaneda, Nicolás. "La escuela sin religión." In idem, *Escritos y Discursos*, vol. 3, 185–247. Buenos Aires: Compañía Sudamericana de Billetes de Banco, 1910.

Avni, Haim. *Argentina and the Jews: A History of Jewish Immigration*. Tuscaloosa: University of Alabama Press, 1991.

Ayrolo, Valentina. *El abrazo reformador. Las reformas eclesiásticas en tiempos de construcción estatal. Córdoba y Cuyo en el concierto iberoamericano (1813–1840)*. Rosario: Prohistoria, 2017.

Azara, Félix de. *Voyages dans l'Amérique méridionale*, edited by C. Walckenaer, 2 vols. Paris: Dentu, 1809.

Babini, Nicolás. *La informática en la Argentina*. Buenos Aires: Letra Buena, 1991.

Bacon, Francis. "Of Atheism." In *Essays*, edited by M. A. Scott, 71–76. New York: Charles Scribner's Sons, 1908.

Bailey, Gauvin Alexander. *Art on the Jesuit Missions in Asia and Latin America, 1542–1773*. Toronto: University of Toronto Press, 1999.

Bailey, Gauvin Alexander. *The Spiritual Rococo. Decor and Divinity from the Salons of Paris to the Missions of Patagonia*. Abingdon: Routledge, 2016.

Banchoff, Thomas and José Casanova, eds. *The Jesuits and Globalization. Historical Legacies and Contemporary Challenges.* Washington: Georgetown University Press, 2016.

Banchoff, Thomas and José Casanova. "The Jesuits and Globalization." In *The Jesuits and Globalization*, edited by Thomas Banchoff and José Casanova, 1–24. Washington: Georgetown University Press, 2016.

Barbour, Ian. *Religion in an Age of Science: The Gifford Lectures 1990–1991*, vol. 1. San Francisco: HarperSanFrancisco, 1990.

Barbour, Ian. "On Typologies for Relating Science and Religion." *Zygon* 37, no. 2 (2002): 345–59.

Barcia, Pedro L. "Composición y temas de *Las fuerzas extrañas*." In Leopoldo Lugones, *Las fuerzas extrañas*, edited by Pedro L. Barcia, 9–45. Buenos Aires: Ediciones del 80, 1984.

Barcia, Pedro L. "Lugones y los saberes ocultos." In Leopoldo Lugones, *Estudios esotéricos*, edited by Pedro L. Barcia, 9–86. Buenos Aires: Docencia, 2018.

Barcón Olesa, José. "A propósito de una conferencia del Dr. Nicolai." *Nosotros* 64, no. 240 (May 1929): 289–91.

Bard, Leopoldo. "Educación especial y profilaxis antivenérea en los centros obreros." *La semana médica* 27, no. 17 (1920): 562–65.

Bard, Philip, C. H. Best et al. "The Argentine Citizens Declaration." *Science* 99, no. 2565 (25 February 1944): 166.

Barnett, Richard. "Dr Jacob de Castro Sarmento and Sephardim in Medical Practice in 18th-Century London." *Transactions of the Jewish Historical Society of England* 27 (1982): 84–114.

Barón, Máximo. *Octavio F. Mossotti. En el amanecer de la ciencia Argentina.* Buenos Aires: ECA, 1981.

Barón, Máximo. "La primera cátedra de física experimental en la Universidad de Buenos Aires." *Quipu* 3, no. 3 (1986): 349–400.

Barrancos, Dora. *La escena iluminada. Ciencias para trabajadores (1890–1930).* Buenos Aires: Plus Ultra, 1996.

Barrancos, Dora. "El otro rostro de la modernidad: socialistas y ciencia esotérica (1890–1930)." *Estudios sociales* (Santa Fe) 21, no. 40 (2011): 101–26.

Barry, Jonathan and Colin Jones, eds. *Medicine and Charity Before the Welfare State.* London: Routledge, 1991.

Basso, Nidia and Norberto Terragno. "History about the Discovery of the Renin-Angiotensin System." *Hypertension* 38 (2001): 1246–49.

Bastero Montserrat, Juan Jesús. "Longinos Navás S.J. An approach to his Life and Entomological Work." In *Overview and Strategies of Ephemeroptera and Plecoptera*, edited by J. Alba-Tercedor and A. Sánchez Ortega, 581–84. Gainesville, FL: Sandhill Crane Press, 1991.

Baubérot, Jean. "The Two Thresholds of Laïcization." In *Secularism and Its Critics*, edited by Rajeev Bhargava, 94–136. Oxford: Oxford University Press, 1998.

Baubérot, Jean. *Histoire de la laïcité en France*, 3rd ed. Paris: PUF, 2005.

Baubérot, Jean. "Los umbrales de la laicización en la Europa latina y la recomposición de lo religioso en la modernidad tardía." In *La modernidad religiosa: Europa latina y América latina en perspectiva comparada*, edited by Jean-Pierre Bastian, 94–110. Mexico: FCE, 2004.

Beaumont, John Barber. *Travels in Buenos Ayres and the Adjacent Provinces of the Río de la Plata*. London: James Ridgway, 1828.
Beck, Eugenio (Guillermo Furlong). "Un benemérito de las ciencias en el Río de la Plata. Bartolomé Doroteo Muñoz (1831–1931)." *Revista de la Sociedad Amigos de la Arqueología* 5 (1931): 53–80.
Beddall, Barbara. "'Un naturalista original': Don Félix de Azara, 1746–1821." *Journal of the History of Biology* 8, no. 1 (1975): 15–66.
Beddall, Barbara. "Isolated Spanish Genius: Myth or Reality: Félix de Azara and the Birds of Paraguay." *Journal of the History of Biology* 16, no. 2 (1983): 225–58.
Bell, Stephen. *A Life in Shadow. Aimé Bonpland in Southern South America, 1817–1858*. Stanford: Stanford University Press, 2010.
Benedetti, Edmund. "Radioastronomía solar y actividades del Observatorio de San Miguel." *Boletín de la Asociación Argentina de Astronomía* 13 (1968): 6–9.
Bernaola, Omar A. *Enrique Gaviola y el Observatorio Astronómico de Córdoba*. Buenos Aires: Saber y Tiempo, 2001.
Bernard, Claude. *La science expérimentale*. Paris: J. B. Baillière, 1878.
Berthelot, Marcellin. "La science et la morale." *Revue de Paris* (1st February 1895): 449–69.
Bertoni, Lilia A. *Patriotas, cosmopolitas y nacionalistas. La construcción de la nacionalidad argentina a fines del siglo XIX*. Buenos Aires: FCE, 2001.
Besio Moreno, Nicolás. "El sistema filosófico de Agustín Álvarez." In Agustín Álvarez, *¿Adónde vamos?*, 7–27. Buenos Aires: La Cultura Argentina, 1915.
Besio Moreno, Nicolás. *Las fundaciones matemáticas de Belgrano*. Buenos Aires: Instituto Nacional Belgraniano, 1995.
Biagini, Hugo, ed. *El movimiento positivista argentino*. Buenos Aires: Editorial de Belgrano, 1985.
Bianchi, Susana. *Catolicismo y peronismo. Religión y política en la Argentina, 1943–1955*. Tandil: IEHS, 2001.
Bianchi, Susana. *Historia de las religiones en la Argentina. Las minorías religiosas*, 2nd ed. Buenos Aires: Sudamericana, 2009.
"Bibliografía del Dr. Gallardo." *Anales de la Sociedad Científica Argentina* 133 (1942): 169–84.
Biermat, Carolina. "La eugenesia argentina y el debate sobre el crecimiento de la población en los años de entreguerras." *Cuadernos del Sur* no. 34 (2005): 251–73.
Bischoff, Efraín U. *Leopoldo Lugones. Un cordobés rebelde*. Córdoba: Editorial Brujas, 2005.
Blackwell, Richard J. *Galileo, Bellarmine, and the Bible*. Notre Dame: University of Notre Dame Press, 1991.
Blair, Ann and Kaspar von Greyerz, eds. *Physico-Theology. Religion and Science in Europe, 1650–1750*. Baltimore: Johns Hopkins University Press, 2020.
Blancke, Stefaan. "Catholic Responses to Evolution, 1859–2009: Local Influences and Mid-Scale Patterns." *Journal of Religious History* 37 (2013): 353–68.
Blanco, José M. *El Credo de Ameghino*. Buenos Aires: R. Herrando, 1916.
Blanco, José M. *La evolución antropológica y Ameghino*. Buenos Aires: Alfa y Omega, 1916.
Blanco, José M. "Las ideas preconcebidas y la ciencia." *Estudios* 13 (1917): 423–35.
Blanco, José M. "Las bolas de Parodi ¿serán bolas?" *Estudios* 20 (1921): 31–35.
Blanco, José M. "Ameghino juzgado por Boule. *Les hommes fossiles*." *Estudios* 20 (1921): 419–26.

Blanco, José M. "Dinámica cerebral." *Estudios* 24 (1923): 5–16, 83–89, 161–67 and 246–54.
Blanco, José M. "La antigüedad del hombre." *Estudios* 25 (1923): 91–103, 161–74.
Blanco, José M. "¿Es posible la evolución del hombre?" *Estudios* 27 (1924): 431–46; 28 (1925): 42–53, 81–90, 344–57; 29 (1925): 40–51.
Blanco, José M. "Hablan los hombres de ciencia del país sobre las asendereadas teorías de Ameghino." *Estudios* 22 (1922): 428–45.
Blanco, José M. *La eugenia naturalista y la campaña profiláctica contra la lúe*. Buenos Aires: Amorrortu, 1923.
Boman, Eric. "Encore l'homme tertiaire dans l'Amérique du Sud." *Journal de la Société des Américanistes*, new series, 11 (1919): 657–64.
Boman, Eric. "Los vestigios de industria humana encontrados en Miramar (República Argentina) y atribuidos a la época terciaria." *Revista Chilena de Historia y Geografía* 39, no. 42 (1921): 330–52.
Bonomo, Mariano. "El hombre fósil de Miramar." *Intersecciones en Antropología* 3 (2002): 69–87.
Borutta, Manuel. *Antikatholizismus: Deutschland und Italien im Zeitalter der europäischen Kulturkämpfe*. Göttingen: Vandenhoeck & Ruprecht, 2010.
Bosca, Roberto. *La iglesia nacional peronista. Factor religioso y poder político*. Buenos Aires: Sudamericana, 1997.
Bosch, Beatriz. "Alejo Peyret, administrador de la colonia San José." In *III Congreso de Historia Argentina y Regional, Santa Fe-Paraná, 1975*, 53–72. Buenos Aires: Academia Nacional de la Historia, 1977.
Botana, Natalio R. *El orden conservador. La política argentina entre 1880 y 1916*. Buenos Aires: Hyspamérica, 1986.
Bougainville, Louis Antoine de. *Voyage autour du monde ... nouvelle edition augmentée*, 2 vols. Neuchâtel: Imprimerie de la Société Typographique, 1772.
Bourgignon d'Anville, Jean-Baptiste. "Le Paraguay, où les RR. PP. de la Compagnie de Jesus ont répandu leurs Missions" (map). *Lettres édifiantes et curieuses* 21 (1734): 278–79.
Bowler, Peter J. *The Eclipse of Darwinism. Anti-Darwinian Evolution Theories in the Decades around 1900*. Baltimore: The Johns Hopkins University Press, 1983.
Bowler, Peter J. *Life's Splendid Drama*. Chicago: University of Chicago Press, 1996.
Brabo, Francisco Javier. *Inventarios de los bienes hallados a la expulsión de los jesuitas y ocupación de sus temporalidades por decreto de Carlos III*. Madrid: Rivadeneyra, 1872.
Braun Menéndez, Eduardo. *Universidades no oficiales e institutos privados de investigación*. Buenos Aires, 1945.
Braun Menéndez, Eduardo. *Bases para el progreso de las ciencias en la Argentina*. Buenos Aires: Ateneo del Club Universitario de Buenos Aires, 1946.
Braun Menéndez, Eduardo. "¿Se cumplen los fines de la universidad?" *Ciencia e Investigación* 11, no. 9 (September 1955): 385–87.
Braun Menéndez, Eduardo. "Una facultad monstruosa." *Ciencia e Investigación* 11, no. 12 (December 1955): 529–31.
Braun Menéndez, Eduardo. "Los títulos habilitantes y la universidad." *Ciencia e Investigación* 12, no. 1 (January 1956): 1–4.
Braun Menéndez, Eduardo. "Las etapas para la creación de una universidad privada." *Ciencia e Investigación* 13, no. 3 (March 1957): 97–99.

Braun Menéndez, Eduardo. "La ley universitaria." *Ciencia e Investigación* 14, no. 7 (July 1958): 289–90.

Braun Menéndez, Eduardo, Juan C. Fasciolo, Luis F. Leloir, Juan M. Muñoz and Alberto C. Taquini, *Renal Hypertension*, translated by Lewis Dexter. Springfield, IL: Charles C. Thomas Pub., 1946.

Braun Menéndez, Eduardo and Irvine Page. "Suggested Revision of Nomenclature-Angiotensin." *Science* 127, no. 3292 (31 January 1958): 242.

Brignardello, Luisa A. *El movimiento estudiantil argentino*. Buenos Aires: Macchi, 1972.

Brixia, Fortunatus a. *Philosophia sensuum mechanica*. Vol. 4. Venice: Ex Typographia Remondiniana, 1755.

Broers, Michael. "Revolution as Vendetta: Patriotism in Piedmont, 1794–1821." *The Historical Journal* 33, no. 3 (1990): 573–97.

Brooke, John H. "Why Did the English Mix Their Science and Their Religion?" In *Science and Imagination in Eighteenth-Century British Culture*, edited by Sergio Rossi, 57–78. Milan: Unicopli, 1987.

Brooke, John H. "Science and the Secularization of Knowledge: Perspectives on Some Eighteenth-Century Transformations." *Nuncius* 4, no. 1 (1989): 43–65.

Brooke, John H. "Scientific Thought and Its Meaning for Religion: the Impact of French Science on British Natural Theology, 1827–1859." *Revue de Synthèse* 110, no. 1 (1989): 33–59.

Brooke, John H. *Science and Religion. Some Historical Perspectives*. Cambridge: Cambridge University Press, 1991.

Brooke, John H. "Science and Secularization." In *The Cambridge Companion to Science and Religion*, edited by Peter Harrison, 103–23. Cambridge: Cambridge University Press, 2010.

Brooke, John H. "Afterword. The Instantiation of Historical Complexity." In *Rethinking History, Science, and Religion. An Exploration of Conflict and the Complexity Principle*, edited by Bernard Lightman, 235–42. Pittsburgh: University of Pittsbugh Press, 2019.

Brooke, John H. and Ronald L. Numbers, eds. *Science and Religion around the World*. Oxford: Oxford University Press, 2011.

Brooke, John H. and Geoffrey Cantor. *Reconstructing Nature. The Engagement of Science and Religion*. Edinburgh: T&T Clark, 2000.

Bruce, Steve. *Secularization. In Defence of an Unfashionable Theory*. Oxford: Oxford University Press, 2011.

Bruch, Carlos. "La obra entomológica del doctor Ángel Gallardo." *Revista de la Sociedad Entomológica Argentina* 6 (1934): 234–42.

Brugier, Eduardo. *Elementos de cosmografía*, 7th ed. Buenos Aires: Estrada, 1933.

Brunetière, Ferdinand. "Charles Darwin." *Revue politique et littéraire*, 3rd series, 2, no. 17 (29 April 1882): 518–24

Brunetière, Ferdinand. *L'évolution des genres dans l'histoire de la littérature*, 2 vols. Paris: Hachette, 1890–1892.

Brunetière, Ferdinand. "Après une visite au Vatican." *Revue des Deux Mondes*, 4th period, 127, no. 1 (1st January 1895): 97–118.

Bruno, Cayetano. *Historia de la Iglesia en la Argentina*, 12 vols. Buenos Aires: Don Bosco, 1966–1981.

Bruno, Cayetano. *Historia argentina*. Buenos Aires: Don Bosco, 1977.

Bruno, Cayetano. *Vida y acción del Padre Luis María Etcheverry Boneo*. Buenos Aires: Asociación Cristo Sacerdote, 1996.
Bruno, Paula. *Pioneros culturales de la Argentina*. Buenos Aires: Siglo XXI, 2011.
Bubello, Juan P. "De 'Jesús no es Dios' a 'Jesús…es el verdadero fundador del socialismo'. Ocultismo y política en el espiritismo kardecista argentino (1870–1930)." *Melancolía* 1 (2016): 51–74.
Buchbinder, Pablo. *Historia de las universidades argentinas*. Buenos Aires: Sudamericana, 2005.
Buchrucker, Cristián. *Nacionalismo y peronismo. La Argentina en la crisis ideológica mundial (1927–1955)*. Buenos Aires: Sudamericana, 1987.
Buckland, William. *Reliquiae Diluvianae*. London: John Murray, 1823.
Buenos Aires & Pacific Railway Company. *History and Characteristics, 1882–1933*. Buenos Aires, 1933.
Buisseret, David. "Jesuit Cartography in Central and South America." In *Jesuit Encounters in the New World: Jesuit Chronicles, Geographers, Educators and Missionaries in the Americas (1549–1767)*, edited by Joseph A. Gagliano and Charles E. Ronan, 113–62. Rome: Institutum Historicum Societatis Iesu, 1997.
Bunge, Carlos O. *Principios de psicología individual y social*. Madrid: Daniel Jorro, 1903.
Bunge, Carlos O. "Viaje a través de la estirpe." In idem, *Viaje a través de la estirpe y otras narraciones*, 7–91. Buenos Aires: Biblioteca de La Nación, 1908.
Bunge, Carlos O. "La enseñanza de la tradición y la leyenda." *El Monitor de la Educación Común* 36, no. 458 (1911): 263–81.
Bunge, Carlos O. "Notas de psicología social." In idem, *Estudios filosóficos*, 117–51. Buenos Aires: La Cultura Argentina, 1919.
Burchardt, Marian, Monika Wohlrab-Sahr and Matthias Middell. "Multiple Secularities beyond the West: An Introduction." In *Multiple Secularities beyond the West. Religion and Modernity*, edited by M. Burchardt, M. Wohlrab-Sahr and M. Middell, 1–15. Boston: De Gruyter, 2015.
Burdick, Michael A. *For God and the Fatherland. Religion and Politics in Argentina*. New York: The State University of New York Press, 1995.
Burmeister, Hermann. *Geschichte der Schöpfung*, 8th ed. Leipzig: Otto Wigand, 1872.
Burmeister, Hermann. *Description physique de la République Argentine. Tome troisième. Animaux vertébrés. Partie 1. Mammifères vivants et éteints*. Buenos Aires: Coni; Paris: Savy, 1879.
Burnouf, Émile-Louis. *La science des religions*. Paris: Maisonneuve, 1876.
Buschiazzo, Mario J. "El templo y convento de Santo Domingo en Buenos Aires." *Anales del Instituto de Arte Americano e Investigaciones Estéticas* 4 (1951): 62–75.
Bussolini, Juan A. "Filosofía y física de los corpúsculos." *Mundo atómico* 1, no. 1 (September–October 1950): 78.
Bussolini, Juan A. "A propósito de un Consejo Nacional de Investigaciones Científicas y Técnicas." *Estudios* no. 477 (August–September 1956): 27–31.
Bussolini, Juan A. "Las ciencias del espíritu, de la materia y de la vida." *Estudios* no. 543 (May 1963): 201–10.
Bustos [y Ferreira], Zenón, ed. *Anales de la Universidad Nacional de Córdoba. Segundo período*, 3 vols. Córdoba: Casa editora Domenici and Tipografía La Industrial, 1901–1910.

Bustos y Ferreyra, Zenón. "Pastoral con motivo de la agitación antirreligiosa en Córdoba." *Revista Eclesiástica del Arzobispado de Buenos Aires* 18 (1918): 489–94.
Cabanel, Patrick. "Le grand exil des congrégations enseignantes au début du XXe siècle. L'exemple des Jesuites." *Revue d'histoire de l'Église de France* 81, no. 206 (1995): 207–17.
Cabanel, Patrick and Jean-Dominique Durand, eds. *Le grand exil des congrégations religieuses françaises, 1901–1914*. Paris: Le Cerf, 2005.
Cabanis, Pierre J.-G. *Rapports du physique et du morale de l'homme*, 8th ed. Paris: J. B. Baillière, 1844.
Cabrera, Ángel. *El pensamiento vivo de Ameghino*. Buenos Aires: Losada, 1944.
Caillet-Bois, Julio and Roberto F. Giusti. "Manuel José de Lavardén." In *Historia de la literatura argentina*, edited by Rafael A. Arrieta, vol. 1, 239–49 (Buenos Aires: Peuser, 1958).
Caimari, Lila. *Perón y la iglesia católica*. Buenos Aires: Sudamericana, 1994.
Caimari, Lila. "Whose Criminals Are These? Church, State, and *Patronatos* and the Rehabilitation of Female Convicts (Buenos Aires, 1890–1940)." *The Americas* 54, no. 2 (1997): 185–208.
Caldcleugh, Alexander. *Travels in South America during the Years 1819–20–21*, 2 vols. London: Murray, 1825.
Calendario astronómico para la parte austral de América del Sur. Año 1911. Buenos Aires: Coni, 1910.
Calleri Damonte, Giacomo C. F. "Carlo Giuseppe Ferraris: un farmacista di adozione biellese." Article published 6 May 1998. https://doczz.it/doc/1276479/carlo-giuseppe-ferraris–un-farmacista-di-adozione-biellese.
Calvo, Nancy. "'Cuando se trata de la civilización del clero': Principios y motivaciones del debate sobre la reforma eclesiástica porteña de 1822." *Boletín del Instituto de Historia Argentina y Americana*, 3rd series, 24 (2001): 73–104
Calvo, Nancy, Roberto Di Stefano and Klaus Gallo, eds. *Los curas de la revolución*. Buenos Aires: Emecé, 2002.
Camacho, Horacio H. *Las ciencias naturales en la Universidad de Buenos Aires*. Buenos Aires: Eudeba, 1971.
Campbell, Colin. *Toward a Sociology of Irreligion*. London: Macmillan, 1971.
Canclini, Arnoldo. *400 años de protestantismo argentino: historia de la presencia evangélica en la Argentina*. Buenos Aires: Facultad Internacional de Educación Teológica, 2004.
Cantón, Eliseo. *La Facultad de medicina y sus escuelas*, 4 vols. Buenos Aires: Coni, 1921.
Cantor, Geoffrey. "What Shall We Do with the Conflict Thesis?" In *Science and Religion. New Historical Perspectives*, edited by Thomas Dixon, Geoffrey Cantor and Stephen Pumfrey, 283–98. Cambridge: Cambridge University Press, 2010.
Cantor, Geoffrey and Chris Kenny. "Barbour's Fourfold Way." *Zygon* 36, no. 4 (2001): 765–81.
Cañizares Esguerra, Jorge. *How to Write the History of the New World: Histories, Epistemologies, and Identities in the Eighteenth-Century Atlantic World*. Stanford: Stanford University Press, 2001.
Canovai, José. "Oración fúnebre." *Estudios* 67 (1941): 70–77.
Cappelletti, Enrique M. *El gran cometa de 1882. Lijeros estudios sobre sus apariencias y curso hechos en el Colegio de Santa Fe (R. A.)*. Buenos Aires: Biedma, 1883.

Caraman, Philip. *The Lost Paradise, an Account of the Jesuits in Paraguay, 1607–1768*. London: Sidgwick & Jackson, 1975.
Carbia, Rómulo. *La revolución de Mayo y la Iglesia*. Buenos Aires: Huarpes, 1945.
Cárdenas Eduardo J. and Carlos M. Payá. *La Argentina de los hermanos Bunge, 1901–1907*. Buenos Aires: Sudamericana, 1995.
Carlson, A. J. Review of *Human Physiology*, by Bernardo A. Houssay et al., translated by Juan T. Lewis and Olive M. Lewis (New York: McGraw-Hill, 1951). *Science* 113, no. 2946 (15 June 1951): 700.
Carranza, Ángel J. "Benjamín Villafañe (nacido en Tucumán el 30 de marzo de 1819)." *Revista Nacional* 12 (1890): 259–85.
Casanova, José. *Public Religions in the Modern World*. Chicago: Chicago University Press, 1994.
Casanova, José. "The Secular, Secularizations, Secularisms." In *Rethinking Secularism*, edited by Craig Calhoun, Mark Juergensmeyer and Jonathan VanAntwerpen, 54–74. Oxford: Oxford University Press, 2011.
Casares, Tomás D. "La naturaleza de la vida según Hans Driesch." *Criterio* no. 28 (13 September 1928): 337–40.
Casas, Santiago de. *Relox universal de péndola y en él nueva idea de la estructura del universo*. Madrid: Herederos de la Viuda de Juan García Infanzón, 1758.
Casella, Edgardo. "Nuestros 'hombres de ciencia'." *Revista del Círculo Médico Argentino y Centro de Estudiantes de Medicina* 27 (1927): 230.
Castellani, Leonardo. *Lugones*. Buenos Aires: Biblioteca Nacional, 2011.
Castelfranco, Diego. "La América del Sud (1876–1880) y las perspectivas católicas sobre el vínculo entre la ciencia y el catolicismo en la Buenos Aires de fines del siglo XIX." *Boletín del Instituto de Historia Argentina y Americana Dr. Emilio Ravignani* 47 (2017): 63–100.
Castellanos, Alfredo R. "La biblioteca científica del Padre Larrañaga." *Revista Histórica* (Montevideo) 16 (1949): 589–626.
Castellanos, Alfredo R. "Contribución al estudio de las ideas del Pbro. Dámaso Larrañaga." *Revista Histórica* (Montevideo) 17 (1951): 1–118.
Castex, Mariano, ed. *Peces y aves del Paraguay Natural Ilustrado, 1767*. Buenos Aires: Compañía General Fabril, 1968.
Castro Sarmento, Jacob de. "Observationes astronomicae variae factae in Paraquaria, Regione Americae Australis, ab anno 1706 [1700] ad annum 1730." *Philosophical Transactions of the Royal Society* 45 (1748): 667–74.
[Castro Sarmento, Jacob de] "Observationes aliquae astronomicae a Reverendo P. P. Suarez e S. J. in Paraquaria habitae, et per D. Suarez M. D. Cum Soc. Regali communicatae." *Philosophical Transactions of the Royal Society*, 46 (1749–1750): 8–10.
Català Gorgues, Jesús I. "La polémica sobre el hombre terciario y su expresión en la Valencia de comienzos del siglo XX." *Asclepio* 64, no. 1 (2012): 63–96.
Cereijido, Marcelino. *La nuca de Houssay*. Buenos Aires: FCE, 1990.
Cerviño, Pedro. "El tridente de Neptuno es el cetro del mundo." In Nicolás Besio Moreno, *Las fundaciones matemáticas de Belgrano*, 159–73. Buenos Aires: Instituto Nacional Belgraniano, 1995.

Cerviño, Pedro. "Prolusión académica. Discurso en que se procura que para ser buen piloto es necesaria la astronomía." In Nicolás Besio Moreno, *Las fundaciones matemáticas de Belgrano*, 173–78. Buenos Aires: Instituto Nacional Belgraniano, 1995.

Cerviño, Pedro. "Discurso o memoria sobre la importancia de la Academia establecida por el Consulado de Buenos Aires." In Nicolás Besio Moreno, *Las fundaciones matemáticas*, 182–94. Buenos Aires: Instituto Nacional Belgraniano, 1995.

Chadwick, Owen. *The Secularization of the European Mind in the Nineteenth Century*. Cambridge: Cambridge University Press, 1975.

Chakrabarti, Pratik and Michael Workboys. "Science and Imperialism since 1870." In *Modern Science in National, Transnational and Global Contexts*, edited by Hugh R. Slotten, Ronald L. Numbers and David N. Livingstone, 9–31. Vol. 8, *The Cambridge History of Science*. Cambridge: Cambridge University Press, 2020.

Charle, Christophe and Eva Telkes. *Les professeurs de la Faculté des sciences de Paris. Dictionnaire biographique (1901–1939)*. Paris: Institut National de la Recherche Pédagogique, 1989.

Charlton, Donald G. *Secular Religions in France, 1815–1870*. London: Oxford University Press, 1963.

Chiaramonte, José Carlos. *La Ilustración en el Río de la Plata*. Buenos Aires: Sudamericana, 2007.

Chorroarín, Luis José. "Curso de Lógica." In *La enseñanza de la filosofía en la época colonial*, translated by Juan Chiabra, vol. 2, 1–171. Buenos Aires: Coni, 1911.

Cibotti, Ema. "La masonería frente a las leyes laicas." *Todo es Historia* 34, no. 405 (2001): 77–82.

"Ciencia prehistórica. Algunas observaciones a los que defienden la existencia de los hombres prehistóricos." *La Cruz* (Sevilla) 2 (1875): 616–26.

"Ciencia y religión." *F.V.D.* 6, no. 66 (6 August 1926).

"Ciencia y fe." *F.V.D.* 13, no. 149 (September 1933): 357–58.

Clark, John. *La pensée de Ferdinand Brunetière*. Paris: Nizet, 1954.

Clossey, Luke. *Salvation and Globalization in the Early Jesuit Missions*. Cambridge: Cambridge University Press, 2008.

Colbert, Ricardo. "La evolución de Dios." *La escuela positiva* 4, no. 6 (1898): 253–55.

Colegio Nacional de Buenos Aires, Departamento de astronomía. "El Observatorio." Accessed 18 September 2020. https://astro.cnba.uba.ar/?page_id=24.

Colvin, F. J. "Los católicos y la eugenesia. Males de la civilización." *Iatría* 1, no. 1 (April 1930): 5–10.

Comisión Organizadora. *Jornadas Universitarias de Tucumán*. San Miguel de Tucumán: Universidad Nacional de Tucumán, 1957.

Compagnon, Olivier. *Jacques Maritain et l'Amérique du Sud*. Villeneuve-d'Ascq: Presses Universitaires du Septentrion, 2003.

Comstock, George C. "Biographical Memoir Benjamin Apthorp Gould, 1824–1896." *Memoirs of the National Academy of Sciences* 17 (1924): 153–80.

"Como se funda un museo. Una de las glorias del Colegio San José: el doctor Francisco P. Moreno." *F.V.D.* 5, no. 55 (September 1955).

Condorcet, Marquis of. *Outlines of a Historical View of the Progress of the Human Mind*. London: J. Johnson, 1795.

Conforte, Anna J. "Cultura científica, magnetología y espiritismo durante fines del siglo XIX en Buenos Aires: un análisis de Ovidio Rebaudi." *Anuario de Investigaciones de la Facultad de Psicología, UBA* 24 (2018): 25–33.

Congreso Nacional, Cámara de Diputados. *Diario de sesiones... Año 1878*, vol. 1. Buenos Aires: Imprenta y Librería de Mayo, 1879.

Congreso Nacional, Cámara de Diputados. *Diario de sesiones... Año 1883*, vol. 1. Buenos Aires: Imprenta La Universidad, 1884.

Congreso Nacional, Cámara de Diputados. *Diario de sesiones... Año 1884*, vol. 2. Buenos Aires: Stiller and Laass, 1885.

Congreso Nacional, Cámara de Diputados. *Diario de sesiones... Año 1906*, vol. 1. Buenos Aires: Talleres Gráficos de la Penitenciaría Nacional, 1907.

Congreso Nacional. Cámara de Diputados. *Diario de sesiones... Año 1918*, vol. 2. Buenos Aires: Rosso, 1918.

Congreso Nacional, Cámara de Diputados. *Diario de sesiones... Año 1926*, vol. 3. Buenos Aires: Imprenta de la Cámara de Diputados, 1926.

Congreso Nacional, Cámara de Diputados. *Diario de sesiones... Año 1927*, vol. 2. Buenos Aires: Imprenta y Encuadernadora de la Cámara de Diputados, 1927.

Congreso Nacional, Cámara de Diputados. *Diario de sesiones... Año 1936*, vol. 5. Buenos Aires: Imprenta del Congreso Nacional, 1937.

Congreso Nacional, Cámara de Diputados. *Diario de sesiones... Año 1958*, vol. 7. Buenos Aires: Imprenta del Congreso, 1959.

Congreso Nacional, Cámara de Senadores. *Diario de sesiones... Período de 1879*. Buenos Aires: Imprenta de la Honorable Cámara de Diputados, 1931.

Congreso Nacional, Cámara de Senadores. *Diario de sesiones... Año 1975*, vol. 6. Buenos Aires: Imprenta del Congreso de la Nación, 1976.

Constituciones argentinas. Compilación histórica, edited by Natalia Monti. Buenos Aires: Infojus, 2015.

Contreras Roqué, Julio R., Bárbara Gasparri, Adrián Giacchino and Yolanda Davies. *Pedro Scalabrini (1848–1916). Educador y naturalista*. Buenos Aires: Fundación Azara and Universidad Maimónides, 2019.

Corbière, Emilio J. *La masonería. Política y sociedades secretas*. Buenos Aires: Sudamericana, 1998.

[Cornoldi, Giovanni M.] "Esame critico della storia del conflitto fra la religione e la scienza di Guglielmo Draper." *La Civiltà Cattolica* 28 (1877): 142–56.

[Cornoldi, Giovanni M.] "Examen crítico de la historia de los conflictos entre la religión y la ciencia de Guillermo Draper." *La Ciencia Cristiana* 2 (1877): 512–29.

Cornoldi, Giovanni M. *Examen crítico de los conflictos entre la religión y la ciencia de Guillermo Draper*. Granada: Librería Católica de San José, 1878.

Cowie, Helen. "A Creole in Paris and a Spaniard in Paraguay: Geographies of Natural History in the Hispanic World (1750–1808)." *Journal of Latin American Geography* 10, no. 1 (2011): 175–97.

Cox, Jeffrey. "Master Narratives of Long-term Religious Change." In *The Decline of Christendom in Western Europe, 1750–2000*, edited by Hugh McLeod and Werner Ustorf, 201–17. Cambridge: Cambridge University Press, 2003.

Crane, Richard F. *Passion of Israel. Jacques Maritain, Catholic Conscience, and the Holocaust*. Eugene, OR: Wipf & Stock, 2014.

Cueto, Marcos. "The Rockefeller Foundation's Medical Policy and Scientific Research in Latin America: the Case of Physiology." *Social Studies of Science* 20 (1990): 229–54.
Cueto, Marcos. "Laboratory Styles in Argentine Physiology." *Isis* 85 (1994): 228–46.
Cushner, Nicholas P. *Jesuit Ranches and the Agrarian Development of Colonial Argentina 1650–1767.* Albany: State University of New York Press, 1983.
Cuvier, Georges. *Recherches sur les ossemens fossiles*, 3rd ed., vol. 5, pt. 1. Paris and Amsterdam: G. Dufour and E. d'Ocagne, 1823.
Cuvier, Georges. *Discours sur les révolutions du globe*, edited by Paul Bory. Paris: Berche and Tralin, 1881.
Daino, Leonardo. "Exégesis histórica de los hallazgos arqueológicos de la costa bonaerense." *Prehistoria Bonaerense* (1979): 95–145.
"Datos biográficos del Iltmo. y Rvmo. Sr. Dr. D. Fortunato Devoto." *Revista Eclesiástica del Arzobispado de Buenos Aires* 28 (1928): 109–13.
Dallas, Ernesto. "Dr. Juan Brèthes. Bio-bibliografía." *Revista de la Sociedad Entomológica Argentina* 2 (1928–1929): 103–12.
Damianovich, Horacio. "Ángel Gallardo y su teoría de la cariocinesis." *Anales de la Sociedad Científica Argentina* 133 (1942): 102–31.
Darwin, Charles. *Journal of Researches*. London: Henry Colburn, 1840.
Darwin, Charles. *De l'origine des espèces*, translated by Clémence-Auguste Roger. Paris: Masson, 1862.
Daum, Andreas. "Naturwissenschaften und Öffentlichkeit in der deutschen Gesellschaft. Zu den Anfängen einer Populärwissenschaft nach der Revolution von 1848." *Historische Zeitschrift* 267, no. 1 (1998): 57–90.
Davie, Grace. *The Sociology of Religion*. London: Sage, 2007.
Davie, Grace. "Religion and *laïcité*." In *Modern France: Society in Transition*, edited by Malcolm Cook and Grace Davie, 140–154. London: Routledge, 1999.
"De spiritismo." *Acta Apostolicae Sedis* 9, no. 1 (1917): 268.
Debans, Camille. *Discurso sobre el espiritismo...por un médico incrédulo*, translated by Miguel Navarro Viola. *La Biblioteca Popular de Buenos Aires* 13 (1878): 126–73.
Deckmann Fleck, Eliane C. "'Da mística às luzes'—medicina experimental nas reduções jesuítico-guaranis da Província Jesuítica do Paraguai." *Revista Complutense de Historia de América* 32 (2006): 153–78.
Deckmann Fleck, Eliane C., ed. *As artes de curar em um manuscrito jesuítico inédito do Setecentos. O* Paraguay Natural Ilustrado *do padre José Sánchez Labrador (1771–1776)*. São Leopoldo: Oikos-Editorial Unisinos, 2015.
Deckmann Fleck, Eliane C. and Franz Obermeier. "O libro de medicina, cirugía, e botica: um manuscrito anônimo de Matérica médica rioplatense da primeira metade do século XVIII." *Antíteses* 11, no. 21 (2018): 132–56.
Deckmann Fleck, Eliane C. and Roberto Poletto. "Transcrição do *Inventário formado por Lorenzo Infante Boticário em la Ciudad de Córdoba de los bienes medicinales, Julio de 1772.*" *Antiguos Jesuitas en Iberoamérica* 1, no. 1 (2013): 162–247.
[Decree of President Roque Sáenz Peña, 14 July 1911] *Revista Eclesiástica del Arzobispado de Buenos Aires* 11 (1911): 593.
Del Bello, Juan C., Osvaldo Barsky and Graciela Giménez. *La Universidad privada argentina*. Buenos Aires: Zorzal, 2007.

Denevan, William M. "The Pristine Myth. The Landscape of the Americas in 1492." *Annals of the Association of American Geographers* 82, no. 3 (1992): 369–85.

Denevan, William M. "The 'Pristine Myth' Revisited." *Geographical Record* 101, no. 4 (2011): 576–91.

Derisi, Octavio M. *La Universidad Católica Argentina en el recuerdo: a los 25 años de su fundación.* Buenos Aires: UCA, 1983.

De Ruschi Crespo, María I. "La Reforma Universitaria de 1918 y la Iglesia: algunas precisiones. Actuación de los Centros Católicos de Estudiantes." *Archivum* 25 (2006): 13–40.

Destutt de Tracy, Antoine L. C. *Élémens d'idéologie. Première partie. Idéologie proprement dite,* 3rd ed. Paris: Courcier, 1816.

Devoto, Fernando J. "Atilio Dell'Oro Maini." *Prismas* 9 (2005): 187–204.

Devoto, Fernando J. *Nacionalismo, fascismo y tradicionalismo en la Argentina moderna.* Buenos Aires: Siglo XXI, 2005.

[Devoto, Fortunato] "El V Congreso Científico Internacional de Católicos." *Revista Eclesiástica del Arzobispado de Buenos Aires* 1 (1901): 49–55.

Devoto, Fortunato. "Introducción." In Observatorio Astronómico de la Universidad Nacional de La Plata, *Calendario astronómico para el año 1909,* i–xxxv. Buenos Aires: Coni, 1908.

Devoto, Fortunato. "Observations de la comète 1913f (Delavan), faites à l'Observatoire de Paris (équatorial tour de l'Ouest de 0^m, 305 d'ouverture)." *Comptes rendus hebdomadaires des séances de l'Académie des sciences* 160 (January–June 1915): 128–29.

Devoto, Fortunato. "A 'Patria y Hogar.' En defensa del Centro de Estudios Religiosos." *Ichtys* 2, no. 25 (1923): 343–46.

Diario de sesiones de la Primera Asamblea de los Católicos Argentinos. Buenos Aires: Igon Hermanos, 1885.

Diccionario Histórico de la Compañía de Jesús. Biográfico-temático. Edited by Charles E. O'Neill and Joaquín M. Domínguez, 4 vols. Rome: Institutum Historicum Societatis Iesu; Madrid: Universidad Pontificia de Comillas, 2001.

Dictionary of Minor Planets, 5th ed. Edited by Lutz D. Schmadel. Berlin: Springer, 2007.

Dictionary of Scientific Biography. Edited by Charles C. Gillispie, 16 vols. New York: Charles Scribner's Sons, 1981.

Diderot, Denis. "Supplément au Voyage de Bougainville." In idem, *Le neveu de Rameau et autres dialogues philosophiques,* 281–336. Paris: Gallimard, 1972.

Di Pasquale, Mariano. "La présence de l'idéologie à Buenos Aires. Un élan philosophique dans le processus de professionalisation de la médecine, 1820–1840." *Cahiers des Amériques latines* 76 (2014): 111–29.

Di Pasquale, Mariano. "Vitalismo, *idéologie* y fisiología en Buenos Aires. La polémica entre Cosme Argerich y Crisóstomo Lafinur en *El Americano,* 1819." *Revista Ciencias de la Salud* 13 (2015): 13–28.

Discursos pronunciados en la manifestación liberal de la juventud universitaria de Buenos Aires. Buenos Aires: El Nacional, 1883.

Di Stefano, Roberto. *Ovejas negras. Historia de los anticlericales argentinos.* Buenos Aires: Sudamericana, 2010.

Di Stefano, Roberto. "Pensar la Iglesia: el Río de la Plata entre la reforma y la romanización (1820–1834)." *Anuario de Historia de la Iglesia* 19 (2010): 221–39.

Di Stefano, Roberto. "Por una historia de la secularización y de la laicidad en la Argentina." *Quinto Sol* 15, no. 1 (2011): 1–31.
Di Stefano Roberto and Diego Mauro. "Our Lady of Luján: Mass Mobilization and Identity in Argentina." In *Marian Devotions, Political Mobilizations, and Nationalism in Europe and America*, edited by Roberto Di Stefano and Ramón Solans, 279–313. Switzerland: Palgrave Macmillan, 2016.
Di Stefano, Roberto and José Zanca, eds. *Fronteras disputadas: religión, secularización y anticlericalismo en la Argentina (siglos XIX y XX)*. Buenos Aires: Imago Mundi, 2016.
Di Stefano Roberto and Loris Zanatta. *Historia de la Iglesia en la Argentina*. Buenos Aires: Grijalbo, 2000.
Dittrich, Lisa. *Antiklerikalismus in Europa: Öffentlichkeit und Säkularisierung in Frankreich, Spanien und Deutschland (1848–1914)*. Göttingen: Vandenhoeck & Ruprecht, 2014.
Dizionario Biografico degli Italiani, 100 vols. Roma: Istituto della Enciclopedia italiana, 1960–2020.
Dobrizhoffer, Martin. *An Account of the Abipones, an Equestrian People of Paraguay*, translated by Sarah Coleridge, 3 vols. London: Murray, 1822.
Dodds, James. *Records of the Scottish Settlers in the River Plate and their Churches*. Buenos Aires: Grant and Sylvester, 1897.
Doello-Jurado, Martín. "Ángel Gallardo (1867–1934). Su actuación en el Museo de Buenos Aires." *Anales del Museo Argentino de Ciencias Naturales* 38 (1934–1936): ix–xliv.
Domingorena, Horacio C. *Artículo 28. Universidades privadas en la Argentina. Sus antecedentes*. Buenos Aires: Editorial Americana, 1959.
Donoso, Armando. "La conversión de Brunetière." *Revista de Filosofía. Cultura, ciencias, educación* 9 (1919): 206–22.
Dorn, Georgette M. "Sarmiento, the United States, and Public Education." In *Sarmiento and His Argentina*, edited by Joseph T. Criscenti, 77–89. Boulder: Lynne Rienner, 1993.
Douglas, Mary. "The Effects of Modernization on Religious Change." *Daedalus* 111, no. 1 (Winter 1982): 1–19.
Draper, John W. *History of the Conflict between Religion and Science*. New York: Appleton, 1875.
Draper, John W. *Les conflits de la science et de la religion*. Paris: G. Baillière, 1875.
Draper, John W. *Historia de los conflictos entre la religión y la ciencia*, translated by Augusto T. Arcimís, with a prologue by Nicolás Salmerón. Madrid: Imprenta de Aribau, 1876.
Driesch, Hans. "Idealismo y vitalismo." *Arx* (Córdoba) 1, no. 1 (1924): 47–69.
Driesch, Hans. *Lebenserinnerungen: Aufzeichnungen eines Forschers und Denkers in entscheidender Zeit*. Munich: Ernst Reinhardt 1951.
Drozdek, Adam. "Noël-Antoine Pluche and the Spectacularity of Nature." *Roczniki Theologie* 67, no. 4 (2020): 41–59.
Duncan, Julian S. "British Railways in Argentina." *Political Science Quarterly* 52, no. 4 (1937): 559–82.
Duprey, Jacques. *Voyage aux origines françaises de l'Uruguay*. Montevideo: Instituto Histórico y Geográfico del Uruguay, 1952.
Durelli, Augusto J. *Carta abierta al Sr. Decano de la Facultad de Ciencias Exactas, Físicas y Naturales. Sobre el concepto de Patria*. Buenos Aires, 1931.
Durelli, Augusto J. *Discursos al iniciar y dar término al período presidencial 1931–32 en el Centro de Estudiantes de Ingeniería*. Buenos Aires, 1932.

Durelli, Augusto J. *Contribution à l'étude du béton traité. Essai photoelastique.* Saint-Amand (Cher): R. Bussière, 1936.
Durelli, Augusto J. "La unidad entre los católicos." *Sur* 8, no. 47 (August 1938): 72–80.
Durelli, Augusto J. "Toma de posición." *Criterio* no. 556 (27 October 1938): 217–20.
Durelli, Augusto J. *Essai sur les "mentalités" contemporaines: bourgeoisisme, capitalisme, nationalisme, christianisme, communisme,* 2nd ed. Paris: Librairie Montsouris, 1939.
Durelli, Augusto J. *El nacionalismo frente al cristianismo.* Buenos Aires: Losada, 1940.
Durelli, Augusto J. *La libération de la liberté.* Montreal: Editions de l'Arbre, 1944.
Durelli, Augusto J. *Forma y sentido de la resistencia universitaria de octubre de 1945.* Buenos Aires, 1945.
Durelli, Augusto J. *La mochila del coronel.* Buenos Aires: Acción Democrática de Ingenieros, Técnicos y Agrimensores, 1946.
Durelli, Augusto J. *La investigación científico-técnica.* Buenos Aires: El Ateneo, 1945.
Durelli, Augusto J. "Democracia y cristianismo." *Orden Cristiano* 5, no. 103 (January 1946): 352–62.
Durelli, Augusto J. "La relación entre el estado y los empleados." *Orden Cristiano* 5, no. 105 (March 1946): 532–34.
Durelli, Augusto J. *Del universo de la universidad al universo del hombre.* Buenos Aires: Palumbo, 1947.
Duruy, Victor. *Histoire sainte d'après la Bible.* Paris: Hachette, 1845.
Echeverría, Esteban. *Dogma socialista de la Asociación de Mayo.* Montevideo: Imprenta del Nacional, 1846.
Echeverría, Esteban. "Avellaneda." In idem, *Obras completas,* edited by Juan M. Gutiérrez, vol. 1, 283–444. Buenos Aires: Imprenta y Librería de Mayo, 1870.
Echeverría, Esteban. "Manual de enseñanza moral para las escuelas primarias del Estado Oriental." In idem, *Obras completas,* edited by Juan M. Gutiérrez, vol. 4, 327–411. Buenos Aires: Imprenta y Librería de Mayo, 1873.
Eggers-Lecour, Conrado. "Einstein, la inteligencia más madura y el espíritu más joven del siglo." *El Hogar* 21, no. 809 (1925): 7, 64.
Egido, Teófanes. "La expulsión de los jesuitas de España." In *La Iglesia en la España de los siglos XVII y XVIII,* 745–92. Vol. 4, *Historia de la Iglesia de España,* edited by Ricardo García-Villoslada. Madrid: La Editorial Católica, 1979.
Einsenstadt, Shmuel. "Multiples Modernities." *Daedalus* 129, no. 1 (2000): 1–29.
"El Consorcio de Médicos Católicos. Su programa cultural." *Iatría* 1, no. 1 (April 1930): 5.
"El Instituto Católico de Ciencias." *Ciencia e Investigación* 9, no. 9 (August 1953): 337–38.
"El nuevo rector de la Universidad." *Archivos de la Universidad de Buenos Aires* 7 (April–May 1932): 11–13.
Elsen, Sydney. "Frederic Harrison and Herbert Spencer: Embattled Unbelievers." *Victorian Studies* 12, no. 1 (1968): 35–56.
"En busca del rayo supersónico." *Argentina. Revista mensual* 1, no. 3 (1949): 77–81.
Enciclopedia della persona nel XX secolo. Edited by Antonio Pavan. Naples: Edizioni Scientifiche Italiane, 2009.
Engstrand, Iris W. *Spanish Scientists in the New World. The Eighteenth-Century Expeditions.* Seattle: University of Washington Press, 1981.
"En la Academia Literaria del Plata. Conferencia del doctor Rey Pastor. La teoría de la relatividad." *Estudios* 23 (1922): 219–24.

Ennis, Juan A. "Las novedosas ciencias del lenguaje y la política de sus usos: Vicente Fidel López en la *Revista de Buenos Aires* (1863–1869)." *Boletín de la Sociedad Española de Historiografía Lingüística* 12 (2018): 53–74.

"En torno al artículo 28." *Criterio* no. 1307 (8 May 1958): 332.

Ernesto Tornquist & Co. *The Economic Development of the Argentine Republic in the Last Fifty Years.* Buenos Aires, 1919.

E. S. H. "Declaration of Citizens of Argentina." *Science* 98, no. 2552 (26 November 1943): 467–68.

Esteves Balado, Luis and Juan Carlos Arizabalo. "La herencia en las enfermedades mentales." *Iatría* 9 (November 1939): 3–14.

Esteyneffer, Juan de. *Florilegio medicinal.* Mexico: Herederos de Juan Joseph Guillena Carrascoto, 1712.

Estrada, José M. *El génesis de nuestra raza. Refutación de una lección del Dr. D. Gustavo Minelli sobre la misma materia.* Buenos Aires: Imprenta de la Bolsa, 1862.

Estrada, José M. "La iglesia y el estado." *Revista Argentina* 10 (1871): 193–260.

Estrada, José M. *Lecciones sobre la historia de la República Argentina*, 2 vols. Buenos Aires: Librería del Colegio, 1896.

Estrada, José M. "La libertad y el liberalismo." In idem, *Obras completas*, edited by Alberto Estrada, vol. 12, 159–81. Buenos Aires: Compañía Sudamericana de Billetes de Banco, 1905.

Estrada, José M. "El naturalismo y la educación." In idem, *Obras completas*, edited by Alberto Estrada, vol. 12, 207–29. Buenos Aires: Compañía Sudamericana de Billetes de Banco, 1905.

Estrada, Marcos. "La casa de Altolaguirre." *Genealogía* 12 (1957): 139–51.

Estrada, Santiago. "Olegario V. Andrade." In idem, *Miscelánea*, vol. 1, 107–68. Barcelona: Henrich, 1889.

Estrada, Santiago. "La neurosis de los hombres célebres en la República Argentina." In idem, *Miscelánea*, vol. 1, 299–322. Barcelona: Henrich, 1889.

Eyzaguirre, Edgardo C. "Ciencia y religión." *F.V.D.* 12, no. 135 (May 1932): 126.

"Excmo. y Revmo. Monseñor Fortunato J. Devoto. Falleció en Buenos Aires el 29 de junio de 1941." *Revista Eclesiástica del Arzobispado de Buenos Aires* 41 (1941): 457–60.

Ewe, Herbert. "Hermann Burmeister." In idem, *Bedeutende Persönlichkeiten Vorpommerns*, 33–40. Weimar: Hermann Böhlaus Nachfolger, 2001.

Fabre, Pierre-Antoine, José Eduardo Franco and Carlos Fiolhais. "The Dynamics of Anti-Jesuitism in the History of the Society of Jesus." *Jesuit Historiography Online.* Last modified November 2018. https://referenceworks.brillonline.com/entries/jesuit-historiography-online/the-dynamics-of-anti-jesuitism-in-the-history-of-the-society-of-jesus-COM_192530.

Falcao Espalter, Mario. "Cartas científicas de Larrañaga. Edición y notas." *Revista del Instituto Histórico y Geográfico del Uruguay* 2, no. 1 (1921): 295–339.

Farrall, Lyndsay A. "Controversy and Conflict in Science: A Case Study. The English Biometric School and Mendel's Laws." *Social Studies of Science* 5 (1975): 269–301.

Farré, Luis. *Cincuenta años de filosofía en Argentina.* Buenos Aires: Peuser, 1958.

Farro, Máximo and Irina Podgorny. "Frente a la tumba del sabio." *Ciencia Hoy* 8, no. 47 (1998): 28–37.

Favaro, Edmundo. *Dámaso Antonio Larrañaga. Su vida y su época*. Montevideo: Universidad de la República, 1950.
Federación de agrupaciones para la defensa y progreso de la universidad democrática y autónoma. *Avasallamiento de la universidad argentina*. Buenos Aires, 1947.
Feijóo, Benito J. *Teatro Crítico Universal*, 1st ed., 9 vols. Madrid: Imprenta de Lorenzo F. Mojados, Imprenta de Francisco del Hierro, 1726–1740.
Feijóo, Benito J. *Cartas eruditas y curiosas*, 1st ed., 5 vols. Madrid: Imprenta de los herederos de Francisco del Hierro and Joaquín Ibarra, 1742–1760.
Feingold, Mordechai. "Jesuits: Savants." In *Jesuit Science and the Republic of Letters*, edited by Mordechai Feingold, 1–45. Cambridge, MA: The MIT Press, 2003.
Fernández, Ángela M. "Pedagogía y positivismo a fines del siglo XIX: Pedro Scalabrini, Alfredo Ferreira y *La Escuela Positiva*." *Estudios Ibero-Americanos* 29, no. 1 (2003): 73–92.
Fernández, Juan M. *Conclusiones publico-historico-dogmatico-scholastico-phisico-theologicae ex praecipuis Sacrae Theologiae tractatibus depromptae*. Buenos Aires: Apud Typographiam Regiam Parvulorum Orphanorum, 1803.
Fernández de Agüero, Juan M. *Principios de Ideología*, edited by Jorge R. Zamudio Silva, 3 vols. Buenos Aires: Universidad de Buenos Aires, Facultad de Filosofía y Letras, Instituto de Filosofía, 1940.
Fernández Verano, Alfredo. "Ensayo de sistematización de las doctrinas eugénicas." *La semana médica* 27, no. 50 (1920): 800–12.
Ferns, Henry S. *Britain and Argentina in the Nineteenth Century*. Oxford: Oxford University Press, 1960.
Ferns, Henry S. *Argentina*. New York: Praeger, 1969.
Ferrándiz Alejandra and Enrique Lafuente. "El pensamiento eugénico de Marañón." *Asclepio* 51, no. 2 (1999): 133–48.
Ferrari, Joseph A. *Philosophia peripatetica*. Vol. 3. Venice: apud Thomas Bettinelli, 1754.
Ferreira, José A. "Evolución y educación religiosa." *Revista Positiva* (Mexico) no. 54 (26 March 1905), 165–89.
Ferreira, José A. "La concepción religiosa de Comte y Spencer." In idem, *Ensayos de ética*, 135–55. Buenos Aires: Imprenta Ferrari, 1944.
Ferrer, Gaspar. "Manuel Demetrio Pizarro." In *Controversias políticas del ochenta*, 131–52. Buenos Aires: Club de Lectores, 1964.
Figueroa, Marcelo F. "Félix de Azara and the Birds of Paraguay. Making Inventories and Taxonomies at the Boundaries of the Spanish Empire, 1784–1802." In *Global Scientific Practice in an Age of Revolutions, 1750–1850*, edited by Patrick Manning and Daniel Rood, 147–62. Pittsburgh: University of Pittsburgh Press, 2016.
Finlen, Paula, ed. *Athanasius Kircher. The Last Man Who Knew Everything*. New York: Routledge, 2004.
Finocchiaro, Maurice A. *Defending Copernicus and Galileo. Critical Reasoning in the Two Affairs*. Dordrecht: Springer, 2010.
Fletcher, John E. *A Study of the Life and Works of Athanasius Kircher, 'Germanus Incredibilis.'* Leiden: Brill, 2011.
Floria, Carlos A. and Marcelo Montserrat. "La política desde *Criterio* (1928–1977)." *Criterio* 50, no. 1777–1778 (24 December 1977): 762–89.

Foglia, Virgilio. "The History of Bernardo A. Houssay's Research Laboratory, Instituto de Biología y Medicina Experimental: The First Twenty Years, 1944–1963." *Journal of History of Medicine and Allied Sciences* 35, no. 4 (1980): 380–96.
Foster, David W. *The Argentine Generation of 1880. Ideology and Cultural Texts.* Columbia and London: University of Missouri Press, 1990.
Fox, Robert. "Science, Celebrity, Diplomacy: The Marcellin Bethelot Centenary, 1927." *Revue d'histoire des sciences* 69, no. 1 (2016): 77–115.
Fradkin, Raúl O. and Jorge Gelman. *Juan Manuel de Rosas. La construcción de un liderazgo político.* Buenos Aires: Edhasa, 2017.
Franceschi, Gustavo. "Ferdinand Brunetière." *Revista Eclesiástica del Arzobispado de Buenos Aires* 7 (1907): 147–56.
Franceschi, Gustavo. *El espiritualismo en la literatura francesa contemporánea.* Buenos Aires: Agencia General de Librería y Publicaciones, 1917.
Franceschi, Gustavo. "Ciencia y fe." *Criterio* no. 375 (9 May 1935): 37–39.
Franceschi, Gustavo. "La salud de la raza." *Criterio* no. 494 (19 August 1937): 373–75.
Franceschi, Gustavo. "El pensamiento 'científico' de la Edad Media." *Criterio* no. 831 (3 February 1944): 101–5.
Franceschi, Gustavo. "La era atómica." *Criterio* no. 941 (28 March 1946): 277–81.
Franceschi, Gustavo. "Los fenómenos físicos sobrenaturales ante la medicina." *Criterio* no. 972 (31 October 1946): 411–16.
Fraser, John F. *The Amazing Argentine. A Land of Enterprise.* New York: Funk & Wagnalls, 1914.
Frederick, Bonnie. "A State of Conviction, a State of Feeling: Scientific and Literary Discourses in the Works of Three Argentine Writers, 1879–1908." *Latin American Literary Review* 19 (1991): 48–61.
Frías, Félix. "La religión y la ciencia." In idem, *Escritos y discursos*, edited by Pedro Goyena, vol. 1, 19–34. Buenos Aires: Imprenta y Librería de Mayo, 1884.
Frías, Félix. "Los derechos de los frailes." In idem, *Escritos y discursos*, edited by Pedro Goyena, vol. 3, 108–38. Buenos Aires: Librería de Mayo, 1884.
Frías, Luis R. "Manuel D. Pizarro." In *La Argentina del ochenta al Centenario*, edited by Gustavo Ferreri and Ezequiel Gallo, 199–214. Buenos Aires: Sudamericana, 1980.
Frizzi de Longoni, Haydée E. *Rivadavia y la reforma eclesiástica.* Buenos Aires, 1947.
Frondizi, Risieri. "La Enseñanza libre." In Federación Universitaria de Buenos Aires, *La Reforma Universitaria, 1918–1958*, 257–65. Buenos Aires: FUBA, 1959.
Funeral cívico de homenaje a la memoria del sabio naturalista Dr. D. Florentino Ameghino en La Plata. La Plata: Taller de Impresiones Oficiales, 1911.
Funes, Gregorio. *Examen crítico de los discursos sobre una constitución religiosa.* Buenos Aires: Hallet, 1825.
Funes, Gregorio. *Plan de estudios para la Universidad de Córdoba.* Córdoba: Imprenta de la Universidad, 1832.
Furlong, Guillermo. "Buenaventura Suárez (1679–1750)." In idem, *Glorias santafecinas. Buenaventura Suárez, Francisco Javier Iturri, Cristóbal Altamirano. Estudios biobibliográficos*, 79–140. Buenos Aires: Editorial Surgo, 1929.
Furlong, Guillermo. *El Padre José Quiroga.* Buenos Aires: Peuser, 1930.
Furlong, Guillermo. "Un médico colonial: Segismundo Aperger." *Estudios* 54 (1936): 117–48.

Furlong, Guillermo. *Cartografía jesuítica del Río de la Plata*, 2 vols. Buenos Aires: Peuser, 1936.

Furlong, Guillermo. *Matemáticos argentinos durante la dominación hispánica*. Buenos Aires: Huarpes, 1945.

Furlong, Guillermo. *Médicos argentinos durante la dominación hispánica*. Buenos Aires: Huarpes, 1947.

Furlong, Guillermo. *Nacimiento y desarrollo de la filosofía en el Río de la Plata, 1536–1810*. Buenos Aires: Kraft, 1952.

Furlong, Guillermo. "Cartografía colonial." In *El momento histórico del Virreinato del Río de la Plata*, 175–200. Vol. 4, bk. 2, *Historia de la Nación Argentina*, 3rd ed., edited by Ricardo Levene. Buenos Aires: El Ateneo, 1961.

Furlong, Guillermo. "Emiliano J. Mac Donagh († 1º de Agosto de 1961)." *Anales de la Academia Argentina de Geografía* 5 (1961): 9–23.

Furlong, Guillermo. *Misiones y sus pueblos guaraníes*. Buenos Aires: Imprenta Balmes, 1962.

Furlong, Guillermo. *Ángel Gallardo*. Buenos Aires: ECA, 1964.

Furlong, Guillermo. *Los jesuitas y la cultura rioplatense*. Buenos Aires: Biblos, 1994.

Furlong, Guillermo. *Historia del Colegio de la Inmaculada Concepción*. Vol. 5. Santa Fe: Colegio de la Inmaculada Concepción, 2011.

Gabardini, Malvina A. de. *Revista "La escuela positiva" (Corrientes, 1895–1899). Introducción e índices*. Resistencia: Editorial Gómez Lestarri, 1995.

Gallardo, Ángel. "Essai d'interprétation des figures karyokinétiques." *Anales del Museo Nacional de Buenos Aires* 5 (1896): 11–22.

Gallardo, Ángel. "Problemas biológicos. Algunas reflexiones sobre la especificidad celular y la teoría física de la vida, de Bard." *Revista de Derecho, Historia y Letras* 4 (1899): 540–65.

Gallardo, Ángel. *La interpretación dinámica de la división celular*. Buenos Aires: Coni, 1902.

Gallardo, Ángel. "Les mathématiques et la biologie." In *Compte rendu du deuxième Congrès international des mathématiciens*, edited by Ernest Duporcq, 395–403. Paris: Gauthier-Villars, 1902.

Gallardo, Ángel. *Las investigaciones modernas sobre la herencia*. Buenos Aires: La ciencia médica, 1908.

Gallardo, Ángel. "Sur l'épreuve statistique de la loi de Mendel." *Comptes rendus hebdomadaires des séances de l'Académie des sciences* 146 (1908): 361–62.

Gallardo, Ángel. "Prologue." In Martín Gil, *Cosas de arriba*, i–v. Córdoba: La Italia, 1909.

Gallardo, Ángel. "Recientes contribuciones matemáticas al estudio de las leyes de la herencia biológica." *Anales de la Sociedad Científica Argentina* 68 (1909): 185–208.

Gallardo, Ángel. [Note on Darwin]. In "Darwin, Science, and the Church," edited by Elmer J. Kneale. *Illinois State Register*, 23 August 1914.

Gallardo, Ángel. "El instinto de las hormigas." *Revista de Filosofía. Cultura, ciencias, educación* 2 (1915): 1–20.

Gallardo, Ángel. "Creencia y ciencia." *Tribuna Universitaria* 3 (1916): 103–7.

Gallardo, Ángel. *Zoología*, 8th ed. Buenos Aires: Estrada, [1917].

Gallardo, Ángel. "Los estudios biológicos en la República Argentina." In *Primera Reunión Nacional de la Sociedad Argentina de Ciencias Naturales. Tucumán, 1916*, 1–12. Buenos Aires: Coni, 1919.

Gallardo, Ángel. *Memorias para mis hijos y mis nietos*. Buenos Aires: Academia Nacional de la Historia, 1982.
Gallardo, Guillermo. *La política religiosa de Rivadavia*. Buenos Aires: Theoria, 1962.
Gallardo, Jorge E. *Conflicto con Roma (1923–1926). La polémica por Monseñor de Andrea*. Buenos Aires: El Elefante Blanco, 2004.
Galles, Carlos. "La obra de Camilo Meyer por la cultura científica argentina." In *La ciencia en la Argentina. Perspectivas históricas*, edited by Miguel de Asúa, 134–44. Buenos Aires: CEAL, 1993.
Gallo, Ezequiel. "Society and Politics, 1880–1916." In *Argentina since Independence*, edited by Leslie Bethell, 79–111. Cambridge: Cambridge University Press, 1993.
Gallo, Klaus. *The Struggle for an Enlightened Republic: Buenos Aires and Rivadavia*. London: Institute for the Study of the Americas, 2006.
Gallo, Klaus. *Bernardino Rivadavia. El primer presidente argentino*. Buenos Aires: Edhasa, 2012.
Gálvez, Lucía. *Delfina Bunge. Diarios íntimos de una época brillante*. Buenos Aires: Planeta, 2000.
Gálvez, Manuel. *Recuerdos de la vida literaria*. Vol. 1. Buenos Aires: Taurus, 2002.
García, Juan Agustín. *Nuestra incultura*. Buenos Aires: Claridad, 1922.
García Belsunce, César A. *Buenos Aires y su gente, 1810–1830*. Buenos Aires: Emecé, 1976.
García Belsunce, César A. *Pertenencias extrañas. Libros en Buenos Aires en 1815*. Buenos Aires: Academia Nacional de la Historia, 2013.
García Camarero, Ernesto and Enrique, eds. *La polémica de la ciencia española*. Madrid: Alianza, 1970.
García de Loydi, Ludovico. *Una luz en la manzana de las luces: Chorroarín*. Buenos Aires: Municipalidad de la Ciudad de Buenos Aires, Secretaría de Cultura, 1973.
García de Tagle, Gregorio, Dámaso Larrañaga and Melchor Fernández. *Theses ex universa philosophia* (Buenos Aires: Apud Typographiam Regiam Parvulorum Orphanorum, 1792). In Juan C. Zuretti, "Tesis sobre filosofía y ciencias, defendidas en 1792 en el Real Colegio de San Carlos de Buenos Aires." *Revista de la Universidad de Buenos Aires* 44 (1948): 515–53.
García Pintos, Salvador. "¿Qué hay de positivo en materia de eugenesia?" *Iatría* 11 (April 1941): 3–27.
Gallo, Ezequiel. "Society and Politics, 1880–1916." In *Argentina since Independence*, edited by Leslie Bethell, 79–111. Cambridge: Cambridge University Press, 1993.
Gargurevich, Eduardo. "La reacción anti-inmigrante en la literatura argentina de los ochenta." *Revista de Crítica Literaria Latinoamericana* 20, no. 39 (1994): 91–107.
Garro, Juan M. *Bosquejo histórico de la Universidad de Córdoba*. Buenos Aires: Biedma, 1882.
Garzón Maceda, Félix. *La medicina en Córdoba: apuntes para su historia*, 3 vols. Buenos Aires: Rodríguez Giles, 1916–1917.
Gascoigne, John. *Cambridge in the Age of Enlightenment. Science, Religion, and Politics from the Restoration to the French Revolution*. Cambridge: Cambridge University Press, 1988.
Gasparini, Sandra. *Espectros de la ciencia. Fantasías científicas de la Argentina del siglo XIX*. Buenos Aires: Santiago Arcos, 2012.
Gaudant, Jean. "Albert Gaudry (1827–1908) et les 'Enchaînements du Monde animal'." *Revue d'histoire des sciences* 44, no. 1 (1991): 117–28.

Gaulin, Morgan. "Refonder la philosophie en 1860: Ernest Renan critique de Vacherot." *Nineteenth-Century French Studies* 38 (2009): 52–66.

Gerbi, Antonello. *The Dispute of the New World; the History of a Polemic, 1750–1900*, revised, edited and translated by Jeremy Moyle. Pittsburgh: University of Pittsburgh Press, 1973.

Gez, Juan W. *El Dr. Juan Crisóstomo Lafinur*. Buenos Aires: Cabaut, 1907.

Ghio, José M. *La iglesia católica en la política argentina*. Buenos Aires: Prometeo, 2007.

Ghirardi, Enrique. *La Democracia Cristiana*. Buenos Aires: CEAL, 1983.

Gicklhorn, Renée. *Missionsapotheker: Deutsche Pharmazeuten in Lateinamerika des 17. und 18. Jahrhunderts*. Stuttgart: Wissenschaftliche Verlagsgesellschaft, 1973.

Gil, Martín. "La bancarrota de la ciencia." *Caras y caretas* no. 792 (6 December 1913): 106–7.

Gilbert, Isidoro. *La Fede. Alistándose para la revolución. La Federación Juvenil Comunista 1921–2005*. Buenos Aires: Sudamericana, 2009.

Gillette, Aaron. "Agostino Gemelli and the Latin Eugenics Movement." *Römische Quartalschrift für Christliche Altertumskunde und Kirchengeschichte* 109, no. 1–2 (2014): 92–102.

Gillham, Nicholas. "The Battle between the Biometricians and the Mendelians: How Sir Francis Galton Caused his Disciples to Reach Conflicting Conclusions about the Hereditary Mechanism." *Science & Education* 24 (2013): 61–75.

Gillispie, Charles C. *Genesis and Geology*. New York: Harper & Row, 1951.

Giménez, Ángel. *Páginas de historia del movimiento social en la República Argentina*. Buenos Aires: Imprenta La Vanguardia, 1927.

Glick, Thomas. *Einstein y los españoles. Ciencia y sociedad en la España de entreguerras*. Madrid: CSIC, 1986.

Glick, Thomas F. "Marañón, Intersexuality and the Biological Construction of Gender in Spain." *Cronos* 8 (2005): 121–38.

Glick, Thomas and David M. Quinlan. "Félix de Azara: The Myth of the Isolated Genius in Spanish Science." *Journal of the History of Biology* 8, no. 1 (1975): 67–83.

Goldish, Matt. "Newtonian, Converso, and Deist: The Lives of Jacob (Henrique) de Castro Sarmento." *Science in Context* 10, no. 4 (1997): 651–75.

Gómez, José Valentín. *Conclusiones ex universa philosophia*. Buenos Aires: Apud Regiam Parvulorum Orphanorum Typographiam, 1802.

Gómez, Leila, ed. and Nicholas F. Calloway, trans. *Darwinism in Argentina: Major Texts (1845–1909)*. Lewisburg: Bucknell University Press, 2012.

Gondra, Luis R. *Las ideas económicas de Manuel Belgrano*. Buenos Aires: Rosso, 1923.

Góngora, Mario. "Estudios sobre el Galicanismo y la 'Ilustración católica' en América española." *Revista Chilena de Historia y Geografía* no. 125 (1957): 96–151.

González, Joaquín V. *La paz por la ciencia. Discurso en la colación de grados y títulos de la Universidad Nacional de La Plata, el 15 de agosto de 1914*. La Plata: Christmann and Crespo, 1914.

González Bollo, Hernán. *La teodicea estadística de Alejandro E. Bunge, 1880–1943*. Buenos Aires: UCA-Imago Mundi, 2012.

Goodman, David C. *The "Conflict Thesis" and Cosmology*, 30–49. Unit 1, *Science and Belief: from Copernicus to Darwin*. Milton Keynes: The Open University Press, 1974.

Gorriti, Juan Ignacio. *Reflecciones sobre las causas morales de las convulsiones interiores en los nuevos estados americanos.* Valparaíso: Imprenta del Mercurio, 1836.
Goudin, Antonio. *Philosophia iuxta inconcusa, tutissimaque Divi Thomae Dogmata.* Vol. 3. Venice: Typis Dominici Lovisa, 1729.
Gould, Benjamin A. "Address of the President." In *Proceedings of the American Association for the Advancement of Science…Eighteenth Meeting…1869*, 1–37. Cambridge, MA: Joseph Lovering, 1870.
Goyena, Pedro. "José Manuel Estrada." In idem, *Crítica literaria*, 135–50. Buenos Aires: La Cultura Argentina, 1917.
Gran Enciclopedia Argentina, edited by Diego Abad de Santillán, 8 vols. Buenos Aires: Ediar, 1956–63.
Greco, M. A., H. Grassi Gallegos and J. R. Seibold. "Solar Physics at the Observatorio Nacional de Física Cósmica, San Miguel, Argentina." *Solar Physics* 14 (1970): 503–7.
Gregory, Brad S. *The Unintended Reformation. How a Religious Revolution Secularized Society.* Cambridge, MA: Harvard University Press, 2012.
Gregory, Frederick. *Scientific Materialism in Nineteenth Century Germany.* Dordrecht: Reidel, 1977.
Grimoult, Cédric. "Albert Gaudry dans l'histoire de la paléontologie évolutioniste." *Physis. Rivista internazionale di storia della scienza* 36, no. 1 (1999): 73–97.
Groussac, Paul. *Los que pasaban.* Buenos Aires: Taurus, 2001.
Guasti, Niccolò. "From Expulsions to Canonical Suppression (1759–1773)." In *The Oxford Handbook of the Jesuits*, edited by Ines G. Županov, 918–49. New York: Oxford University Press, 2019.
Guido y Spano, Carlos. *Autobiografía.* Buenos Aires: Kapeluz, 1969.
Gutiérrez, Juan M. "D. Juan Manuel de Lavardén." In Idem, *Estudios biográficos y críticos*, 35–128. Buenos Aires: Imprenta del Siglo, 1865.
Gutiérrez, Juan M. *Noticias históricas sobre el orijen y desarrollo de la enseñanza pública superior en Buenos Aires.* Buenos Aires: Imprenta del Siglo, 1868.
Gutiérrez, Juan M. "Don Félix de Azara. Su mérito, sus servicios, su juicio sobre las misiones del Paraná y del Uruguay." *Revista de Buenos Aires* 18 (1869): 167–92.
Gutiérrez, Juan M. "Las restauraciones religiosas en 1835–1841–1875." *Revista del Río de la Plata* 11 (1875): 399–433.
Gutiérrez Lanza, Mariano. *La teoría de la relatividad.* Havana: Imprenta Molina, 1929.
Gvirtz, Silvina, M. Aisenstein, J. Cornejo and A. Valerani. "The Natural Sciences in the Schools: Tension in the Modernization Process of Argentine Society (1870–1960)." *Science and Education* 10, no. 6 (2001): 545–58.
Häfner, Claudia. *Heimischwerdung am La Plata. Von der Deutschen Evangelischen La Plata Synode zur Iglesia Evangélica del Río de la Plata.* Münster: LIT Verlag, 2008.
Hagood, Jonathan. "Arming and Industrializing Perón's 'New Argentina': The Transfer of German Scientists and Technology after World War II." *Icon* 11 (2005): 63–78.
Hahn, Oscar. "Estudio." In idem, *El cuento fantástico hispanoamericano en el siglo XIX*, 9–84. Mexico: Premia, 1978.
Hale, Frederick. "Appropriating the Closure of Jesuit Missions: Fritz Hochwälder's *Das heilige Experiment.*" *Acta Theologica* 28, no. 1 (2008): 58–75.
Halperin Donghi, Tulio. *Historia de la Universidad de Buenos Aires.* Buenos Aires: Eudeba, 1962.

Halperin Donghi, Tulio. *Una nación para el desierto argentino*. Buenos Aires: Prometeo, 2005.
Halperin Donghi, Tulio. *Revolución y Guerra. Formación de una elite dirigente en Argentina*, 3rd ed. Buenos Aires: Siglo XXI, 2014.
Halperin Donghi, Tulio, Iván Jaksic, Gwen Kirkpatrick and Francine Masiello, eds. *Sarmiento: Author of a Nation*. Berkeley: University of California Press, 1994.
Harris, Jonathan. "Rivadavia and Benthamite 'Discipleship'." *Latin American Research Review* 33, no. 1 (1998): 129–49.
Harris, Steven J. "Transposing the Merton Thesis: Apostolic Spirituality and the Establishment of the Jesuit Scientific Tradition." *Science in Context* 3, no. 1 (1989): 29–65.
Harris, Steven J. "Confession-Building, Long-Distance Networks, and the Organization of Jesuit Science." *Early Science and Medicine* 1, no. 3 (1996): 287–318.
Harris, Steven J. "Mapping Jesuit Science: the Role of Travel in the Geography of Knowledge." In *The Jesuits. Cultures, Sciences, and the Arts, 1540–1773*, edited by John W. O'Malley, Gauvin A. Bailey, Steven J. Harris, and T. Frank Kennedy, 212–40. Toronto: University of Toronto Press, 1999.
Harris, Steven J. "Jesuit Scientific Activity in the Overseas Missions, 1540–1773." *Isis* 96, no. 1 (2005): 71–79.
Harrison, Frederic and Herbert Spencer. *The Nature and Reality of Religion. A Controversy*. New York: Appleton, 1885.
Harrison, Peter. *"Religion" and the Religions in the English Enlightenment*. Cambridge: Cambridge University Press, 2002.
Harrison, Peter. *The Territories of Science and Religion*. Chicago: University of Chicago Press, 2015.
Harrison, Peter. "The Modern Invention of 'Science-and-Religion': What Follows?" *Zygon* 51, no. 3 (2016): 742–47.
Harrison, Peter. "Science and Secularization." In *Narratives of Secularization*, edited by Peter Harrison, 47–70. London: Routledge, 2018.
Harrison, Peter. "Conflict, Complexity, and Secularization in the History of Science and Religion." In *Rethinking History, Science, and Religion. An Exploration of Conflict and the Complexity Principle*, edited by Bernard Lightman, 221–42. Pittsburgh: University of Pittsbugh Press, 2019.
Hartog, Marcus. "Dynamic Interpretation of Cell Division." *Nature* 67, no. 1724 (13 November 1902): 42–43.
Hayes, Carlton. *Nationalism: A Religion*. New York: Macmillan, 1960.
Head, Brian W. *Ideology and Social Science. Destutt de Tracy and French Liberalism*. Dordrecht: Martinus Nijhoff, 1985.
Hentschke, Jens R. "Argentina's Escuela Normal de Paraná and Its Disciples: Mergers of Liberalism, Krausism, and Comtean Positivism in Sarmiento's Temple for Civilizing the Nation, 1870–1916." *Iberian and Latin American Studies* 17 no. 1 (2011): 1–31.
Hernández, Pablo. *El extrañamiento de los jesuitas del Río de la Plata*. Madrid: Victoriano Suárez, 1908.
Herr, Richard. *The Eighteenth-Century Revolution in Spain*. Princeton: Princeton University Press, 1958.
Herrero Ducloux, Enrique. "Profesor Juan Brèthes, 1871–1928." *Revista de la Facultad de Agronomía* 18, no. 1 (1928): 5–12.

Hsia, Ronnie Po-chia. "Introduction. Catholic Global Missions and the Expansion of Europe." In *A Companion to Early Modern Catholic Global Missions*, edited by Ronnie Po-chia Hsia, 1–13. Leiden: Brill, 2017.

Hodge, John E. "Benjamin Apthorp Gould and the Founding of the Argentine National Observatory." *The Americas* 28, no. 2 (1971): 152–75.

Hodge, John E. "Charles Dillon Perrine and the Transformation of the Argentine National Observatory." *Journal of the History of Astronomy* 8 (1977): 12–25.

Hodges, W. H., ed. *History of the Anglican Church St. John the Baptist (pro-cathedral). Buenos Aires, 1831–1931*. Buenos Aires: W. F. Story, 1932.

Holmberg, Eduardo L. "Arácnidos." In *Informe oficial de la Comisión científica agregada al Estado Mayor General de la Expedición al Río Negro*, vol. 1, 117–68. Buenos Aires: Ostwald y Martínez, 1881.

Holmberg, Eduardo L. *Carlos Roberto Darwin*. Buenos Aires: El Nacional, 1882.

Holmberg, Eduardo L. *Dos partidos en lucha*, edited by Sandra Gasparini. Buenos Aires: Corregidor, 2005.

Holmberg, Eduardo L. "Viajes a las Sierras del Tandil y de la Tinta (1880)." In idem, *Excursiones bonaerenses*, edited by Juan C. Chebez and Bárbara Gasparri, 51–208. Buenos Aires: Albatros, 2009.

Holmberg, Luis. *Holmberg, el último enciclopedista*. Buenos Aires, 1952.

"Homenaje a Darwin." *Anales del Círculo Médico Argentino* 5 (1882): 546–47.

Hosne, Ana C. *The Jesuit Missions to China and Peru, 1570–1610*. Abingdon: Routledge, 2013.

Houssay, Bernardo. "Concepto de la Universidad." In idem, *Escritos y Discursos*, 41–64. Buenos Aires: El Ateneo, 1942.

Houssay, Bernardo. "Ángel Gallardo y el porvenir de las ciencias en la Argentina." In idem, *Escritos y discursos*, 491–514. Buenos Aires: El Ateneo, 1942.

Houssay, Bernardo. "La ciencia necesita un ambiente de libertad." *Ciencia e Investigación* 2, no. 3 (March 1946): 97–98.

Hubeñak, Florencio. *Historia de la Universidad Católica Argentina*. Buenos Aires: Universidad Católica Argentina, 2016.

Humboldt, Alexander von. *Kosmos*. Vol. 3. Stuttgart and Tübingen: Gotta'scher Verlag, 1850.

Humboldt, Alexander von. *Personal Narrative of the Equinoctial Regions of America: During the Years 1799–1804*, translated by Thomasina Ross, 3 vols. (London: Henry Bohn, 1852–1853).

Hurtado de Mendoza, Diego. "Los primeros años de la AAPC (1933–1945)." *Ciencia e Investigación* 56, no. 2 (2004): 35–40.

Hurtado de Mendoza, Diego. "Autonomy, even Regional Hegemony: Argentina and the 'Hard Way' toward Its First Research Reactor (1945–1958)." *Science in Context* 18, no. 2 (2005): 285–308.

Hurtado de Mendoza, Diego and Analía Busala. *Los ideales de universidad "científica" (1931–1959)*. Buenos Aires: Libros del Rojas, 2002.

Hurtado de Mendoza, Diego and Analía Busala. "De la 'movilización industrial' a la 'Argentina científica': la organización de la ciencia durante el peronismo (1943–1955)." *Revista da Sociedade Brasileira de História da Ciência* 4, no. 1 (2006): 17–33.

Ibarguren, Carlos. *La literatura y la Gran Guerra*. Buenos Aires: Agencia General de Librería y Publicaciones, 1920.

Ibarguren, Carlos. *La historia que he vivido*. Buenos Aires: Dictio, 1977.

Iglesias, Fidel. "The Platine Clergy, Independence, and the Rivadavian Reforms, 1806–27." *Journal of Religious and Theological Information* 3, no. 2 (2000): 119–51.
Ignace of Loyola. "The Spiritual Exercises." In idem, *Personal Writings*, 279–360. London: Penguin, 2004.
Imbruglia, Girolamo. *The Jesuit Missions of Paraguay and a Cultural History of Utopia (1568–1789)*. Leiden: Brill, 2017.
"Ingeniero Augusto J. Durelli. Su visita a nuestro país." *Ciencia y Técnica* 124, no. 620 (July 1957): 234.
Ingenieros, José. "Los reptiles burgueses. I. Los que van al santuario." *La Montaña* 1, no. 2 (15 April 1897): 5.
Ingenieros, José. "La ciencia oficial y la Facultad de Ciencias Herméticas." *La Montaña* 1, no. 11 (1st September 1897): 6–7.
Ingenieros, José. "La religión de la raza." *Caras y caretas* no. 839 (31 October 1914): 47.
Ingenieros, José. "Los fundamentos de la psicología biológica." *Revista de Filosofía. Cultura, ciencias, educación* 1, no. 1 (1915): 442–71.
Ingenieros, José. *La locura en Argentina*. Buenos Aires: Editorial Buenos Aires, 1920.
Ingenieros, José. "La Universidad del Porvenir." *Ateneo* no. 3 (1920): 63–102.
Inman, Samuel G. "Parana, Exponent of North American Education." *Bulletin of the Pan American Union* 53 (1921): 463–74.
"Inventario de los documentos de la donación Segurola, recibidos por el Director de la Biblioteca Pública." *Revista de la Biblioteca Nacional* 4, no. 13 (1940): 15–73.
Iriarte, Tomás de. *Memorias. Rivadavia, Monroe y la guerra argentino-brasileña*. Buenos Aires: Sociedad Impresora Americana, 1945.
Isabelle, Arsène. *Voyage à Buenos Ayres et à Porto-Alègre*. Paris: J. Morlent, 1835.
Isern, Juan. *La formación del clero secular de Buenos Aires y la Compañía de Jesús*. Buenos Aires: San Miguel, 1936.
Ivereigh, Austen. "Franceschi y el movimiento católico integral, 1930–1943." *Criterio* no. 2081 (14 November 1991): 623–30; no. 2082 (28 November 1991): 660–68.
Ivereigh, Austen. *Catholicism and Politics in Argentina, 1810–1960*. New York: St. Martin's Press, 1995.
Jackson, Robert. "Mortality Crises in the Jesuit Missions of Paraguay, 1730–1740." *World History Review* 1, no. 2 (2004): 2–23.
Jaim Etcheverry, Guillermo, ed. *Retratos. Eduardo Braun Menéndez (1903–1959)*. Buenos Aires, 1989.
Jaim Etcheverry, Guillermo. "La concepción universitaria de Eduardo Braun Menéndez." *Medicina* 60 (2000): 149–54.
Jakob, Christfried. *Atlas des gesunden und kranken Nervensystems*. Munich: Lehmann, 1895.
Jakob, Christfried. "Del mecanismo al dinamismo del pensamiento. Estudio histórico-crítico de psicología orgánica." *Anales de la Facultad de Derecho y Ciencias Sociales* 18 (1898): 195–238.
Jakob, Christfried. *Das Menschenhirn*. Munich: Lehmann, 1911.
Jakob, Christfried. "El cultivo artificial del órgano del pensamiento." *Revista del Jardín Zoológico*, 2nd epoch, 10, no. 37 (1914): 9–22.
Jakob, Christfried. *Filosofía de la naturaleza. Un curso de conferencias*. Buenos Aires: Kraft, 1920.

Jakob, Christfried. "La religión de la naturaleza y el porvenir del hombre." *Humanidades* (La Plata) 22 (1930): 107–19.
Jakob, Christfried. *Folia neurobiológica argentina. Atlas*, 3 vols. Buenos Aires: Aniceto López, 1939–41.
Jakob, Christfried. *Folia neurobiológica argentina*, 5 vols. Buenos Aires: Aniceto López, 1941–46.
Jolís, José. *Ensayo sobre la historia natural del Gran Chaco*, translated by M. L. Acuña. Resistencia: Universidad Nacional del Nordeste, Facultad de Humanidades, Instituto de Historia, 1972.
Jovellanos, Gaspar Melchor de. "Oración pronunciada en el Instituto Asturiano sobre el estudio de las ciencias naturales." In idem, *Obras publicadas e inéditas*, ed. Cándido Nocedal, vol. 1, 335–42. Madrid: Rivadeneyra, 1858.
Kapusta, Pawel. "Darwin from *Humani generis* to the Present." In *Darwin and Catholicism. The Past and Present Dynamics of a Cultural Encounter*, edited by Louis Caruana, 27–42. London: T&T Clark, 2009.
Katra, William H. *The Argentine Generation of 1837: Echeverría, Alberdi, Sarmiento, Mitre*. London: Associated University Press, 1996.
Keller, Héctor. "Notas sobre medicina y magia entre los guaraníes de Misiones, Argentina. Un enfoque etnobotánico." *Suplemento Antropológico* 42, no. 2 (2007): 345–83.
Kelly, Alfred. *Descent of Darwin. The Popularization of Darwinism in Germany, 1860–1914*. Chapel Hill: The University of North Carolina Press, 1981.
Kennedy, Emmet. *A Philosophe in the Age of Revolution. Destutt de Tracy and the Origins of "Ideology."* Memoirs of the American Philosophical Society no. 129. Philadelphia: American Philosophical Society, 1978.
Kennedy, Emmet. "The Secularism of Destutt de Tracy's 'Ideology'" (2001). Accessed 17 September 2020. https://www.geisteswissenschaften.fu-berlin.de/v/grammaire_generale/Actes_du_colloque/Textes/Kennedy/Emmet_Kennedy.pdf.
Kennedy, John J. *Catholicism, Nationalism, and Democracy in Argentina*. Notre Dame: University of Notre Dame Press, 1958.
Kennedy, T. Frank. "Music and the Jesuit Mission in the New World." *Studies in the Spirituality of the Jesuits* 39, no. 3 (2007).
Kilgore, W. J. "Latin American Philosophy or the Place of Alejandro Korn." *Journal of Inter-American Studies* 2, no. 1 (1960): 77–82.
King, John. *Sur. A Study of the Argentine Literary Journal*. Cambridge: Cambridge University Press, 1986.
Kippenberg, Hans G. *Discovering Religious History in the Modern Age*, translated by Barbara Harshav. Princeton: Princeton University Press, 2002.
Kirwan, Richard. "On the Primitive State of the Globe and Its Subsequent Catastrophe." *Transactions of the Royal Irish Academy* 6 (1797): 233–308.
Klappenbach, Miguel A. *Larrañaga y el Viejo Museo*, Publicación Extra del Museo Nacional de Historia Natural y Antropología, no. 53. Montevideo: MNHN, 2004.
Knott, Kim. *The Location of Religion. A Spatial Analysis*. London: Equinox, 2001.
Knott, Kim. "The Secular Sacred: In between or both/and?" In *Social Identities between the Sacred and the Secular*, edited by Abby Day, Giselle Vincett and Christopher R. Cotter, 145–60. Farnham: Ashgate, 2013.

Kong, Lily and Orlando Woods. *Religion and Space. Competition, Conflict, and Violence in the Contemporary World*. London: Bloomsbury, 2016.
Korn, Alejandro. *Locura y crimen*. Buenos Aires: Imprenta de la Penitenciaría, 1883.
Korn, Alejandro. *La libertad creadora*. Buenos Aires: Losada, 1944.
Korn, Alejandro. "La evolución de las ciencias en la República Argentina." In idem, *Obras completas*, vol. 3, 603–7. Buenos Aires: Claridad, 1949.
Korn, Alejandro. *El pensamiento argentino*. Buenos Aires: Nova, 1961.
Kragh, Helge. "The Vortex Atom: A Victorian Theory of Everything." *Centaurus* 44 (2002): 32–114.
Kraglievich, Lucas. "Las teorías de Ameghino. Sobre una titulada réplica." In idem, *Obras de Geología y Paleontología*, edited by Alfredo J. Torcelli and Carlos Marelli, vol. 1, 65–76. La Plata: Taller de Impresiones Oficiales, 1940.
Kress, Lee B. "Argentine Liberalism and the Church under Julio Roca, 1880–1886." *The Americas* 30 (1974): 319–40.
Kuethe, Allan J. and Kenneth J. Andrien. *The Spanish Atlantic World in the Eighteenth Century: War and the Bourbon Reforms, 1713–1796*. Cambridge: Cambridge University Press, 2014.
Laclau, Narciso C. "El vitalismo de Hans Driesch." *Revista de Filosofía. Cultura, ciencias, educación* 23, no. 2 (March 1926): 297–326.
"La encíclica *Casti connubii*." *Criterio* no. 150 (8 January 1931): 85–86.
"La esterilización de los anormales mentales." *Iatría* 5, no. 1 (1934): 16–19.
"La eugenesia." *Criterio* no. 218 (5 May 1932): 130.
"La fe y los sabios." *F.V.D.* 7, no. 77 (July 1927).
Lafinur, Juan Crisóstomo. "Otro comunicado." *El Americano* no. 29, 8 October 1819, 7–11.
Lafinur, Juan Crisóstomo. *Curso filosófico dictado en Buenos Aires en 1819*. Buenos Aires: Instituto de Filosofía de la Facultad de Filosofía y Letras de la Universidad de Buenos Aires, 1938.
Lafuente, Antonio. "Las Academias militares y la inversión en ciencia en la España ilustrada (1750–1760)." *Acta Hispanica ad Medicinae Scientarumque Historiam Illustrandam* 2 (1982): 193–209.
Lafuente, Antonio. "Enlightenment in an Imperial Context: Local Science in the Late-Eighteenth-Century Hispanic World." *Osiris*, 2nd series, 15 (2000): 155–73.
Lafuente, Antonio, Alberto Elena and M. L. Ortega, eds. *Mundialización de la ciencia y cultura nacional*. Madrid: Universidad Autónoma-Doce Calles, 1993.
Lafuente Machain, Ricardo de. *El barrio de la Recoleta*. Buenos Aires: Municipalidad de Buenos Aires, Instituto Municipal de Extensión Artística, 1945.
"La iniciativa privada en los estudios superiores." *Ciencia e Investigación* 3, no. 2 (February 1947): 45–46.
"La juventud argentina de Córdoba a los hombres libres de Sud América." In *La Reforma Universitaria. Tomo I. El movimiento argentino*, edited by Gabriel del Mazo, 1–5. Lima: Universidad Nacional Mayor de San Marcos, 1967.
Lalouette, Jacqueline. "La querelle de la foi et de la science et le banquet Berthelot." *Revue historique* 300, no. 608 (1998): 825–44.
Lalouette, Jacqueline. *La république anticléricale, XIXe-XXe siècles*. Paris: Éditions du Seuil, 2002.

Lamarca, Emilio. "Necesidad de la acción social." *Revista Eclesiástica del Arzobispado de Buenos Aires* 10 (1910): 3–30.
Lambert, Robert A. "The Argentine Declaration." *Science* 101, no. 2629 (18 May 1945): 502.
La nacionalidad y la obra de Ameghino, 2nd ed. Buenos Aires: Imprenta de El Pueblo, 1917.
Landi Dessy, Jorge. "Charles Dillon Perrine y el desarrollo de la astrofísica en la República Argentina." *Boletín de la Academia Nacional de Ciencias* 48 (1970): 219–40.
Lanning, John Tate. *Academic Culture in the Spanish Colonies*. London: Oxford University Press, 1940.
Lanning, John Tate. "Tradition and Enlightenment in the Spanish Colonial Universities." *Cahiers d'histoire mondiale* 10 (1967): 705–21.
"La radio en el Congreso." In *XXXII Congreso Eucarístico Internacional*, vol. 1, 100–106. Buenos Aires: Comité ejecutivo del Congreso, 1935.
Laromiguière, Pierre. *Leçons de philosophie*, 2 vols. Paris: Brunot-Labbé, 1815–18.
[Larrañaga, Dámaso] "Note sur le Megaterium de Cuvier, l'Hydromis, et une variété nouvelle de Maïs." *Bulletin des Sciences (de la Société Philomatique de Paris)* 1823, 83.
Larrañaga, Dámaso. *Escritos*, 5 vols. Montevideo: Instituto Histórico y Geográfico del Uruguay, 1922–30.
Larrañaga, Dámaso. "Memoria geológica sobre la reciente formación del Río de la Plata deducida de sus conchas fósiles." In idem, *Escritos*, vol. 3, 7–20 (1924).
Larrañaga, Dámaso. "Tierra." In idem, *Escritos*, vol. 3, 29–36 (1924).
Larrañaga, Dámaso. "Oración inaugural." In idem, *Escritos*, vol. 3, 135–146 (1924).
Larrañaga, Dámaso. *Diario de Historia Natural: 1808–1814*, edited by Ariadna Islas. Montevideo: Biblioteca Artigas, 2015.
Larrañaga, Dámaso. *Diario de Historia Natural: 1813–1824*, edited by Ariadna Islas. Montevideo: Biblioteca Artigas, 2017.
Larsen, Juan Mariano. "Filología americana: la lengua quichua y el doctor López." *Revista de Buenos Aires* no. 84 (1869): 409–31.
Larson, Carolyne R. "'Argentine Man': Human Evolution and Cultural Citizenship in Argentina, 1911–1940." In *Making Citizens in Argentina*, edited by Benjamin Bryce and David M. K. Sheinin, 43–61. Pittsburgh: The University of Pittsburgh Press, 2017.
Lascano, Julio R. *Los estudios superiores en la historia de Buenos Aires*. Buenos Aires: Municipalidad de la Ciudad de Buenos Aires, 1981.
"Las universidades libres." *Ciencia e Investigación* 3, no. 1 (January 1947): 1–2.
Laura, Patricio. "Augusto J. Durelli." *Anales de la Academia Nacional de Ciencias Exactas, Físicas y Naturales* 39 (1987): 247–55.
Lavardén, Manuel José de. *Nuevo aspecto del comercio en el Río de la Plata*, edited by Enrique Wedovoy. Buenos Aires: Raigal, 1955.
Lavilla, Esteban O. and Guillermo Wilde, eds. *Los anfibios y reptiles del Paraguay Ilustrado de Joseph Sánchez Labrador (Rávena, 1776)*. Tucumán: Fundación Miguel Lillo, 2020.
Lederkremer, Rosa M. de and Eduardo G. Gros. "Venancio Deulofeu, 1902–1984." *Advances in Carbohydrate Chemistry and Biochemistry* 46 (1988): 11–15.
Lee, Lois. *Recognizing the Non-Religious. Reimagining the Secular*. Oxford: Oxford University Press, 2015.
Lee, Lois. "Feeling Rational: Affinity and Affinity Narrations in British Science-Non-Religion Relations." In *Science, Belief and Society: International Perspectives on Religion, Non-*

Religion and the Public Understanding of Science, edited by Stephen H. Jones, Tom Kaden and Rebecca Catto, 173–195. Bristol: Bristol University Press, 2019.

Leguizamón, Onésimo. *Memoria presentada al Congreso Nacional de 1875 por el Ministro de Justicia, Culto e Instruccion Publica.* Buenos Aires: Imprenta Americana, 1875.

Leguizamón, Onésimo. "Discurso del doctor Onésimo Leguizamón." *Anales del Círculo Médico Argentino* 8 (1885): 219–40.

Lehner, Ulrich L. *The Catholic Enlightenment.* Oxford: Oxford University Press, 2016.

Leon, Sharon M. *An Image of God. The Catholic Struggle with Eugenics.* Chicago: The University of Chicago Press, 2013.

Leonard, Virginia W. *Politicians, Pupils, and Priests. Argentine Education since 1943.* New York: Peter Lang, 1989.

Lértora Mendoza, Celina. *La enseñanza de la filosofía en tiempos de la colonia.* Buenos Aires: Fecic, 1979.

Lértora Mendoza, Celina. "Introducción de las teorías newtonianas en el Río de la Plata." In *Mundialización de la ciencia y cultura nacional*, edited by Antonio Lafuente, Alberto Elena and M. L. Ortega, 307–23. Madrid: Universidad Autónoma-Doce Calles, 1993.

Lértora Mendoza, Celina. "La enseñanza elemental y universitaria." In *Nueva Historia de la Nación Argentina*, edited by Academia Nacional de la Historia, vol. 3, 369–402. Buenos Aires: Planeta, 1999.

Lértora Mendoza, Celina. "Nollet y la difusión de Newton en el Río de la Plata." In *The Spread of the Scientific Revolution in the European Periphery*, edited by Celina Lértora Mendoza, Efthymios Nicolaïdis and Jan Vandersmissen, 123–36. Turnhout: Brepols, 2000.

Lessl, Thomas M. *Rhetorical Darwinism. Religion, Evolution and the Scientific Identity.* Waco: Baylor University Press, 2012.

Levene, Ricardo, ed. *Historia de la Nación Argentina*, 3rd ed., 15 vols. Buenos Aires: El Ateneo, 1961.

Levine, Alex and Adrian Novoa, eds. *¡Darwinistas! The Construction of Evolutionary Thought in Nineteenth Century Argentina.* Leiden: Brill, 2012.

Lewis, Juan T. *Antecedentes, títulos y trabajos presentados a la Universidad Nacional del Litoral al hacerse cargo de la Cátedra de Fisiología de la Escuela de Medicina de Rosario.* Buenos Aires: Frascoli & Bindi, 1929.

Lewis, Juan T. *Comentarios sobre el proyecto ministerial de ley universitaria. Un anteproyecto de ley universitaria.* Rosario, 1932.

Lewis, Juan T. *Influencia de la investigación científica sobre la sociedad.* Buenos Aires: Las Ciencias, 1934.

Lewis, Juan T. "Objeto y fin de la Universidad." *Universidad* (Universidad del Litoral) 1 (October 1935): 37–63.

Lewis, Juan T. "La Universidad no debe callar." *Mundo Médico* 10, no. 108 (1945): 21–23.

Lewis, Juan T. [Answer to a survey] *Orden Cristiano* 5, no. 97 (1 October 1945): 22.

Lewis, Juan T. "El hombre de ciencia." *Revista de la Asociación Médica Argentina* 62, no. 631–32 (15 and 30 June 1948): 279–80.

Lewis, Juan T. "El Instituto de Investigaciones Médicas de Rosario." *Ciencia e Investigación* 5, no. 1 (1949): 1–2, 37–39.

Lewis, Juan T. "La vocación intelectual." *Criterio* no. 1201–1202 (24 December 1953): 990–93.

Lewis, Juan T. "El magisterio de la Iglesia." *Criterio* no. 1230 (24 February 1955): 127–29.
Lewis, Jimmy. "El doctor Juan T. Lewis." *Rosario. Su historia y región* no. 83 (March 2010): 8–10.
Lezaun, Secundino. "Esterilización masculina." *Iatría* 5, no. 1 (1934): 3–15.
Lida, Miranda. "Los orígenes del catolicismo de masas en la Argentina, 1900–1934." *Jahrbuch für Geschichte Lateinamerikas* 46 (2009): 345–70.
Lida, Miranda. *Monseñor Miguel de Andrea. Obispo y hombre de mundo (1877–1960)*. Buenos Aires: Edhasa, 2013.
Lida, Miranda. *Historia del catolicismo en la Argentina entre el siglo XIX y el XX*. Buenos Aires: Siglo XXI, 2015.
Lida, Miranda and Mariano Fabris, eds. *La revista* Criterio *y el siglo XX argentino. Religión, cultura y política*. Rosario: Prohistoria, 2019.
Lightman, Bernard. "Unbelief." In *Science and Religion around the World*, edited by John H. Brooke and Ronald L. Numbers, 252–77. Oxford: Oxford University Press, 2011.
Lightman, Bernard. "Does the History of Science and Religion Change Depending on the Narrator? Some Atheist and Agnostic Perspectives." *Science and Christian Belief* 24 (2012): 149–68.
Lightman, Bernard. "The Victorians: Tyndall and Draper." In *The Warfare between Science and Religion. The Idea That Wouldn't Die*, edited by Jeff Hardin, Ronald Numbers and Ronald A. Benzley, 65–83. Baltimore: Johns Hopkins University Press, 2018.
Lightman, Bernard. "Introduction." In *Rethinking History, Science, and Religion. An Exploration of Conflict and the Complexity Principle*, edited by Bernard Lightman, 3–16. Pittsburgh: University of Pittsburgh Press, 2019.
Lightman, Bernard and Michael S. Reidy. *The Age of Scientific Naturalism*. London: Pickering and Chatto, 2014.
Lindberg, David C. and Ronald L. Numbers, eds. *God and Nature. Historical Essays on the Encounter between Christianity and Science*. Berkeley: University of California Press, 1986.
Linnaeus, Carolus. *Miscellaneous Tracts Relating to Natural History, Husbandry, and Physick*, edited by Benjamin Stillingfleet, 3rd ed. London: R. and J. Dodsley, 1762.
Littera annua [of the Jesuit province of *Paraquaria*] for the years 1635–1637. In *Cartas anuas de la Provincia del Paraguay, Chile y Tucumán, de la Compañía de Jesús (1615–1637)*, edited by Carlos Leonhardt, 385–775. Buenos Aires: Peuser, 1929.
Livingstone, David N. "Science, Region, and Religion: The Reception of Darwinism in Princeton, Belfast, and Edinburgh." In *Disseminating Darwinism: The Role of Place, Race, Religion, and Gender*, edited by Ronald L. Numbers and John Stenhouse, 7–38. Cambridge: Cambridge University Press, 1999.
Livingstone, David N. *Putting Science in Its Place. Geographies of Scientific Knowledge*. Chicago: The University of Chicago Press, 2003.
Livingstone, David N. "Science, Religion, and the Cartographies of Complexity." *Historically Speaking* 8 (2007): 15–16.
Livingstone, David N. "Which Science? Whose Religion?" In *Science and Religion around the World*, edited by John H. Brooke and Ronald L. Numbers, 278–96. Oxford: Oxford University Press, 2011.

Livingstone, David N. "Scientific Inquiry and the Missionary Enterprise." In *Participating in the Knowledge Society. Researchers beyond the University Walls*, edited by Ruth Finnegan, 50–64. Basingstoke: Palgrave Macmillan, 2015.
Llambías, Héctor A. "Límites de la libertad." *Criterio* no. 554 (13 October 1938): 171–72.
Llanas Jubero, Eduardo. *Los seis días de la Creación*. Barcelona: Fidel Giró, 1889.
Lóizaga, Ludovico L. *Tristán Achával Rodríguez. Su vida y su obra*. Buenos Aires: L. J. Rosso, 1930.
López, José A. "Observatorio Astronómico Félix Aguilar." In *Astronomía*, edited by Simón Gershanik and Luis A. Milone, 201–12. Vol. 7, *Evolución de las ciencias en la República Argentina, 1923–1972*. Buenos Aires: Sociedad Científica Argentina, 1979.
López, Octavio V. "Cristianismo y eugenesia." *Anales de biotipología, eugenesia y medicina social* 1, no. 9 (1 August 1933): 6–7.
López, Vicente F. *Les races aryennes du Pérou. Leur langue, leur religión, leur histoire*, translated by Gaston Maspero. Paris: Franck, 1871.
López, Vicente F. *Historia de la República Argentina*, 10 vols. Buenos Aires: La Facultad, 1911.
López, Vicente F. "Introducción." In José María Ramos Mejía, *La neurosis de los hombres célebres en la República Argentina*, 79–88. Buenos Aires: La Cultura Argentina, 1915.
López Silvestre, Federico A. "Dios y la naturaleza. El desarrollo de la teología natural en España y en Galicia en el s. XVIII." *Semata. Ciencias sociais e Humanidades* 14 (2002): 417–53.
"Los sabios cristianos." *F.V.D.* 12, no. 139 (September 1932): 250; 12, no. 140 (October 1932): 288; 12, no. 141 (November 1932): 325.
[Love, George T.] *Five Years' Residence in Buenos Ayres, during the Years 1820 to 1825…by an Englishman*. London: G. Hebert, 1825.
Löwith, Karl. *Meaning in History*. Chicago: The University of Chicago Press, 1949.
Lozano, Pedro. *Historia de la conquista del Paraguay, Río de la Plata y Tucumán*, edited by Andrés Lamas, 5 vols. Buenos Aires: Imprenta Popular, 1873–75.
Lozano, Pedro. *Descripción Corográfica del Gran Chaco Gualamba*, edited by Radamés Altieri. Tucumán: Universidad Nacional de Tucumán, Instituto de Antropología, 1941.
Lubac, Henri de. *The Drama of Atheist Humanism*, translated by Edith M. Riley, Anne E. Nash and Mark Sebanc. San Francisco: Ignatius Press, 1995.
Ludueña, Gustavo A. "Popular Epistemologies and 'Spiritual Science' in Early Twentieth-Century Buenos Aires." In *Handbook of Religion and the Authority of Science*, edited by James R. Lewis and Olav Hammer, 609–31. Leiden: Brill, 2010.
Lue, Thomas. "L'apologétique de Ferdinand Brunetière et le positivisme: un bricolage idéologique 'génereux et acueillant'." *Revue des sciences philosophiques et théologiques* 87, no. 1 (2003): 101–26.
Lugones, Leopoldo. *El tamaño del espacio. Un ensayo de psicología matemática*. Buenos Aires: El Ateneo, 1921.
Lugones, Leopoldo. "Los Mundos." *Nosotros* 3, no. 7 (1938): 165–74.
Lugones, Leopoldo. *Strange Forces*, translated by Gilbert Alter-Gilbert. Pittsburgh: Latin America Literary Review Press, 2011.
Lugones, Leopoldo. *Estudios esotéricos*, edited by Pedro L. Barcia. Buenos Aires: Docencia, 2018.
Luna, Félix. *Alvear*. Buenos Aires: Hyspamerica, 1986.

Lustig, Andrew J. "Erich Wasmann, Ernst Haeckel, and the Limits of Science." *Theory in Bioscience* 121, no. 3 (2002): 252–59.
Luys, Jean-Bernard. *Traité clinique et pratique des maladies mentales*. Paris: Delahaye and Lecrosnier, 1881.
Lynch, John. *The Spanish American Revolutions, 1808–1826*, 2nd ed. New York: Norton, 1986.
Lynch, John. *Argentine Caudillo. Juan Manuel de Rosas*. Wilmington, DE: Scholarly Resources Books, 2001.
Lynch, John. *Edad Moderna. Crisis y recuperación, 1598–1808*. Vol. 5, *Historia de España*. Barcelona: Crítica, 2005.
MacCormack, Sabine. "Grammar and Virtue: The Formulation of a Cultural and Missionary Program by the Jesuits in Early Colonial Peru." In *The Jesuits II. Cultures, Sciences and the Arts, 1540–1773*, edited by John W. O'Malley, Gauvin A. Bailey, Steven J. Harris, and T. Frank Kennedy, 576–601. Toronto: University of Toronto Press, 2006.
Mac Donagh, Emiliano J. "Ensayo sobre la regulación orgánica en Hans Driesch." *Signo* no. 8 (1922): 246–59.
Mac Donagh, Emiliano J. "Sobre un estudio de Darwin por su *Voyage*." *Arx* 1, no. 1 (1924): 1–14.
Mac Donagh, Emiliano J. "La difícil doctrina de Hans Driesch." *Criterio* no. 29 (20 September 1928): 363–65.
Mac Donagh, Emiliano J. "Apología de un plural inusitado." *Criterio* no. 33 (18 October 1928): 83–84.
Mac Donagh, Emiliano J. Review of *Los orígenes de la vida*, by D. L. Saint-Ellier. *Criterio* no. 38 (22 November 1928): 251.
Mac Donagh, Emiliano J. "El peligro de un buen ejemplo." *Criterio* no. 44 (3 January 1929): 9–12.
Mac Donagh, Emiliano J. "A favor de los símiles." *Criterio* no. 56 (28 May 1929): 403–5.
Mac Donagh, Emiliano J. "La libertad intelectual del investigador." *Criterio* no. 428 (14 May 1936): 36–39; no. 429 (21 May 1936): 63–65.
Mac Donagh, Emiliano J. "Darwin no es para los niños." *Criterio* no. 1169 (14 August 1952): 561–65.
Mac Donagh, Emiliano J. "Carlos Darwin y el Origen de las especies." *Anales de la Academia Argentina de Geografía* 3 (1959): 20–32.
Mac Donagh, Emiliano J. *150 Años de Evolución Científica Argentino-Británica*. La Plata: CICBA, 1960.
Machoni, Antonio. *Arte y vocabulario de la lengua lule y tonocote*. Madrid: Herederos de Juan García Infanzón, 1732.
Maeder, Ernesto J. *Misiones del Paraguay. Construcción jesuítica de una sociedad cristiano-guaraní (1610–1768)*. Resistencia: Contexto, 2013.
Maeder, Ernesto J. "Libros, bibliotecas, control de lecturas e imprentas rioplatenses en los siglos XVI al XVIII." *Teología* 77 (2001): 5–24.
Maeder, Ernesto J. and A. S. Bolsi. "La población de las misiones guaraníes." *Estudios Paraguayos* 2, no. 1 (1974): 111–37.
Magendie, François. *Précis élémentaire de physiologie*, 2 vols. Paris: Méquignon-Marvis, 1816–1817.
Mallimaci, Fortunato. *El catolicismo integral en la Argentina (1930–1945)*. Buenos Aires: Biblos, 1988.

Mallimaci, Fortunato. "Catolicismo y liberalismo. Las etapas del enfrentamiento por la definición de la modernidad religiosa en América Latina." In *La modernidad religiosa. Europa latina y América Latina en perspectiva comparada*, edited by Jean-Pierre Bastian, translated by Dulce M. López Vega, 19–44. Mexico: FCE, 2004.

Mallimaci, Fortunato. "Laïcité de subsidiarité en Argentine: entre élargissement des droits sociaux et recherche mutuelle de légitimité." In *Laïcité, laïcités. Reconfigurations et nouveaux défis*, edited by Jean Baubérot, Micheline Milot and Philippe Portier, 221–52. Paris: Éditions de la Maison des sciences de l'homme, 2014.

Mallimaci, Fortunato. *El mito de la Argentina laica. Catolicismo, política y estado*. Buenos Aires: Capital Intelectual, 2016.

"Manifiesto de los demócratas cristianos en auspicio de la fórmula presidencial Tamborini-Mosca." *Orden Cristiano* no. 104 (February 1946): 412–15.

Manuel, Frank E. *The Prophets of Paris*. Cambridge, MA: Harvard University Press, 1962.

[Mariño, Cosme] "*La Unión* y el espiritismo." *Constancia* 6, no. 7 (30 July 1883): 193–96.

[Mariño, Cosme] "El arzobispo señor Aneiros y el Espiritismo." *Constancia* 10, no. 130 (30 January 1887): 1–2.

Mariño, Cosme. *El espiritismo en la Argentina*. Buenos Aires: Constancia, 1963.

Marion, Corinne. "Auguste Nicolas. Un laïc apologiste au XIXe siècle." *Communio* no. 231–232 (2014): 121–32.

Maritain, Jacques. *Théonas*. Paris: Nouvelle librairie nationale, 1921.

Maritain, Jacques. "À propos de la question juive." *La Vie Spirituelle* 2 (1921): 305–10.

Maritain, Jacques. "De la métaphysique des physiciens." *La Revue universelle* no. 10 (15 August 1922): 426–45.

Maritain, Jacques. "Nota sobre la cuestión judía." *Criterio* no. 33 (9 August 1934): 356–57.

Maritain, Jacques. *Science and Wisdom*, translated by Bernard Wall. London: Geoffrey Bles, 1940.

Márquez Miranda, Fernando. *Ameghino. Una vida heroica*. Buenos Aires: Nova, 1951.

Martin, David. *A General Theory of Secularization*. New York: Harper, 1978.

Martin, David. *On Secularization. Towards a Revised Theory*. Aldershot: Ashgate, 2005.

Martin, David. "Does the Advance of Science Mean Secularisation?" *Science and Christian Belief* 19 no. 1 (2007): 3–14.

Martin, Eric C. "Science and Ideology." In *Internet Encyclopedia of Philosophy*. Article revised in 2020. https://iep.utm.edu/sci-ideo/.

Martín Martín, Carmen and José L. Valverde, eds. *La farmacia en la América colonial: el arte de preparar medicamentos*. Granada: Universidad de Granada y Hermandad Farmacéutica Granadina, 1995.

Martínez, Ignacio. *Una nación para la Iglesia argentina*. Buenos Aires: Academia Nacional de la Historia, 2013.

Martínez, Ignacio and Diego Mauro. "Ctéato y Éurito. Iglesia, religión y poder político en la Argentina en el siglo XX." In *Fronteras disputadas: religión, secularización y anticlericalismo en la Argentina (siglos XIX y XX)*, edited by Roberto Di Stefano and José Zanca, 1–34. Buenos Aires: Imago Mundi, 2016.

Martínez, José M. "La arquitectura de *Las fuerzas extrañas* de Leopoldo Lugones: analogía universal y taxonomías de lo fantástico." *Bulletin of Hispanic Studies* 96, no. 5 (2019): 533–52.

Martínez de Aldunate, Francisco J. and José Elías del Carmen Pereyra. *Conclusiones sobre toda la filosofía* (Buenos Aires: Imprenta de los Niños Expósitos, 1790). In Enrique Martínez Paz, "Una tesis de filosofía del siglo XVIII de la Universidad de Córdoba." *Revista de la Universidad de Córdoba* 6 (1919): 228–86.

Massini Ezcurra, José M. *Los Argerich*. Buenos Aires: Instituto Amigos del Libro Argentino, 1955.

Mather, Kirtley F. and Harry Grundfest. "The Argentine Citizens Declaration." *Science* 99, no. 2566 (3 March 1944): 176.

Matignon, Ambroise. *Los muertos y los vivos. Conferencias acerca de las comunicaciones de ultratumba*. Barcelona: Tipográfica Católica, 1872.

Maton, George W. "Description of Seven New Species of Testacea." *Transactions of the Linnean Society* 10 (1811): 325–33.

Mauro Diego. "*I popolari* en la Argentina. Luigi Sturzo y el antifascismo católico de entreguerras." *Anuario IEHS* no. 29–30 (2011): 267–88.

Mauro, Diego and José A. Zanca, eds. *La reforma universitaria cuestionada*. Rosario: HyA ediciones, 2018.

Mavor, William. *Natural History for the Use of Schools*. London: Printed for R. Phillips, 1800.

Mayo, Carlos A. *Los betlemitas en Buenos Aires: convento, economía y sociedad*. Sevilla: Publicaciones de la Excma. Diputación Provincial de Sevilla, 1991.

Mayordomo, Alejandro. "La recepción en España de la obra de J. W. Draper. Una perspectiva del conflicto entre religión y ciencia." *Historia de la educación. Revista Interuniversitaria* 4 (1985): 139–58.

McGeagh, Robert. "Catholicism and Sociopolitical Change in Argentina: 1943–1973." PhD Diss., University of New Mexico, 1974.

McGeagh, Robert. *Relaciones entre el poder político y eclesiástico en la Argentina*. Buenos Aires: Itinerarium, 1987.

McLeod, Hugh. *Secularization in Western Europe, 1848–1914*. Basingstoke: Macmillan, 2000.

McLeod, Hugh and Werner Ustorf, eds. *The Decline of Christendom in Western Europe, 1750–2000*. Cambridge: Cambridge University Press, 2003.

McMullin, Ernan. "Galileo's Theological Venture." In *The Church and Galileo*, edited by Ernan McMullinn, 88–116. Notre Dame: University of Notre Dame Press, 2005.

Medan, Diego. *Pedro Arata. Un viaje a los cimientos de la Argentina*. Buenos Aires: Eudeba, 2017.

Medina, José Toribio. *La Inquisición en las Provincias del Plata*. Santiago de Chile: Imprenta Elzeviriana, 1899.

Megill, Allan. "Recounting the Past: 'Description,' Explanation and Narrative in Historiography." *American Historical Review* 94, no. 3 (1989): 627–53.

Mejía, Jorge M. "Educación y libertad." *Criterio* no. 1316 (25 September 1958): 683–87.

Melià, Bartomeu. *El guaraní conquistado y reducido. Ensayos de etnohistoria*. Asunción: Centro de Estudios Antropológicos, Universidad Católica, 1986.

Memoria de la Penitenciaría Nacional correspondiente al año 1894 acompañada de un retrospecto. Buenos Aires: Penitenciaría Nacional, 1895.

Méndez, Jesús. "Church-State Relations in Argentina in the Twentieth Century: A Case Study of the Thirty-second International Eucharistic Congress." *Journal of Church and State* 27 (1985): 223–43.

Menéndez Pelayo, Marcelino. *Historia de los heterodoxos españoles*. Vol. 3. Madrid: Librería católica de San José, 1881.

Menéndez Pelayo, Marcelino, ed. *Antología de poetas hispano-americanos*, 4 vols. Madrid: Sucesores de Rivadeneyra, 1893–95.

Métraux, Alfred. "Le shamanisme chez les Indiens de l'Amérique du Sud tropicale." *Acta Americana* 2 (1944): 197–219; 320–41.

Métraux, Alfred. "The Guaraní." In *The Tropical Forest Tribes*, edited by Julian H. Stewart, 69–94. Vol. 3, *Handbook of South-American Indians*. Smithsonian Institution, Bureau of American Ethnology, Bulletin 143. Washington: Government Publishing Office, 1948.

Meyer, Camilo. *Conferencias de astronomía popular dadas en 1912 y 1913*. Buenos Aires: Kidd, 1916.

Meyer Arana, Alberto. *La caridad en Buenos Aires*, 2 vols. Buenos Aires, 1911.

Mignone, Emilio F. *Política y universidad. El estado legislador*. Buenos Aires: Lugar Editorial, 1998.

Míguez, Eduardo. *Bartolomé Mitre. Entre la nación y la historia*. Buenos Aires: Edhasa, 2018.

Míguez, Eduardo. "Reforma y primitivismo. Tierra y fiscalidad en el Río de la Plata, de la colonia a la independencia." In *Cambio institucional y fiscalidad. Mundo hispánico, 1760–1850*, edited by Michel Bertrand and Zacarías Moutoukias, 287–303. Madrid: Casa de Velázquez, 2018.

Míguez Bonino, José. *Faces of Latin American Protestantism*, translated by Eugene L. Stockwell. Grand Rapids: William B. Eermans, 1997.

Millé, Andrés. *La Recoleta de Buenos Aires. Una visión del siglo XVIII*. Buenos Aires: Talleres Gráficos de Domingo Taladriz, 1952.

Miller, Richard A. "New Spectroheliograph at Manila Observatory." *Applied Optics* 4, no. 9 (1965): 1085–89.

Millones Figueroa, Luis and Domingo Ledesma, eds. *El saber de los jesuitas. Historias naturales y el nuevo mundo*. Frankfurt: Vervuert; Madrid: Iberoamericana, 2005.

Minelli, Gustavo. *Curso de Historia Universal. Discurso preliminar*. Buenos Aires: Imprenta La Tribuna, 1862.

Ministerio de Educación, Secretaría de Cultura. *Junta Nacional de Intelectuales. Antecedentes de su creación, decretos y reglamento interno*. Buenos Aires, 1949.

Ministerio de Educación y Justicia. *La Revolución Libertadora y la Universidad, 1955–1957*. Buenos Aires: Ministerio de Educación y Justicia, 1957.

Ministerio de Educación y Justicia. *Antecedentes sobre reglamentación legal de las universidades privadas*. Buenos Aires, 1958.

Ministerio de Justicia e Instrucción Pública. *Ley Universitaria. Mensaje y proyecto de ley*. Buenos Aires: Talleres Gráficos de la Penitenciaría Nacional, 1932.

Mitre, Bartolomé. Letter to Diego Barros Arana, 20 October 1975. In *Correspondencia literaria, años 1859–1881*, 48–78. Vol. 20, *Archivo del General Mitre*. Buenos Aires: Biblioteca de La Nación, 1912.

Molinari, José L. "Hospitales coloniales." In idem, *Historia de la medicina argentina. Tres conferencias*, 71–110. Buenos Aires: López, 1937.

Molinari, José L. "La reforma de las instituciones médicas por la Asamblea de 1813 y en la época de Rivadavia." *Historia* (Buenos Aires) 9, no. 32 (1963): 119–35.

"Monseñor Dr. Fortunato J. Devoto. Nuevo Obispo Auxiliar de la Arquidiócesis." *Revista Eclesiástica del Arzobispado de Buenos Aires* 27 (1927): 711.

Monti, Daniel P. *Presencia del protestantismo en el Río de la Plata durante el siglo XIX.* Buenos Aires: La Aurora, 1969.

Montserrat, Marcelo. "Sarmiento, propulsor de la ciencia." *Ciencia e Investigación* 42, no. 5 (1988): 277–83.

Montserrat, Marcelo. *Ciencia, historia y sociedad en la Argentina del siglo XIX.* Buenos Aires: CEAL, 1993.

Montserrat, Marcelo. "The Evolutionist Mentality in Argentina: an Ideology of Progress." In *The Reception of Darwinism in the Iberian World*, edited by Thomas Glick, Miguel Ángel Puig-Samper and Rosaura Ruiz, 1–27. Dordrecht: Springer, 2001.

Montenegro, Pedro. *Materia medica misionera.* Buenos Aires: Biblioteca Nacional, 1945.

Moore, James R. *The Post-Darwinian Controversies. A Study of the Protestant Struggle to Come to Terms with Darwin in Great Britain and America, 1870–1900.* Cambridge: Cambridge University Press, 1979.

More, Thomas. *Utopia*, ed. Peter Marshall. New York: Simon and Schuster, 1988.

Morello, Gustavo. "'It Takes Two to Tango': The Religious and Secular in Argentina's Political Dance, 1860–1960." In *Laicidad and Religious Diversity in Latin America*, edited by Juan M. Vaggione and José M. Morán Faúndes, 57–72. Cham: Springer, 2017.

Moreno, Francisco P. *Viaje a la Patagonia Austral.* Vol. 1. Buenos Aires: Imprenta de La Nación, 1879.

[Moreno, Manuel]. *Impugnación a la respuesta dada al mensaje del gobierno.* Buenos Aires: Imprenta del Estado, 1827.

[Moreno, Mariano]. "El editor a los habitantes de esta América." In *Del Contrato Social...por el ciudadano de Ginebra Juan Jacobo Rousseau*, edited by Mariano Moreno. Buenos Aires: Real Imprenta de Niños Expósitos, 1810.

Mörner, Magnus. *The Political and Economic Activities of the Jesuits in the La Plata Region; the Hapsburg Era.* Stockholm: Victor Pettersons Bokindustri Aktiebolag, 1953.

Mörner, Magnus. "Introduction." In *The Expulsion of the Jesuits from Latin America*, edited by Magnus Mörner, 3–30. New York: Alfred A. Knopf, 1965.

Mörner, Magnus. "The Role of the Jesuits in the Transfer of Secular Baroque Culture to the Río de la Plata Region." In *The Jesuits. Cultures, Sciences, and the Arts, 1540–1773*, edited by John W. O'Malley, Gauvin A. Bailey, Steven J. Harris, and T. Frank Kennedy, 305–16. Toronto: University of Toronto Press, 1999.

Mornet, Daniel. *Les sciences de la nature en France au XVIIIe siècle.* Paris: Armand Colin, 1911.

Mossotti, Ottaviano F. "Mémoire sur le climat de la ville de Buenos-Ayres." *Comptes rendus de la Académie des Sciences* 1835, pt. 1, 283.

Mossotti, Ottaviano F. "Observations of the Transit of Mercury over the Sun's Disc, in May 1832; and of the Comet of Encke in June 1832: at Buenos Ayres." *Monthly Notices of the Royal Astronomical Society* 3 (1833–36): 37–38.

Mossotti, Ottaviano F. "On the Forces Which Regulate the Internal Constitution of Bodies." In *Scientific Memoirs*, edited by Richard Taylor, vol. 1, 448–69. London: Richard and John E. Taylor, 1837.

Mossotti, Ottaviano F. and Thomas Henderson. "Places of Encke's Comet from Observations in Buenos Ayres." *Memoirs of the Royal Astronomical Society* 1835, 244–50.

Mühn, Juan B. "Primera exposición entomológica argentina, 19 al 25 de septiembre de 1928." *Estudios* 36 (1928): 151–56.

Mulhall, Michael G. and Edward T. *Handbook of the River Plate*, 5th ed. Buenos Aires: Mulhall; London: Trübner, 1885.

Mulleady, Ricardo T. *Breve historia de la telefonía argentina, 1886–1956*. Buenos Aires: Kraft, 1957.

Muniaín, Claudia and Jesús Ortea. "The Taxonomic Status and Redescription of *Polycera marplatensis* Franceschi, 1928 (Nudibranchia: Polyceratidae) from Argentina." *The Veliger* 61 (1998): 142–47.

Municipalidad de la Capital, *Documentos y planos relativos al período edilicio colonial de la ciudad de Buenos Aires*, 5 vols. Buenos Aires: Peuser, 1910.

Muratori, Ludovico. *Il cristianesimo felice nelle missioni de' Padri della Compagnia di Gesù nel Paraguai*. Venice: Giambattista Pasquali, 1752.

Navarro, Jaume. "The Conflicting Circulation of the Conflict Thesis." *Zygon* 54, no. 4 (2019): 1107–24.

Navarro Viola, Alberto. Review of *La vida y el transformismo moderno*, by Pedro S. Alcacer. *Anuario Bibliográfico de la República Argentina* 4 (1883): 215–16.

Navarro Viola, Alberto. Review of *Locura y crimen*, by Pedro S. Alcacer. *Anuario Bibliográfico de la República Argentina* 5 (1884): 284.

Navarro Viola, Alberto. Review of *Locura y crimen*, by Alejandro Korn. *Anuario Bibliográfico de la República Argentina* 5 (1884): 285.

Nebel, Graco (José M. Blanco). "¡Ameghino filósofo!" *Estudios* 13 (1917): 277–79.

New Catholic Encyclopedia, 2nd ed., 15 vols. Detroit: Thomson/Gale; Washington, D.C.: The Catholic University of America, 2003.

Newman, Carlos and Barry Poulson. "Purely Animal: Pastoral Production and Early Argentine Economic Growth, 1825–1865." *Explorations in Economic History* 35, no. 3 (1998): 325–45.

Newson, Linda A. "Medical Practice in Early Colonial Spanish America: A Prospectus." *Bulletin of Latin American Research* 25, no. 3 (2006): 367–91.

Nicolai, Georg F. "La vida." *Revista de Filosofía. Cultura, ciencias, educación* 16, no. 4 (July 1922): 65–76.

Nicolai, Georg F. "La base biológica del relativismo científico y sus complementos absolutos." *Revista de la Universidad Nacional de Córdoba* 12, no. 1–3 (January–March 1925): 1–182.

Nicolai, Georg F. "La ciencia y la moral." *Revista de Filosofía. Cultura, ciencias, educación* 26, no. 3 (May 1927): 301–17.

Nicolai, Georg F. "Las definiciones de la vida." *Revista de Filosofía. Cultura, ciencias, educación* 26, no. 4 (July 1927): 1–21.

Nicolas, Auguste. *Études philosophiques sur le christianisme*, 4 vols. Paris: A. Vaton, 1842–1845.

Nicolas, Auguste. *Estudios filosóficos sobre el cristianismo*, 3 vols. Madrid: Rivadeneyra, 1845.

Nicolau, Juan C. "Historia de la Sociedad Científica Argentina en el siglo XIX (1872–1900)." *Anales de la Sociedad Científica Argentina* 231, no. 1 (2002): 5–72.

Nicolau, Juan Carlos. *Ciencia y técnica en Buenos Aires, 1800–1860*. Buenos Aires: Eudeba, 2005.

Noakes, Richard. *Physics and Psychics. The Occult and the Sciences in Modern Britain*. Cambridge: Cambridge University Press, 2019.

Noel, Charles. "Clerics and Crown in Bourbon Spain, 1700–1808: Jesuits, Jansenists, and Enlightened Reforms." In *Religion and Politics in Enlightened Europe*, edited by James E. Bradley and Dale K. Van Kley, 119–53. Notre Dame: University of Notre Dame Press, 2001.

Nongbri, Brent. *Before Religion. A History of a Modern Concept.* New Haven: Yale University Press, 2013.

Noonan, John T., Jr. *Contraception. A History of Its Treatment by Catholic Theologians and Canonists.* Cambridge, MA: Harvard University Press, 1965.

Normandin, Sebastian. "Claude Bernard and *An Introduction to the Study of Experimental Medicine:* 'Physical Vitalism,' Dialectic, and Epistemology." *Journal of the History of Medicine and Allied Sciences* 62, no. 4 (2007): 495–528.

"Nota al Exmo. Sr. Ministro de R. E. y Culto sobre nombramientos en la curia." *Revista Eclesiástica del Arzobispado de Buenos Aires* 23 (1923): 379.

Novoa, Adriana. "The Rise and Fall of Spencer's Evolutionary Ideas in Argentina, 1870–1910." In *Global Spencerism. The Communication and Appropriation of a British Evolutionist*, edited by Bernard Lightman, 173–91. Leiden: Brill, 2015.

Novoa, Adriana and Alex Levine. *From Man to Ape. Darwinism in Argentina, 1870–1920.* Chicago: Chicago University Press, 2010.

"Nuevas autoridades en observatorios nacionales." *Revista Astronómica* 29 (1956): 81–87.

Numbers, Ronald L. "Science, Secularization, and Privatization." In idem, *Science and Christianity in Pulpit and Pew*, 129–36. New York: Oxford University Press, 2007.

Numbers, Ronald L. "Simplifying Complexity: Patterns in the History of Science and Religion." In *Science and Religion. New Historical Perspectives*, edited by Thomas Dixon, Geoffrey Cantor and Stephen Pumfrey, 263–82. Cambridge: Cambridge University Press, 2010.

Numbers, Ronald L. "Revisiting the Battlefields of Science and Religion. The Warfare Thesis Today." In *Rethinking History, Science, and Religion*, edited by Bernard Lightman, 183–90. Pittsburgh: University of Pittsburgh Press, 2019.

[Nysten, Pierre-H.] *Dictionnaire de médecine*, edited by Émile Littré and Charles Robin, 13th ed. Paris: Baillière, 1873.

"Obituary. Mr. M. J. Petty." *British Medical Journal* no. 4810 (10 March 1953): 623.

Observatorio Nacional de Física Cósmica. San Miguel, 1970.

Ochoa, Javier. "Al margen de la última encíclica. De eugenesia." *Criterio* no. 152 (29 January 1931): 147–48.

Ochoa, Javier. "Al margen de la última encíclica. Las taras de la civilización." *Criterio* no. 157 (5 March 1931): 309–10.

Ochoa, Javier. "Después de la *Casti connubii*. Pornografía y falsa ciencia." *Criterio* no. 200 (31 December 1931): 475–76.

O'Connor, Thomas F. "John A. Zahm, C.S.C.: Scientist and Americanist." *The Americas* 7, no. 4 (1951): 435–62.

Ogilvie, Brian W. *The Science of Describing. Natural History in Renaissance Europe.* Chicago: University of Chicago Press, 2006.

Olcott, Henry Steel. *Old Diary Leaves. The Only Authentic History of the Theosophical Society. Fourth Series, 1887–1892.* London: The Theosophical Publishing Society; Madras: Theosophist Office, 1910.

Oldroyd, David and Sally Newcomb. "Richard Kirwan (1733–1812)." *Earth Sciences History* 31 (2012): 287–314.

Oliver, George. *Collections towards Illustrating the Biography of the Scotch, English and Irish Members of the Society of Jesus*. London: Charles Dolman, 1854.
Olson, Joseph W. *Jesuit Inculturation in the New World. Experiments in Missions of 16th, 17th, and 18th Centuries*. Denver: Outskirts Press, 2008.
Orías, Oscar and Eduardo Braun Menéndez. *The Heart-Sounds in Normal and Pathological Conditions*. Oxford: Oxford University Press, 1939.
Ortiz, Eduardo L. "A Convergence of Interests: Einstein's Visit to Argentina in 1925." *Ibero-Amerikanisches Archiv* 21 (1995): 67–126.
Ortiz, Eduardo L. "La Comisión del Arco de Meridiano. Astronomía, geodesia, oceanografía y geofísica en la Argentina de 1935–1945." *Saber y Tiempo* 19 (2005): 127–87.
Pagden, Anthony. *The Fall of Natural Man. The American Indian and the Origins of Comparative Ethnology*. Cambridge: Cambridge University Press, 1982.
Palacios, Alfredo L. "El ideal de las democracias ibero-americanas." *Nosotros* 17, no. 172 (1923): 5–46.
Pals, Daniel L. *Nine Theories of Religion*, 3rd ed. New York: Oxford University Press, 2015.
Paneth, Tomás. "Ajuste del celóstato del espectroheliógrafo de San Miguel." *Boletín de la Asociación Argentina de Astronomía* 15 (1970): 38.
Papanikas, Dimitri P. "La iglesia de la raza. La iglesia católica española y la construcción de la identidad nacional en Argentina." PhD diss., Universidad Autónoma de Madrid, 2012.
Paquette, Gabriel B. *Enlightenment, Governance, and Reform in Spain and Its Empire, 1759–1808*. London: Palgrave Macmillan, 2008.
Parera, Ricardo G. *Democracia Cristiana en la Argentina*. Buenos Aires: Nahuel, 1967.
Parks, Vincent J. "Professor August J. Durelli: A Short Biography." In *Progress in Experimental Mechanics. Durelli Anniversary Volume. Contributions in Honor of the Sixty-Fifth Birthday of Professor A. J. Durelli*, edited by V. J. Parks, ix–xi. Washington, D.C.: The Catholic University of America, 1975.
Patton, Elda C. *Sarmiento in the United States*. Evansville, IN: University of Evansville Press, 1976.
Paucke, Florian. *Hacia allá y para acá (una estada entre los indios mocovíes, 1749–1767)*, translated by Ernesto Wernicke, 4 vols. Tucumán: Universidad Nacional de Tucumán; Buenos Aires: Institución Cultural Argentino-Germana, 1942–44.
Paul, Harry W. "The Debate over the Bankruptcy of Science in 1895." *French Historical Studies* 5, no. 3 (1968): 299–327.
Paul, Harry W. *The Edge of Contingency. French Catholic Reaction to Scientific Change from Darwin to Duhem*. Gainesville: University Presses of Florida, 1979.
Paul, Harry W. "Religion and Darwinism: Varieties of Catholic Reaction." In *The Comparative Reception of Darwinism*, edited by Thomas F. Glick, 407–36. Chicago: The University of Chicago Press, 1988.
Paz Trueba, Yolanda E. de. "Masonería y sociabilidad en el centro y sur de la campaña bonaerense. Fines del siglo XIX, principios del XX." *Prohistoria* no. 16 (July–December 2011): 1–10.
Pearce, Adrian J. *The Origins of Bourbon Reform in Spanish South America, 1700–1763*. New York: Palgrave Macmillan, 2014.
Peard, Julyan G. *An American Teacher in Argentina. Mary Gorman's Nineteenth-Century Odyssey from New Mexico to the Pampas*. Lewisburg: Bucknell University Press, 2016.

Pearson, Karl. "On the Ancestral Gametic Correlations of a Mendelian Population Mating at Random." *Proceedings of the Royal Society of London. Series B, Containing Papers of a Biological Character* 81 (1909): 225–29.
Peire, Jaime. *El taller de los espejos. Iglesia e imaginario, 1765–1815.* Buenos Aires: Claridad, 2000.
Pelosi Hebe C. and Geraldine Mackintosh. "El ensayo de Universidad Católica (1909–1920)." *Archivum* 16 (1994): 185–95.
Pence, Charles H. and Daniel G. Swaim. "The Economy of Nature: the Structure of Evolution in Linnaeus, Darwin, and the Modern Synthesis." *European Journal for Philosophy of Science* 8 (2018): 435–54.
Penchaszadeh, Pablo and Miguel de Asúa. *El deslumbramiento/L'éblouissement. Aimé Bonpland y Alexander von Humboldt en Sudamérica.* Buenos Aires: MACN-Conicet, 2010.
Pennington, A. Stuart. *The Argentine Republic.* London: Stanley Paul, 1910.
Peña, Ignacio and Guillermo Jaim Etcheverry, eds. *Eduardo Braun Menéndez. Ciencia y conciencia.* Buenos Aires: MINCYT, 2015.
Peretó Juli and Jesús I. Catalá Gorgues. "A Reconciliation with Darwin. Erich Wasmann and Jaime Pujiula's Divergent Views on Evolutionism: Biologists and Jesuits." *Mètode. Science Studies Journal* 7 (2017): 87–93.
Pereyra, Elías del Carmen. *Física*, translated by Juan Chiabra. In *La enseñanza de la filosofía en la época colonial*, vol. 2, 173–333. Buenos Aires: Coni, 1911.
Perrine, Charles D. "Résumé of Observations of Halley's Comet at Córdoba." *Publications of the Astronomical Society of the Pacific* 22 (1910): 211–13.
Peterson, Harold F. *Argentina and the United States, 1810–1960.* New York: SUNY Press, 1964.
Petrus Canisius [pseud.] "Eugenica." *Criterio* no. 425 (26 April 1936): 396–97.
Petrus Canisius [pseud.] "Segundo Congreso Internacional de Médicos Católicos, Viena, 28 de mayo a 3 de junio. Su significación." *Criterio* no. 457 (3 December 1936): 327–29.
Piccirilli, Ricardo. *Rivadavia y su tiempo*, 2nd ed., 3 vols. Buenos Aires: Peuser, 1960.
Pickering, Mary. *Auguste Comte. An Intellectual Biography*, 3 vols. Cambridge: Cambridge University Press, 1993–2009.
Pico, César E. "Discurso sobre el positivismo contemporáneo." *Estudios* 6 (1916): 425–36.
Pico, César E. "Sobre la significación morfo-fisiológica del apéndice ileo-cecal." *Tribuna Universitaria* 3, no. 18 (15 June 1916): 14–23.
Pico, César E. "Las reglas del método experimental y las consecuencias filosóficas de su aplicación incorrecta." *Estudios* 29 (1925): 104–23.
Pico, César E. "Marañón y los excesos eugénicos." *Criterio* no. 4 (29 March 1928): 114.
Pico, César E. "Los fantasmas del vitalismo." *Criterio* no. 37 (15 November 1928): 201–203; no. 38 (22 November 1928): 233–40.
Pico, César E. "¡Otra vez Aníbal Ponce!" *Criterio* no. 38 (22 November 1928): 240.
Pico, César E. "El hombre de ciencia y la filosofía." *Criterio* no. 49 (7 February 1929): 169–71; no. 52 (28 February 1929): 265–68.
Pico, César E. "Teorías del conocimiento." *Número* no. 2 (February 1930): 14–15.
Pico, César E. "Antidemocracia." *Número* no. 17 (May 1931): 44.
Pico, César E. "Autodestrucción de la democracia." *Número* no. 20 (August 1931): 63.
Pico, César E. "Syllabus." *Número* no. 23–24 (December 1931): 78.

Pico, César E. *Carta a Jacques Maritain sobre la colaboración de los católicos con los movimientos de tipo fascista*. Buenos Aires: Adsum, 1937.

Pico, César E. "Materialismo. Ciencia y religión." *Iatría* 8 (October 1937–May 1938): 8–16.

Pico, César E. "Las ciencias. Introducción a su metodología y a su epistemología." *Iatría* 8 (June 1938): 3–11.

Pico, César E. "Las ciencias. Del conocimiento en general." *Iatría* 8 (September 1938): 3–13.

Pico, César E. "Las ciencias. Las evidencias primordiales." *Iatría* 9 (April 1939): 3–18.

Pico, César E. "Absurdos del especialismo." *Sol y Luna* 2 (1939): 32–46.

Pico, César E. "Inteligencia y revolución." In *Los nacionalistas*, edited by María I. Barbero and Fernando Devoto, 96–98. Buenos Aires: CEAL, 1983.

Pimentel, Juan. "The Iberian Vision: Science and Empire in the Framework of Universal Monarchy, 1500–1800." *Osiris*, 2nd series, 15 (2000): 17–30.

Pimentel, Juan. "Baroque Natures: Juan E. Nieremberg, American Wonders and Preterimperial Natural History." In *Science in the Spanish and Portuguese Empires, 1500–1800*, edited by Daniela Bleichmar, Paula De Vos, Kristin Huffine and Kevin Sheehan, 93–114. Stanford: Stanford University Press, 2009.

Piñero, Antonio F. *Localizaciones cerebrales en las alteraciones auditivas y de la visión*. Buenos Aires: La Nación, 1883.

Pizarro, Manuel D. "El gobierno puramente civil." In idem, *La escuela simiana y la Constitución Nacional. Relaciones de la Iglesia y del Estado*, 27–43. Buenos Aires: Imprenta de La Unión, 1883.

Pizarro, Manuel D. "Teología transcendental." In idem, *La escuela simiana y la Constitución Nacional. Relaciones de la Iglesia y del Estado*, 45–72. Buenos Aires: Imprenta de La Unión, 1883.

Pizarro, Manuel D. "La religión y la política." In idem, *Miscelánea*, vol. 1, 237–261. Córdoba: La Minerva, 1897.

Placher, William C. *The Domestication of Transcendence. How Modern Thinking about God Went Wrong*. Louisville, KY: Westminster John Knox Press, 1996.

Planella, Juan. "¿Impostura o verdad?" *Estudios* 13 (July–December 1917): 44–53.

Planella, Juan. "Lo de Giles y a propósito de ello." *Estudios* 22 (January–June 1922): 260–73, 334–47.

Planella, Juan. "Un testimonio más sobre 'lo de Giles'." *Estudios* 23 (July–December 1922): 460–75.

Pluche, Noël-Antoine. *Espectáculo de la naturaleza*, translated by Esteban Terreros, 16 vols. Madrid: Gabriel Ramírez, 1753–1755.

Podgorny, Irina. "Bones and Devices in the Constitution of Paleontology in Argentina at the End of the Nineteenth Century." *Science in Context* 18 (2005): 249–83.

Podgorny, Irina. "Fossil Dealers, the Practices of Comparative Anatomy and British Diplomacy in Latin America, 1820–1840." *British Journal for the History of Science* 46 (2013): 647–74.

Poli, Mario A. "El Seminario Metropolitano de Buenos Aires en la Facultad de Teología (1622–2015)." In Facultad de Teología, *100 años de la Facultad de Teología: memoria, presente, futuro*, 169–93. Buenos Aires: UCA, 2015.

Politis, Gustavo G. and Mariano Bonomo. "Nuevos datos sobre el hombre fósil." In Asociación Paleontológica Argentina, *Vida y obra de Florentino Ameghino*, special issue, 12 (2011): 101–19.

Ponce, Aníbal. "Hans Driesch y los fantasmas del vitalismo." *El Hogar* 24, no. 994 (2 November 1928): 16.
Ponce, Aníbal. Review of *Homenaje de despedida a la tradición de Córdoba docta y santa*, by G. F. Nicolai. *El Hogar* 24, no. 995 (9 November 1928): 8.
Popescu, Oreste. *Studies in the History of Latin American Economic Thought*. London: Routledge, 1997.
Porter, Roy. *The Greatest Benefit to Mankind. A Medical History of Humanity from Antiquity to the Present*. London: Fontana, 1997.
Potash, Robert A. *El ejército y la política en la Argentina*, 2 vols. Buenos Aires: Hyspamérica, 1985.
Poulat, Émile. *Intégrisme et catholicisme intégral. Un réseau secret international antimoderniste: "La Sapinière" (1909–1921)*. Paris: Casterman, 1969.
Poulat, Émile. "La querelle de l'intégrisme en France." *Social Compass* 32 (1985): 343–51.
Prieto, Andrés I. *Missionary Scientist. Jesuit Science in Spanish South America, 1570–1810*. Nashville, TN: Vanderbilt University Press, 2011.
Principe, Lawrence M. "The Warfare Thesis." In *The Warfare between Science and Religion. The Idea That Wouldn't Die*, edited by Jeff Hardin, Ronald Numbers and Ronald A. Benzley, 6–26. Baltimore: Johns Hopkins University Press, 2018.
Pró, Diego F. "Periodización y caracterización de la historia del pensamiento filosófico argentino." In idem, *Historia del pensamiento filosófico argentino. Cuaderno 1*, 143–84. Mendoza: Universidad Nacional de Cuyo, Facultad de Filosofía y Letras, 1973.
Probst, Juan. "Introducción." In *La enseñanza durante la época colonial (1771–1810)*, xi–ccxii. Vol. 18, *Documentos para la Historia Argentina*, ed. Instituto de Investigaciones Históricas, Facultad de Filosofía y Letras, Universidad de Buenos Aires. Buenos Aires: Peuser, 1924.
Probst, Juan. *Juan Baltazar Maziel. El maestro de la generación de Mayo*. Buenos Aires: Instituto de Didáctica de la Facultad de Filosofía y Letras de la Universidad de Buenos Aires, 1946.
[Provincia de Buenos Aires] *Debates de la Convención Constituyente de Buenos Aires, 1870–1873*, 2 vols. Buenos Aires: La Tribuna, 1877.
Provincia de Buenos Aires, Cámara de Diputados. *Diario de sesiones 1916*. La Plata: Taller de Impresiones Oficiales, 1917.
Provine, William B. *The Origins of Theoretical Population Genetics*. Chicago: The University of Chicago Press, 2001.
Puig, Ignacio. *El Observatorio de San Miguel, República Argentina*. Buenos Aires, 1935.
Pujiula, Jaime. "El apéndice vermiforme del ciego." *Estudios* 6 (June 1916): 397–404.
Pyenson, Lewis. *Cultural Imperialism and Exact Sciences. German Expansion Overseas, 1900–1930*. New York: Peter Lang, 1985.
Pyenson, Lewis. "Science and Imperialism." In *Companion to the History of Modern Science*, edited by R. C. Olby, G. N. Cantor, J. R. R. Christie and M. J. S. Hodge, 920–33. London: Routledge, 1990.
Pyenson, Lewis. "Athena's Retinue: Nineteenth-Century Scientists Embedded in the Army." *British Journal for the History of Science* 45, no. 3 (2012): 377–400.
Pyenson, Lewis. "La ciencia en Córdoba en el siglo XIX." In *Universidad Nacional de Córdoba. Cuatrocientos años de historia*, edited by Daniel Saur and Alicia Servetto, vol. 1, 251–81. Córdoba: Universidad Nacional de Córdoba, 2013.

Pyenson, Lewis and Jean-François Gauvin, eds. *The Art of Teaching Physics. The Eighteenth-Century Demonstration Apparatus of Jean Antoine Nollet.* Montreal: Septentrion, 2002.

Quereilhac, Soledad. *Cuando la ciencia despertaba fantasías. Prensa, literatura y ocultismo en la Argentina de entresiglos.* Buenos Aires: Siglo XXI, 2016.

Quesada, Ernesto. *Herbert Spencer y sus doctrinas sociológicas.* Buenos Aires: Talleres Gráficos de la Penitenciaría Nacional, 1907.

Quién es quién en la Argentina, 7th ed. Buenos Aires: Kraft, 1958.

Rabin, Sheila J. "Early Modern Jesuit Science. A Historiographical Essay." *Journal of Jesuit Studies* 1, no. 1 (2014): 88–104.

Raia, Courtenay. *The New Prometheans. Faith, Science, and the Supernatural Mind in the Victorian Fin de Siècle.* New York: Columbia University Press, 2019.

Rainger, Ronald. *An Agenda for Antiquity. Henry Fairfield Osborn and Vertebrate Palaeontology at the American Museum of Natural History, 1890–1935.* Tuscaloosa: The University of Alabama Press, 1991.

Rapport annuel sur l'état de l'Observatoire de Paris pour l'année 1914. Paris: Imprimerie Nationale, 1915.

Rapport annuel sur l'état de l'Observatoire de Paris pour l'année 1916. Paris: Imprimerie Nationale, 1917.

Ravignani, Emilio. "Notas para la historia de las ideas en la Universidad de Buenos Aires. El doctor Carta y la enseñanza de la física experimental." *Revista de la Universidad de Buenos Aires* 13, no. 34 (1916): 70–96.

Redmont, Walter B. *Bibliography of the Philosophy in the Iberian Colonies of America.* The Hague: Martinus Nijhoff, 1972.

Reber, Vera Blinn. *British Mercantile Houses in Buenos Aires, 1810–1880.* Cambridge, MA: Harvard University Press, 1979.

Rectenwald, Michael. *Nineteenth-Century British Secularism: Science, Religion, and Literature.* Basingstoke: Palgrave Macmillan, 2016.

Reggiani, Andrés H. "Dépopulation, fascisme et eugénisme 'Latin' dans l'Argentine des années 1930." *Le Mouvement Social* no. 230 (2010): 7–26.

Regnault, Jules. "La eugénica." *La semana médica* 29, no. 1 (5 January 1922): 22–25.

Rein, Raanan. *Argentine Jews or Jewish Argentines? Essays on Ethnicity, Identity, and Diaspora.* Leiden: Brill, 2010.

Reinke, Johann. *Die Welt als That. Umrisse einer Weltansicht auf wissenschaftlicher Grundlage.* Berlin: Gebrüder Paetel, 1899.

"Religious Broadcasting at the Eucharistic Congress." *Nature* 136, no. 3438 (21 September 1935): 471.

Remond, René. "L'intégrisme catholique. Portrait intelectual." *Études* 370, no. 1 (1989): 95–105.

Remond, René. *Religion and Society in Modern Europe*, translated by Antonia Nevill. Oxford: Blackwell, 1999.

"Renuncia Fortunato Devoto." *Revista Eclesiástica del Arzobispado de Buenos Aires* 11 (1911): 508–12.

República Argentina. *Tercer Censo Nacional levantado el 1º de junio de 1914.* Vol. 2. Buenos Aires: Rosso, 1916.

[Review of] *Les conflits de la science et de la religion*, by John W. Draper. *Revue scientifique de la France et de l'étranger*, 2nd series, 4, no. 50 (5 June 1875): 1170–72.

Richards, Robert J. *The Tragic Sense of Life. Ernst Haeckel and the Struggle over Evolutionary Thought*. Chicago: The University of Chicago Press, 2008.
Richet, Charles. "La science a-t-elle fait banqueroute?" *Revue scientifique* 3 (12 January 1895): 33–39.
Rieznik, Marina. *Los cielos del sur. Los observatorios astronómicos de Córdoba y de La Plata, 1870–1920*. Rosario: Prohistoria, 2011.
Risse, Guenter B. "Medicine in New Spain." In *Medicine in the New World. New Spain, New France, and New England*, edited by Ronald L. Numbers, 12–63. Knoxville: The University of Tennessee Press, 1980.
Risse, Guenter B. *Mending Bodies, Saving Souls. A History of Hospitals*. New York: Oxford University Press, 1999.
Rivero de Olazábal, Raúl. *Por una cultura católica. El compromiso de una generación argentina*. Buenos Aires: Claretiana, 1986.
Rock, David. *Argentina, 1516–1987. From Spanish Colonization to Alfonsín*. Berkeley: University of California Press, 1987.
Rock, David. *Authoritarian Argentina. The Nationalist Movement, its History and its Impact*. Berkeley: University of California Press, 1993.
Rock, David. *The British in Argentina. Commerce, Settlers, and Power, 1800–2000*. Cham: Palgrave Macmillan, 2019.
Rodríguez, Ana María T. "Los médicos católicos argentinos en los años treinta." *Sociedad y Religión* 20, no. 30–31 (2008): 137–70.
Rodríguez, Ángel. *Sobre la teoría de la relatividad propuesta por el Dr. A. Einstein*. Madrid: Imprenta del Asilo de Huérfanos del Sagrado Corazón de Jesús, 1924.
Rodríguez, Julia. *Civilizing Argentina: Science, Medicine, and the Modern State*. Chapel Hill: The University of North Carolina, 2006.
Rojas, Ricardo. *La literatura argentina. Los modernos*, 2nd ed., 2 vols. Buenos Aires: La Facultad, 1925.
Romano, Antonella. "Les jésuites entre apostolat missionnaire et activité scientifique (XVIe–XVIIIe siècles)." *Archivum Historicum Societatis Iesu* 74, no. 147 (2005): 213–36.
Romero, José L. *Las ideas políticas en la Argentina*, 2nd ed. México D.F.: FCE, 1956.
Romero Carranza, Ambrosio. *Itinerario de Monseñor de Andrea*. Buenos Aires, 1957.
Rossi Belgrano, Alejandro and Mariana Rossi Belgrano. *Juan Brèthes (Frère Judulien Marie): primer entomólogo del Museo Nacional*. Buenos Aires, 2018.
Rosso, Giuseppe. "Nicolò Mascardi missionario Gesuita esploratore del Cile e della Patagonia (1624–1674)." *Archivum Historicum Societatis Iesu* 19 (1950): 3–74.
"R. P. José Ubach Medir S.J." *Revista Astronómica* 7 (1935): 337.
Rubiés, Joan-Pau. "The Jesuits and Enlightenment". In *The Oxford Handbook of the Jesuits*, edited by Ines G. Županov, 855–90. New York: Oxford University Press, 2019.
Rudwick, Martin J. S. *Bursting the Limits of Time. The Reconstruction of Geohistory in the Age of Revolution*. Chicago: Chicago University Press, 2005.
Ruiz de Montoya, Antonio. *Conquista Espiritual*, edited by E. J. Maeder. Rosario: EDEHI, 1989.
Ruiz Moreno, Aníbal, ed. *La medicina en "El Paraguay natural."* Tucumán: Universidad Nacional de Tucumán, 1946.
Ruiz Moreno, Aníbal, Vicente A. Risolía and Rómulo D'Onofrio. "El pleito de Bonpland con los bethlemitas por la propiedad de la Quinta de los sauces." *Publicaciones del Instituto de Historia de la Medicina* 17 (1955): 6–20 (mimeo).

Rupke, Nicolaas. "Alexander von Humboldt and Monism." In *Monism. Science, Philosophy, Religion, and the History of a Worldview*, edited by Todd Weir, 79–80. New York: Palgrave Macmillan, 2012.

Russell, Colin A. "The Conflict of Science and Religion." In *Science and Religion. A Historical Introduction*, edited by Gary B. Ferngren, 3–12. Baltimore: John Hopkins University Press, 2002.

Saeger, James S. "*The Mission* and Historical Missions: Film and the Writing of History." *The Americas* 51, no. 3 (1995): 393–415.

"Sagrada Congregación Consistorial. S. E. Revma. Mons. Dr. Fortunato J. Devoto, Obispo Auxiliar del Excmo. Sr. Arzobispo." *Revista Eclesiástica del Arzobispado de Buenos Aires* 33 (1933): 345.

[Saint Ildephonsus] "Opusculum de partu Virginis." In *SS. PP. Toletanorum quotquot extant Opera*, edited by Francisco de Lorenzana, vol. 1, 294–317. Madrid: Joaquín Ibarra, 1782.

Samperio, Manuel J. "Cuestión de paleoantropología argentina." *Estudios* 18 (1920): 350–66.

Sánchez de Loria, Horacio M. *Félix Frías. Acción y pensamiento político*. Buenos Aires: Quorum, 2004.

Sánchez de Loria, Horacio M. *El pensamiento político de José Manuel Estrada. Del liberalismo católico al utramontanismo*. Buenos Aires: Torre de Hércules, 2021.

Sandford, Roscoe F. "The Seventy-fifth Anniversary of the Córdoba Observatory." *Publications of the Astronomical Society of the Pacific* 58, no. 345 (1946): 341–48.

Sanguinetti, Horacio. "Laica o libre. Los alborotos estudiantiles de 1958." *Todo es Historia* no. 80 (January 1974): 8–23.

Sanguinetti, Manuel J. *Chorroarín. El prócer olvidado*. Buenos Aires: Stella, 1984.

Santaló, Luis A. "Agustín Durañona y Vedia." *Anales de la Academia Nacional de Ciencias Exactas, Físicas y Naturales* 33 (1981): 315–17.

Sardá y Salvany, Félix ¿*Qué hay sobre el espiritismo? Cuatro palabras sobre esta secta*. Barcelona: Tipografía Católica, 1872.

Sarmiento, Domingo F. *Darwin en una conferencia seguido de El Congreso de Tucumán*. Buenos Aires: El Nacional, 1882.

Sarmiento, Domingo F. "Memoria sobre ortografía americana." In idem, *Obras*, edited by Augusto Belin Sarmiento, vol. 4, 1–18. Buenos Aires: Félix Lajouane, 1886.

Sarmiento, Domingo F. "Discurso inaugural de la Exposición de Córdoba." In idem, *Obras completas*, edited by Augusto Belin Sarmiento, vol. 21, 316–23. Buenos Aires: Imprenta Mariano Moreno, 1899.

Sarmiento, Domingo F. "Discurso de inauguración del Observatorio Astronómico." In idem, *Obras*, edited by Augusto Belin Sarmiento, vol. 21, 324–328. Buenos Aires: Imprenta Mariano Moreno, 1899.

Sarmiento, Domingo F. "Inauguración del ferro-carril de Concordia (Entre Ríos)." In idem, *Obras*, edited by Augusto Belin Sarmiento, vol. 21, 356–73. Buenos Aires: Imprenta Mariano Moreno, 1899.

Sarmiento, Domingo F. "Alocución aceptando la visita de los jóvenes estudiantes de la Universidad y Colegios." In idem, *Obras*, edited by Augusto Belin Sarmiento, vol. 22, 196–200. Buenos Aires: Imprenta Mariano Moreno, 1899.

Sarmiento, Domingo F. "La inteligencia en la vida argentina." In idem, *Obras*, edited by Augusto Belin Sarmiento, vol. 46, 176–87. Buenos Aires: Imprenta Mariano Moreno, 1900.

Sarmiento, Domingo F. "Lágrimas de cocodrilo." In idem, *Obras*, edited by Augusto Belin Sarmiento, vol. 48, 183–87. Buenos Aires: Imprenta Mariano Moreno, 1900.

Sarmiento, Domingo F. "Las manifestaciones." In idem, *Obras*, edited by Augusto Belin Sarmiento, vol. 48, 324–28. Buenos Aires: Imprenta Mariano Moreno, 1900.

Sarmiento, Domingo F. "Darwin: Lecture Given in the Teatro Nacional, Following the Death of Charles Darwin." In *Darwinism in Argentina. Major Texts (1845–1909)*, translated by Nicholas F. Callaway, 113–38. Lewisburg: Bucknell University Press, 2012.

Sarrailh, Jean. *L'Espagne éclairée de la deuxième moitié du XVIIIe siècle*. Paris: Imprimerie nationale, 1954.

Sarramone, Alberto. *Alexis Peyret*. Buenos Aires: Biblos, 2016.

Sarreal, Julia J. S. *The Guaraní and Their Missions. A Socioeconomic History*. Stanford: Stanford University Press, 2014.

Sarthou, Basilio. *Historia centenaria del Colegio San José de Buenos Aires (1858–1958)*. Buenos Aires, 1960.

Scalabrini, Pedro. *Materialismo, darwinismo, positivismo. Diferencias y semejanzas*. Paraná: La Velocidad, 1889.

Schaefer, Richard. "Andrew Dickson White and the History of a Religious Future." *Zygon* 50 (2015): 7–27.

Schaposchnik, Ana. *The Lima Inquisition: The Plight of Crypto-Jews in Seventeenth-Century Peru*. Madison: University of Wisconsin Press, 2015.

Schneider, Otto. *Geofísica y geodesia*. Vol. 8, *Evolución de las ciencias en la República Argentina, 1923–1972*. Buenos Aires: Sociedad Científica Argentina, 1980.

Sciammarella, César A. "Augusto J. Durelli." *Experimental Techniques* 25, no. 4 (July–August 2001), https://sem.org/Files/about/sem_history25_4.pdf.

"Science and Religion. Museum of Geology." *The Builder* 17 (1859): 35–36.

Scobie, James R. *La lucha por la consolidación de la nacionalidad argentina, 1852–1862*. Buenos Aires: Hachette, 1964.

Secord, James A. *Victorian Sensation: The Extraordinary Publication, Reception, and Secret Authorship of Vestiges of the Natural History of Creation*. Chicago: University of Chicago Press, 2000.

Segarra, Vicente. "La psicología empírica y los fenómenos mediánicos." *Estudios* 18 (January–June 1920): 278–92.

Seibold, Jorge and Tomás Paneth. "El espectroheliógrafo del Observatorio de Física Cósmica de San Miguel." *Boletín de la Asociación Argentina de Astronomía* 15 (1970): 37–38.

Seiguer, Paula. *"Jamás he estado en casa." La Iglesia Anglicana y los ingleses en la Argentina*. Buenos Aires: Biblos, 2017.

Sepp, Anton. *Continuación de las labores apostólicas*, edited by Werner Hoffmann. Buenos Aires: Eudeba, 1973.

Sepp, Anton. *Jardín de Flores Paraquario*, edited by Werner Hoffmann. Buenos Aires: Eudeba, 1974.

Sharp, Lynn L. *Secular Spirituality. Reincarnation and Spiritism in Nineteenth-Century France* (Lanham: Rowan and Littlefield 2006), 207.

Sheets-Pyenson, Susan. *Cathedrals of Science. The Development of Colonial Natural History Museums during the Late Nineteenth Century*. Kingston and Montreal: McGill-Queen's University Press, 1988.

Shiels, Eugene. *King and Church: The Rise and Fall of the Patronato Real*. Chicago: Loyola University Press, 1961.
Shipman, Pat. *The Man Who Found the Missing Link. Eugène Dubois and his Lifelong Quest to Prove Darwin Right*. Cambridge, MA: Harvard University Press, 2002.
Shumway, Nicholas. *The Invention of Argentina*. Berkeley: University of California Press, 1993.
Sierra, Vicente D. *Los jesuitas germanos en la conquista espiritual de Hispano-América*. San Miguel: Facultades de Filosofía y Teología; Buenos Aires: Institución Cultural Argentino-Germana, 1944.
Silva, Ignacio, ed. *Latin American Perspectives on Science and Religion*. London: Pickering and Chatto, 2014.
Simon, W. M. "The 'Two Cultures' in Nineteenth-Century France: Victor Cousin and Auguste Comte." *Journal of the History of Ideas* 26 (1965): 45–58.
Simpson, George G. *The Beginnings of the Age of Mammals in South America. Part 1*. New York: American Museum of Natural History, 1948.
Simpson, George G. *Discoverers of the Lost World*. New Haven: Yale University Press, 1984.
Sisson, Henri-Dominique. *La République Argentine. Description, étude sociale et histoire*. Paris: Plon, 1910.
Sloan, Phillip R. "The Buffon-Linnaeus Controversy." *Isis* 67 (1976): 356–75.
Sloan, Phillip R. "Bringing Evolution to Notre Dame: Father John Zahm, C.S.C. and Theistic Evolutionism." *The American Midland Naturalist* 161 (2009): 189–205.
Smart, Ninian. *The World's Religions*, 2nd ed. Cambridge: Cambridge University Press, 1998.
Smidt, Andrea J. "Bourbon Regalism and the Importation of Gallicanism: The Political Path for a State Religion in Eighteenth-Century Spain." *Anuario de Historia de la Iglesia* 19 (2010): 25–53.
Smidt, Andrea J. "'Luces por la Fe': The Cause of Catholic Enlightenment in 18th-Century Spain." In *A Companion to the Catholic Enlightenment in Europe*, edited by Ulrich L. Lehner and Michael O'Neill Printy, 403–52. Leiden: Brill, 2010.
Smith, Christian. "Introduction. Rethinking the Secularization of American Public Life." In *The Secular Revolution. Power, Interests, and Conflict in the Secularization of American Public Life*, edited by Christian Smith, 1–96. Berkeley: University of California Press, 2003.
Smith, Jonathan Z. *Imagining Religion. From Babylon to Jonestown*. Chicago: The University of Chicago Press, 1982.
Smith, Wilfred C. *The Meaning and End of Religion*. Toronto: Mentor Books, 1964.
Sociedad Científica Argentina. *Congreso Científico Internacional Americano*, 2 vols. Buenos Aires: Coni, 1910.
Sociedad Luz. *Ameghino. Homenaje de la Sociedad Luz en el XXV aniversario de su muerte. 1911-Agosto 6–1936*. Buenos Aires: Federación Gráfica Bonaerense, 1936.
Socolow, Susan M. *The Merchants of Buenos Aires, 1778–1810*. Cambridge: Cambridge University Press, 1978.
Soler, César. "La función crea el órgano." *Criterio* no. 582 (27 April 1939): 393–94.
Soler, Ricaurte. *El positivismo argentino*. Buenos Aires: Paidós, 1968.
Solomon, Harry M. *Robert Dodsley. Creating the New Age of Print*. Carbondale, IL: Southern Illinois University Press, 1996.
Spencer, Frank. *Piltdown: A Scientific Forgery*. Oxford: Oxford University Press, 1990.
Spencer, Frank. *History of Physical Anthropology*, 2 vols. New York: Garland Publishing, 1997.

Spencer, Herbert. *De la educación intelectual, moral y física*. Seville: Administración de la Biblioteca Científico-Literaria; Madrid: Librería de Victoriano Suárez, 1879.
Spencer, Herbert. *Education: Intellectual, Moral and Physical*. New York: Appleton, 1896.
Spenninger, Claus. "A Movement That Never Materialized: The Perception of Scientific Materialism as a Secular Movement in Nineteenth-Century Germany." In *Freethinkers in Europe. National and Transnational Secularities, 1789–1920s*, edited by Carolin Kosuch, 273–96. Berlin: De Gruyter, 2020.
Standaert, Nicolas. *Handbook of Christianity in China*. Vol. 1, *635–1800*. Leiden: Brill, 2001.
Stark, Rodney and Roger Finke. *Acts of Faith. Exploring the Human Side of Religion*. Berkeley: University of California Press, 2000.
Staum, Martin S. *Cabanis. Enlightenment and Medical Philosophy in the French Revolution*. Princeton: Princeton University Press, 1980.
Steele, Arthur R. *Flowers for the King. The Expedition of Ruiz and Pavón and the Flora of Peru*. Durham, NC: Duke University Press, 1964.
Stenhouse, John. "Missionary Science." In *Modern Science in National, Transnational, and Global Context*, edited by Hugh R. Slatten, Ronald L. Numbers and David N. Livingstone, 90–107. Vol. 8, *The Cambridge History of Science*. Cambridge: Cambridge University Press, 2020.
Stepan, Nancy. *"The Hour of Eugenics." Race, Gender, and Nation in Latin America*. Ithaca: Cornell University Press, 1991.
Stepan, Nancy. "Race, Gender and Nation in Argentina: the Influence of Italian Eugenics." *History of European Ideas* 15, no. 4–6 (1992): 749–56.
Stroumsa, Guy G. *A New Science. The Discovery of Religion in the Age of Reason*. Cambridge, MA: Harvard University Press, 2010.
Suárez, Anastasio M. *Asserta ex universa philosophia*. Buenos Aires: Apud Typographiam Regiam Parvulorum Orphanorum, 1792.
Sutton, C. W. (rev. Geoffrey Scott). "Falkner, Thomas." In *Oxford Dictionary of National Biography*, Oxford University Press, 2004. Article published 23 September 2004. doi.org/10.1093/ref:odnb/9124.
Taylor, Charles. *A Secular Age*. Cambridge, MA: Harvard University Press, 2007.
Temkin, Owsei. "The Philosophical Background of Magendie's Physiology." *Bulletin of the History of Medicine* 20 (1946): 10–35.
Terán, Oscar. *Vida intelectual en el Buenos Aires fin-de-siglo*. Buenos Aires: FCE, 2000.
Terán, Oscar. *Historia de las ideas políticas en la Argentina. Diez lecciones iniciales, 1810–1980*. Buenos Aires: Siglo XXI, 2008.
Tessi, Francisco S. *Vida y obra de José Manuel Estrada*. Buenos Aires: Peuser, 1928.
"The Institute of Biology and Experimental Medicine at Buenos Aires." *Science* 99, no. 2575 (5 May 1944): 360–61.
Thibaud, Clément. *La Academia Carolina y la independencia de América. Los abogados de Chuquisaca (1776–1809)*. Sucre: Editorial Charcas, 2010.
Thrower, Norman J. W. *Maps and Civilization. Cartography in Culture and Society*, 3rd ed. Chicago: The University of Chicago Press, 2008.
Thun, Harald. "Félix de Azara, los jesuitas y el guaraní." In *Século das Luzes. Portugal e Espanha, o Brasil e a Região do Rio da Prata*, edited by Werner Thielemann, 475–502. Frankfurt a. M.: Ibero-Amerikanisches Institut, 2006.

Thun, Harald. "El saber médico de los guaraníes y la medicina de los jesuitas. Transmisiones y transformaciones." In *Indigenous Knowledge as a Resource. Transmission, Reception, and Interaction of Knowledge between the Americas and Europe, 1492–1800*, edited by Laura Dierksmeier, Fabian Fechner and Kazuhisa Takeda, 41–73. Tübingen: Tübingen University Press, 2021.

Tognetti, Luis. *La Academia Nacional de Ciencias. Etapa Fundacional: siglo XIX*. Córdoba: Academia Nacional de Ciencias, 2000.

Tognetti, Luis. *La Academia Nacional de Ciencias en el siglo XIX. Los naturalistas, publicaciones y exploraciones*. Córdoba: Academia Nacional de Ciencias, 2004.

Tonda, Américo A. *Rivadavia y Medrano. Sus actuaciones en la reforma eclesiástica*. Santa Fe: Castellvi, 1952.

Tonda, Américo A. *El Dean Funes y la reforma de Rivadavia. Los regulares*. Santa Fe: Castellvi, 1961.

Tonda, Américo A. *Castro Barros*. Buenos Aires: Academia del Plata, 1961.

Tonda, Américo A. *La iglesia argentina incomunicada con Roma (1810–1858)*. Santa Fe: Castellvi, 1965.

Tonda, Américo A. *La eclesiología de los doctores Gorriti, Zavaleta y Agüero*. Rosario: Universidad Católica Argentina, Facultad de Derecho y Ciencias Sociales del Rosario, n. d.

Tonni, Eduardo P., Ricardo C. Pasquali and Mariano Bond. "Ciencia y fraude: el hombre de Miramar." *Ciencia Hoy* 11, no. 62 (2001): 58–62.

Tonni Eduardo P. and Laura H. Zampatti. "El 'hombre fósil' de Miramar. Comentarios sobre la correspondencia de Carlos Ameghino a Lorenzo Parodi." *Revista de la Asociación Geológica Argentina* 68, no. 3 (2011): 436–44.

Torchia Estrada, Juan Carlos. *La filosofía en la Argentina*. Washington, D. C.: Unión Panamericana, 1961.

Torino, Inocencio. "Las teorías evolutivas y la ciencia médica." *Nueva Revista de Buenos Aires* 2 (1881): 241–57.

Torino, Inocencio. "La vida y el transformismo moderno." *Anales del Círculo Médico Argentino* 6 (1882): 34–36.

Torre Revello, José. "La biblioteca que poseía en Potosí don Pedro de Altolaguirre (1799)." *Historia* (Buenos Aires) 1 (1956): 153–62.

Torres, Santiago. "Detalle de los objetos recibidos por el encargado del Museo, en cumplimiento de la resolución ministerial." *Revista de la Biblioteca Nacional* 4, no. 13 (1940): 14.

Triarhou, Lazaros C. and Manuel del Cerro. "Semicentennial Tribute to the Ingenious Neurobiologist Christfried Jakob." *European Neurology* 56 (2006): 176–88, 189–98.

Trincado, Joaquín. *El espiritismo estudiado o política del creador y gobierno del espiritismo*. Buenos Aires: Voz Informativa, 1922.

Truxillo, Manuel María. *Exhortación Pastoral*. Madrid: Viuda de Ibarra, 1786.

Tulchin, Joseph A. "La Reforma Universitaria. Córdoba 1918." *Criterio* no. 1599–1600 (9 and 23 July 1970), accessed 21 September 2020, https://www.researchgate.net/publication/324213193_La_Reforma_Universitaria_Cordoba_1918.

Tulchin, Joseph A. *La Argentina y los Estados Unidos. Historia de una desconfianza*. Buenos Aires: Planeta, 1990.

Turda, Marius and Aaron Gillette. *Latin Eugenics in Comparative Perspective*. London: Bloomsbury, 2016.
Turner, Frank M. *Contesting Natural Authority. Essays in Victorian Cultural Life*. Cambridge: Cambridge University Press, 1993.
[Tyndall, John] "What Knowledge is of Most Worth. Lectures on Education Delivered at the Royal Institution of Great Britain. London, 1855." *The Westminster Review* 141 (July 1859): 1–23.
Ubach, José. *El origen del hombre*. Buenos Aires: Talleres Gráficos Rodríguez Giles, 1917.
Ubach, José. *El eclipse anular del 3 de Diciembre de 1918 en la República Argentina, Chile y Uruguay*. Buenos Aires: R. Herrando, 1918.
Ubach, José. *Observaciones del eclipse anular del 3 de diciembre de 1918 en Buenos Aires (República Argentina)*. Tortosa: Algueró y Baiges, 1919.
Ubach, José. *La teoría de la relatividad en la física moderna. Lorentz, Minkowski, Einstein*. Buenos Aires: Amorrortu, 1920.
Ubach, José. "Observaciones astronómicas del eclipse del 1 de octubre de 1921." *Estudios* 22 (January–June 1922): 19–29.
Ubach, José. "El pasado y presente del problema de la evolución." *Estudios* 22 (January–June 1922): 248–59.
Ubach, José. *Observaciones astronómicas del eclipse del 3 de enero de 1927*. Tortosa: Algueró y Baiges, 1927.
Ubach, José. "El eclipse del día 3 de enero." *Revista Eclesiástica del Arzobispado de Buenos Aires* 27 (3 January 1927): 105–106.
Ubierna, Pablo. *Las humanidades. Notas para una historia institucional*. Gonnet: UNIPE, 2016.
Udaondo, Enrique. *Reseña histórica del templo de Ntra. Sra. del Pilar (Recoleta)*. Buenos Aires: Imprenta Mirau, 1918.
Udaondo, Enrique. *Diccionario biográfico colonial argentino*. Buenos Aires: Huarpes, 1945.
Udías, Agustín. *Searching the Heavens and the Earth. The History of Jesuit Observatories*. Dordrecht: Springer Science, 2003.
"Una proposta ai fautori degli uomini preistorici." *La Civiltà Cattolica* 24 (1873): 265–68.
"Una visita al Observatorio de La Plata." *Caras y Caretas* 13, no. 606 (1910): 77–79.
Ungureanu, James C. "Science and Religion in the Anglo-American Periodical Press, 1860–1900: A Failed Reconciliation." *Church History* 88, no. 1 (2019): 120–49.
Ungureanu, James C. *Science, Religion, and the Protestant Tradition: Retracing the Origins of Conflict*. Pittsburgh: University of Pittsburgh Press, 2019.
"Universities of the Argentine: Retirement of Prof. Bernardo A. Houssay, For. Mem R.S." *Nature* 158, no. 4021 (23 November 1946): 739.
Vallejo, Gustavo. "La hora cero de la eugenesia en la Argentina." *História, Ciências, Saúde. Manghinos* 25, suppl. 1 (August 2018): 15–32.
Van Kley, Dale K. "Jansenism and the International Suppression of the Jesuits." In *Enlightenment, Awakening, and Revolution, 1660–1815*, edited by Stewart J. Brown and Timothy Tackett, 302–8. Vol. 7, *The Cambridge History of Christianity*. Cambridge: Cambridge University Press, 2006.
Van Kley, Dale K. "Plots and Rumors of Plots: The Role of Conspiracy in the International Campaign against the Society of Jesus, 1758–1768." In *The Jesuit Suppression in Global*

Context, edited by Jeffrey D. Burson and Jonathan Wright, 13–39. Cambridge: Cambridge University Press, 2015.
Vicq d'Azyr, Félix. "Essai sur les lieux et les dangers des sépultures." In idem, *Oeuvres complètes*, vol. 6, 257–358. Paris: Badouin, 1805.
Vignati, Milcíades A. *Los restos de industria humana de Miramar. A propósito de los despropósitos del comandante Romero*. Buenos Aires: Oceana, 1919.
Vignati, Milcíades A. "Descripción de los molares humanos fósiles de Miramar." *Revista del Museo de La Plata. Nueva Serie. Sección Antropología* 1, no. 8 (1941): 271–358.
Voltaire (François-Marie Arouet). *Candide, où l'optimisme*. (Geneva), 1759.
Voltaire (François-Marie Arouet). *Relation du Paraguai sous la domination des jesuites*. Basle, 1777.
Walter, Richard J. *Student Politics in Argentina. The University Reform and Its Effects, 1918–1964*. New York: Basic Books, 1968.
Walter, Richard J. *The Socialist Party of Argentina, 1890–1930*. Texas: Institute of Latin American Studies, 1977.
Walter, Richard J. "The Right and the Peronists." In *The Argentine Right. Its History and Intellectual Origins, 1910 to the Present*, edited by Sandra McGee Deutsch and Ronald H. Dolkart, 99–118. Wilmington, DE: SR Books, 1993.
Wargentin, Pehr W. "Series observationum primi satellitis Jovis, ex quibus theoria motuum ejusdem satellitis est deducta." *Acta Societatis Regiae Scientiarum Upsaliensis* 3 (1748): 1–32.
Weir, Todd H. *Secularism and Religion in Nineteenth-Century Germany*. Cambridge: Cambridge University Press, 2014.
Wernick, Andrew. *Auguste Comte and the Religion of Humanity. The Post-Theistic Program of French Social Theory*. Cambridge: Cambridge University Press, 2001.
White, Andrew D. *A History of the Warfare of Science with Theology in Christendom*, 2 vols. New York: Appleton, 1897.
Whitehead, Maurice. "On the Road to Suppression. The Jesuits and Their Expulsion from the Reductions of Paraguay." In *The Jesuit Suppression in Global Context*, edited Jeffrey D. Burson and Jonathan Wright, 83–99. Cambridge: Cambridge University Press, 2015.
Wilde, Eduardo. *La cuestión religiosa en el Congreso Argentino*. Buenos Aires: La Tribuna, 1883.
Wilde, Guillermo, ed. *Saberes de la conversión. Jesuitas, indígenas e imperios coloniales en las fronteras de la cristiandad*. Buenos Aires: Editorial Sb, 2011.
Wilde, Guillermo. "The Political Dimensions of Space-Time Categories in the Jesuit Missions of Paraguay (Seventeenth and Eighteenth Centuries)." In *Space and Conversion in Global Perspective*, edited by Giuseppe Marcocci, Aliocha Maldavsky, Wietse de Boer and Ilaria Pavan, 175–213. Leiden: Brill, 2014.
Wilde, Guillermo. "The Missions of Paraguay: Rise, Expansion and Fall." In *A Companion to Early Modern Catholic Global Missions*, edited by Ronnie Po-chia Hsia, 73–101. Leiden: Brill, 2017.
Willey, Basil. *The Eighteenth-Century Background*. Harmondsworth: Penguin, 1962.
Williner, Gregorio J. "P. Jorge Mühn." *Noticias de la Provincia Argentina* 22, no. 159 (1952): 19–23.
Willink, Abraham. "Vida y obra de Hendrik Weyenbergh." *Boletín de la Academia Nacional de Ciencias* 49 (1972): 51–62.

Wilson, Bryan. *Religion in Secular Society. Fifty Years On*, edited by Steve Bruce. Oxford: Oxford University Press, 2016.

Wilson, Edward B. *The Cell in Development and Inheritance*, 2nd ed. New York: Macmillan, 1911.

Wiseman, Nicholas Patrick S. *On the Connection between Science and Revealed Religion*, 2 vols. London: J. Booker, 1836.

Wiseman, Nicholas Patrick S. *Discursos sobre las relaciones que existen entre la ciencia y la religión revelada*, 2 vols. Madrid: Palacios, 1844.

Wright, Winthrop R. *British-Owned Railways in Argentina. Their Effect on Economic Nationalism, 1854–1948*. Austin: University of Texas Press, 1974.

Yriberry, Arturo J. "Cámaras de niebla en la investigación atómica." *Mundo atómico* 2, no. 5 (1951): 11–13, 83.

Young, Frank and Virgilio G. Foglia. "Bernardo Alberto Houssay, 1887–1971." *Biographical Memoirs of the Royal Society* 20 (1974): 246–70.

Zahm, John A. *La evolución y el dogma*, translated by Miguel Asúa. Madrid: Sociedad Editorial Española, 1905.

Zahm, John A. *Through South America's Southlands*. New York: Appleton, 1916.

Zanatta, Loris. *Del estado liberal a la nación católica. Iglesia y ejército en los orígenes del peronismo, 1930–1943*. Buenos Aires: Universidad Nacional de Quilmes, 1996.

Zanatta, Loris. *Perón y el mito de la nación católica*. Buenos Aires: Sudamericana, 1999.

Zanca, José. *Los intelectuales católicos y el fin de la cristiandad, 1955–1966*. Buenos Aires: FCE, 2006.

Zanca, José. *Cristianos antifascistas. Conflictos en la cultura católica argentina*. Buenos Aires: Siglo XXI, 2013.

Zanca, José. *Los Humanistas universitarios. Historia y memoria (1950–1966)*. Buenos Aires: Eudeba, 2019.

Ziller Camenietzki, Carlos. "Baroque Science between the Old and the New World: Father Kircher and His Colleague Valentin Stansel (1621–1705)." In *Athanasius Kircher. The Last Man who Knew Everything*, edited by Paula Findlen, 311–28. New York: Routledge, 2004.

Zimmerman, Eduardo A. "Racial Ideas and Social Reform: Argentina, 1890–1916." *The Hispanic American Historical Review* 72, no. 1 (1992): 23–46.

Zimmermann, Eduardo A. *Los liberales reformistas. La cuestión social en la Argentina, 1890–1916*. Buenos Aires: Sudamericana, 1995.

Zorrilla de San Martín, Juan. "Discurso [21 July 1915, addressed to the Marian Congregation of Colegio del Salvador]." *Estudios* 9 (July–December 1915): 8–19.

Zuckerman, Phil and John R. Shook. "Introduction: The Study of Secularism." In *The Oxford Handbook of Secularism*, edited by Phil Zuckerman and John R. Shook, 1–17. New York: Oxford University Press, 2017.

Zuelzer, Wolf. *The Case Nicolai. A Biography*. Detroit: Wayne State University Press, 1982.

Zuretti, Juan C. *Nueva historia eclesiástica argentina*. Buenos Aires: Itinerarium, 1972.

Zuretti, Juan C. "La fundación de la primera Universidad Católica." *Universitas* 9, no. 38 (July–September 1975): 89–101.

Index of Names

Achával Rodríguez, Tristán 150 f.
Acosta, Joseph de 25 f., 46
Acquaviva, Claudio 29
Agassiz, Louis 134, 142
Aguilar, Félix 213, 216 f., 267
Alberdi, Juan Bautista 111–113, 197
Alberini, Coriolano 188 f.
Alcacer, Pedro S. 140–145, 147
Alembert, Jean le Rond de 72
Almeida, Teodoro de 76
Altieri, Lorenzo 29, 59
Altolaguirre, Francisco 86
Altolaguirre, Martín 50, 78
Altolaguirre, Martín José 50, 86
Altolaguirre, Pedro 76
Alvear, Marcelo T. de 196, 199, 206 f., 223
Ambrose 68
Ameghino, Carlos 184–185
Ameghino, Florentino 159 f., 175–188, 194, 204, 213, 218, 282 f., 285
Ampère, André-Marie 149
Andrade, Olegario V. 153 f.
Andrea, Miguel de 23, 53, 199, 207, 282
Aneiros, León F. 169
Anquin, Nimio de 230
Anville, Jean-Baptiste Bourguignon de 41
Aperger, Sigismund 32, 34, 48
Aquinas, Thomas 244
Arago, François 94, 149
Aramburu, Pedro E. 262
Aranda, Count of 22
Arata, Pedro 173
Argerich, Cosme 101 f., 108
Aristotle 57, 77, 193, 232
Arriaga, Nilo 268
Artigas, José 64, 67, 120
Atwell, Samuel 87 f.
Augustine 60, 62, 64, 75, 123, 269
Avellaneda, Nicolás 104, 111, 153
Azara, Félix de 47 f., 65, 72, 160

Bacon, Francis 14, 101, 152
Balestra, Juan 158

Bard, Louis 210, 238, 253
Bas, Arturo M. 198, 271
Basil 77, 287
Beaumont, John B. 92 f., 95
Belgrano, Manuel 57, 71 f., 74, 79, 100 f., 159, 204
Benedetti, Edmund 267 f.
Bentham, Jeremy 17, 80 f., 84, 99, 105, 112, 147
Béranger, Pierre 87
Berdiáyev, Nikolái 227, 235
Berg, Karl 203, 209
Bergson, Henri 174
Berkeley, Robert 6, 16, 35, 53, 133, 197, 278
Bernard, Claude 5–7, 9, 125, 144, 152, 187, 235, 279
Berthelot, Marcellin 172 f., 289 f.
Besio Moreno, Nicolás 72 f., 287
Bethancourt, Pedro de 90
Bevans, James 88
Bigourdan, Guillaume 214
Bismarck, Otto von 131
Blaeu, Willem 41
Blanco, José M. 159, 180–187, 189, 191–194, 199 f., 218, 238 f., 273, 283
Blavatsky, Helena 174
Boman, Eric 185
Boneo, Juan A. 215
Bonpland, Aimé 65, 81, 88 f., 96
Bottaro, José M. 199, 215, 239
Bougainville, Louis Antoine de 19
Boule, Marcellin 185 f.
Boyle, Richard 13
Brauly, Édouard 203
Braun Menéndez, Eduardo 247 f., 250, 252–257, 259–266, 271–274, 285 f.
Brescia, Fortunato da 59
Brèthes, Jean 204
Bridarolli, Albino 205
Broca, Paul 126
Brooke, John Hedley 2, 6 f., 13 f., 76, 246, 287 f.

Brugier, Eduard 219
Brunetière, Ferdinand 171–175
Bruno, Giordano 84, 87, 90 f., 106, 114, 116, 120, 231, 260
Bucareli, Francisco de Paula 21
Büchner, Ludwig 128, 138
Buckland, William 130
Buffon (Georges-Louis Leclerc) 4, 26, 29–31, 48, 65 f., 70, 76, 112
Bunge, Alejandro 198
Bunge, Carlos Octavio 162 f.
Burmeister, Hermann 118, 134–137, 142, 154, 183, 203, 209, 280
Burnet, Thomas 63
Bussolini, Juan A. 267–270, 286

Cabanis, Pierre J. G. 93, 99, 102–104
Cabred, Domingo 189
Cafferata, Juan F. 198
Camaño, Joaquín 40
Cambaceres, Eugenio 124
Campanella, Tommaso 19
Campomanes, Count of 21 f., 71
Canning, George 85
Cannon, Walter B. 249
Cantù, Cesare 120
Cappelletti, Enrique M. 201 f.
Carbia, Rómulo 55, 83
Cardiel, José 40, 42
Carta Molino, Pietro 31, 76, 88, 92–94, 96, 108, 235, 257
Casanova, José 8–10, 12, 15 f., 44, 46, 114
Casares, Tomás 232
Castañeda, Francisco de Paula 86, 103
Castex, Mariano 27, 243
Castillo, Ramón S. 223
Castro Sarmento, Jacob de 38 f.
Cauchy, Augustin-Louis 203
Celsus 122
Cerviño, Pedro A. 71–74, 77
Chambers, Robert 287
Charles III (of Spain) 21 f., 51, 78, 82
Charles IV (of Spain) 22
Chomé, Ignace 37
Chorroarín, Luis José 60 f., 64
Cires, Pablo de 50
Civit, Emilio 148

Clairaut, Alexis 72
Clara, Gerónimo 116
Clavigero, Francisco Javier 27
Clement XIV (Pope) 23, 48
Clemente 33 f.
Clossey, Luke 40, 44 f.
Cocito, Ricardo 268
Columbus, Christopher 124
Combes, Émile 203
Comte, Auguste 5, 116, 122, 152, 159–163, 167, 188, 280 f., 284
Condillac, Étienne Bonnot de 71, 99, 101, 105, 112
Condorcet, Nicolas de 5, 14 f., 93, 134
Confucius 132
Coninck, Jean-Raymond 36
Copello, Santiago 223, 260
Copernicus, Nicolaus 4, 58–60, 72, 132, 136, 149
Cornoldi, Giovanni Maria 123, 156
Cortés Funes, Gerónimo 126 f.
Cousin, Victor 112, 122, 280
Crookes, William 164
Cullen, Joaquín 198
Cuvier, Georges 66 f., 69 f., 76

Dale, Henry H. 22 f., 249
Darwin, Charles 4, 68, 110, 120, 125, 128, 131–144, 146 f., 153, 156, 158, 163, 171, 177 f., 181, 187, 212 f., 228, 230, 232 f., 245, 279 f., 282, 285
Darwin, Erasmus 142
Dávila, Juan Francisco 41
Dawson, Charles 184
De Angelis, Pietro 98
De Pauw, Cornelius 26
De Vries, Hugo 213
Deban, Camille 168
Delage, Yves 209
Delisle, Joseph-Nicolas 37
Dell'Oro Maini, Atilio 199, 223, 262, 265, 273
Deluc, Jean-André 66
Derisi, Octavio 266, 273
Descartes, René 235
Destutt de Tracy, Antoine 80, 99 f., 102 f., 107

Deulofeu, Venancio 260
Devoto, Fortunato 195–197, 199, 213–219, 223f., 227, 267, 274, 276, 285
Dickmann, Adolfo 182
Dickmann, Enrique 158, 219
Diderot, Denis 19, 112
Dioscorides 33
Dobrizhoffer, Martin 26, 32
Dolomieu, Déodat de 66
Donoso, Armando 174
Dorrego, Manuel 88, 105
Douglas, Mary 14
Draper, John W. 4f., 9, 17, 110, 116, 119–125, 128, 138, 152–157, 236f., 278
Dreyfus, Alfred 172
Driesch, Hans 228, 230–232
Dubois, Eugène 178
Duilhé de Saint-Projet, François 271
Dujovne, León 231
Dumas, Alexandre 112
Duprat, Luis 202
Durañona y Vedia, Agustín 272f.
Durelli, Augusto J. 247f., 255, 257–259, 263, 270–274, 286

Echeverría, Esteban 104f., 112
Edison, Thomas 135
Einstein, Albert 166, 188, 224–226, 228
Eisenstadt, Shmuel 10
Espinosa, Mariano 213, 215
Esponda, Carlos 268
Estrada, José Manuel 50, 100, 106, 103, 128–132, 147, 153f., 159, 176, 198, 202, 213, 219
Estrada, Santiago 153f.
Etcheverry Boneo, Luis María 260, 273
Euler, Leonhardt 149
Evans, Charles Lovatt 254

Fabre, Henri 20, 204
Falkner, Thomas 26, 32, 35, 45, 58
Faraday, Michael 149, 210
Farrell, Edelmiro 223, 252, 254, 258
Fasciolo, Juan 254
Feijóo, Benito 30f., 58, 76f.
Fernández, Juan 62–64
Fernández, Melchor 62f.

Fernández de Agüero, Juan Manuel 99, 103–105
Fernández Long, Hilario 272
Fernández Verano, Alfredo 238
Ferns, Henry 85, 229
Ferrari, Gaspar S. 64, 162
Ferraris, Carlo G. 92f.
Ferreira, José Alfredo 160–162
Finke, Roger 16
Flexner, Abraham 251
Flourens, Jean-Pierre 140
Fol, Hermann 210
Fontana, Luis J. 89, 153
Franceschelli, Ferdinando 41
Franceschi, Gustavo 174, 236f., 241, 244, 258
Franco, Francisco 20, 224, 234, 236, 247, 257f., 285
Franklin, Benjamin 78, 173
Fraser, John Foster 110
Frías, Félix 48, 112f., 125, 147
Frondizi, Arturo 263, 286
Frondizi, Risieri 264
Frye, Northrop 188
Fulton, Robert 124
Funes, Gregorio 56, 82, 84, 126
Furlong, Guillermo 21, 24, 34, 36–43, 57, 61, 72, 74, 90, 96f., 201, 206, 208, 228, 239

Gache, Samuel 151
Galiani, Ferdinando 71
Galileo 58, 60f., 75, 93, 120, 124–126, 132f., 148, 151, 153, 231
Gallardo, Ángel 187, 196f., 199, 206–213, 218, 245, 251, 274, 276, 285, 289
Gallardo, José León 218
Galton, Francis 211, 242
Gálvez, Manuel 160, 163, 197
García, Juan Agustín 22, 29, 53, 59f., 62f., 77, 155f.
García Pintos, Salvador 240f.
Garibaldi, Giuseppe 128, 151
Gaudry, Jean-Albert 177f.
Gaviola, Enrique 216f.
Gemelli, Agostino 243f.
Genovesi, Antonio 71

Giard, Alfred 209
Gil, Martín 34, 169f., 173, 187, 211
Gilman, Daniel Coit 255
Gmelin, Johann F. 65
Gómez, José Valentín 63, 83, 105, 111, 161
Gómez de Vidaurre, Felipe 27
Góngora, Mario 55, 82
González, Joaquín V. 162, 173f.
Gorriti, Juan Ignacio 55–57
Gould, Benjamin A. 118, 132, 135–137, 150, 216
Goyena, Pedro 48, 113, 129, 144, 146, 148f., 152f., 198
Grammatici, Nicasius 37
Gregory, Brad 13, 280, 282
Gregory XV (Pope) 20
Grote, Friedrich (Federico) 167f., 198, 282
Guénon, René 235
Guevara, José 26–28
Guignard, Léon 209
Gutiérrez, Juan María 48, 74, 88, 94f., 104, 106–109, 112f., 118, 128f.
Gutiérrez Lanza, Mariano 226

Haeckel, Ernst 17, 131f., 140, 142, 158, 164, 177f., 183, 187f., 218, 280, 282
Harris, Steven 21, 29, 44f., 81
Harrison, Frederic 162
Harrison, Peter 6, 12
Hartog, Marcus 210
Hatcher, John Bell 178
Herder, Johann Gottfried von 112, 154
Hernández, Rafael 21, 106, 167, 201
Hessen, Johannes 226
Hieronymus, Georg 279
Hipparchus 135
Hochwälder, Fritz 19
Holmberg, Eduard Ladislaus Kaunitz Baron von 137
Holmberg, Eduardo L. 134, 137–140, 142, 149f., 155, 158, 166, 187, 189, 219, 280
Houssay, Bernardo 212, 247–256, 259f., 262–266, 270, 274, 285, 291
Hrdlička, Aleš 179
Hsia, Ronnie Po-chia 23, 46
Hudson, William 228
Hugo, Victor 112, 159, 200, 267, 280

Humboldt, Alexander von 40, 48, 88f., 150
Huxley, Thomas H. 134, 152f., 203

Ibarguren, Carlos 170, 197, 208
Ignatius of Loyola 28
Ildephonsus 70
Ingenieros, José 88, 93, 164, 174, 179f., 190, 229, 258, 283
Isabelle, Arsène 93

Jacolliot, Louis 138
Jakob, Christfried 159, 189–193
James, Constantin 4, 19, 23, 59, 85, 93, 111, 141, 155, 167, 287
Jansenius (Cornelius Jansen) 22
Jaurès, Jean 179, 283
Jefferson, Thomas 174
Jekhowsky, Benjamin 214
Joffé, Roland 19
Jolís, José 25f., 29f.
Joshua 132
Jouffroy, Théodore 112
Jovellanos, Gaspar Melchor de 71, 77
Justo, Agustín P. 196, 206, 208, 223, 251, 257
Justo, Juan B. 229
Justo, Sara 195

Kelvin, Lord (William Thomson) 210
Kepler, Johannes 138, 148
Kino, Eusebius 35, 43
Kircher, Athanasius 24, 30, 36, 44
Kirwan, Richard 66f., 76
Klimann, Maurice 181
Kögler, Ignaz 37
Korn, Alejandro 61, 143, 145, 160f., 190
Kraglievich, Lucas 183f.

La Hire, Philippe de 37
Laclau, Narciso C. 230f.
Lafinur, Crisóstomo 99–103, 108
Laguna, Andrés 33
Lamarca, Emilio 198, 218, 281f.
Lamarck, Jean-Baptiste 158, 177, 213
Lammenais, Félicité de 112
Laplace, Pierre-Simon 133
Laromiguière, Pierre 102

Larrañaga, Dámaso   52, 62–70, 75, 78 f., 96–98
Larsen, Juan Mariano   154
Lavardén, Manuel José de   73 f., 76
Le Dantec, Félix   209
Legón, Faustino J.   273
Leguizamón, Onésimo   132 f., 135, 137, 140
Leloir, Luis F.   254
Leo XIII (Pope)   171
Lesser, Friedrich Christian   77
Lewis, Juan Traherne   247–256, 259, 262 f., 266, 270–274, 285 f.
Lewis, Olive Mary   248
Lewis, Thomas   249
Lindberg, David   6
Liniers, Jacques (Santiago) de   128
Linnaeus (Carl von Linné)   30, 65, 68–70, 78, 98, 232
Littré, Émile   141
Llanas Jubero, Eduardo   144 f.
Llorente, Juan A.   82
Locke, John   99, 101
Loisel, Gustave   209
López, Octavio   243
López, Vicente   83
López, Vicente Fidel   83, 111, 124, 154
Loreto, marquis of   130
Love, George Thomas   87, 92
Löwith, Karl   5
Lozano, Pedro   26–30
Lugones, Leopoldo   164–166, 170, 197, 225
Luther, Martin   133
Luys, Jean-Bernard   145
Lydekker, Richard   178
Lynch Arribálzaga (brothers, Enrique and Félix)   204

Mac Donagh, Emiliano J.   223 f., 228, 230–234, 242–245, 260, 272, 285
Macaulay, Thomas B.   48
Mach, Ernst   231
Machoni, Antonio   29
Maciel, Baltasar   57, 108
Magendie, François   102, 108
Maignan, Emmanuel   59
Malthus, Thomas R.   139
Mann, Horace   160

Marañón, Gregorio   43, 239
Mariño, Cosme   158, 167–169
Maritain, Jacques   224, 226, 234–236, 244, 249, 257
Markgraf, Georg   33
Márquez de la Plata, José   51
Martínez Zuviría, Gustavo   252
Mascardi, Nicolò   35 f.
Masella, Antonio   91
Maspero, Gaston   154
Matera, Luigi   114
Matignon, Ambroise   169
Maton, William   65
Maurras, Charles   207, 224, 234
Mavor, William Fordyce   68
Medrano, Mariano   83 f., 106
Mejía, Jorge M.   264 f.
Mendel, Gregor   205, 210, 240, 242
Menéndez Pelayo, Marcelino   80, 108
Meyer, Camille   90, 181
Meyerson, Émile   231
Michelet, Jules   280
Mieli, Aldo   5
Miller, Richard   267 f.
Minkowski, Hermann   166, 225
Minnelli, Gustavo   128
Mitre, Bartolomé   48, 111–113
Moigno, François-Marie   141
Moleschott, Jacob   128
Molina, Juan Ignacio   27, 226
Montalembert, Charles de   112, 125, 128
Montenegro, Pedro   32–34
Montesquieu (Charles-Louis de Secondat)   19, 112
More, Thomas   19, 40, 70, 108, 122, 156
Moreau, Alicia   158
Moreno, Francisco P.   125–127, 160, 203, 280
Moreno, Manuel   94, 105
Moreno, Mariano   70
Moreno, Nicolás Besio   287
Mörner, Magnus   21–23
Morton, Samuel   127
Moses   62, 65 f., 129, 138, 145
Mossotti, Ottaviano Fabrizio   81, 94 f., 106 f.
Mounier, Emmanuel   258

Mozzoni, Umberto   265
Muckermann, Hermann   242
Mühn, Juan B.   204 f.
Mullaly (family)   169
Mulleady, Ricardo T.   221
Muñoz, Bartolomé   96–99
Muñoz, Juan M.   254
Murat, Joachim   98
Muratori, Ludovico   28
Musschenbroek, Pieter van   59, 78
Mussolini, Benito   207, 243

Napoleon (Bonaparte)   54, 98, 100, 107
Navarro Viola, Miguel   144 f., 168
Navàs, Longinos   205
Nevares, Nicanor de   182
Newman, John H.   54, 251
Newton, Isaac   5, 58, 61, 72, 100, 133, 135, 138, 148 f.
Nicolai, Georg Friedrich   228, 230 f.
Nicolas, Auguste   47, 129 f., 154
Nieremberg, Eusebius   24, 30, 44
Nieto, Hakhim   38
Noah   130
Nollet, Jean-Antoine   50, 58, 78
Nysten, Pierre-H.   141

Ocampo, Victoria   258
Ochoa, Javier   241
O'Farrell, Santiago   198
O'Gorman, Michael   57
Olcott, Henry Steel   164
Oliveira Salazar, António de   236
Onelli, Clemente   189 f., 192
Ortega y Gasset, José   224
Ortiz, Roberto M.   216, 223, 225

Page, Irvine H.   254
Palacios, Alfredo   129, 164, 167, 173, 207
Paneth, Thomas   268
Parish, Woodbine   85
Parodi, Lorenzo   184–186
Pasqualini, Rodolfo   260
Pasteur, Louis   143, 271
Paucke, Florian   26, 32
Pearson, Karl   210 f.
Pende, Nicola   243

Pennington, Arthur Stuart   179
Peralta y Barnuevo, Pedro de   37
Pereyra, Elías del Carmen   59–63, 77
Pérez Castellano, José Manuel   96
Perón, Juan D.   223, 247, 250, 252, 256, 259–261, 264–270, 274, 285 f.
Perrine, Charles D.   195, 216
Peschke, Heinrich   34
Petroschi, Joannes   41
Petty, Miguel A. (Michael)   221, 240
Peyret, Alejo   121–123
Philolaus   132
Piattoli, Scipione   86
Pico, César E.   223–228, 231, 234–236, 239 f., 243–245
Pinard, Adolphe   238
Piñero, Antonio F.   172 f.
Piso, Willem   33
Pius VII (Pope)   82, 100
Pius IX (Pope)   106, 235
Pius X (Pope)   214, 222
Pius XI (Pope)   215, 221, 223, 241 f.
Pius XII (Pope)   232
Pizarro, Manuel D.   127, 147
Planck, Max   181
Planella, Juan   169 f.
Plato   19, 218
Pliny (the Elder)   70, 135
Pluche, Noël-Antoine   76
Poincaré, Raymond   290
Pombal, Marquis of   22
Ponce, Aníbal   231
Porphyry   122
Poulat, Émile   221 f.
Primo de Rivera, Miguel   239
Ptolemy   132
Puig, Ignacio   111, 217, 267
Pujiula, Javier   180 f., 245

Quatrefages, Jean-Louis   140
Quesada, Ernesto   287
Quesnay, François   71
Quiroga, José   37, 40–42

Ramírez, Pedro P.   76, 223, 252
Ramos Mejía, José María   154, 167, 189–191

Rebaudi, Ovidio 167
Rechberg, Karl 36
Reggini, Horacio C. 272
Reinke, Johann 210
Renan, Ernest 122, 153, 172, 280
Rey Pastor, Julio 225
Ricci, Matteo 46
Richet, Charles 172
Richter, Ronald 269, 286
Riemann, Bernhard 166
Riva, Benito 57–59, 61
Rivadavia, Bernardino 80–85, 87–95, 97–100, 103–108, 110, 236, 276 f.
Robertson, William 26
Robin, Charles 141
Roca, Julio A. 113 f., 116, 139, 147, 200 f., 278 f.
Rodin, Auguste 172
Rodney, Caesar A. 16, 87
Rodríguez, Ángel 225
Rodríguez, Cayetano 59 f., 62
Rodríguez, Martín 80
Rodríguez de Campomanes, Pedro 21
Roger, Clémence 26, 141
Rojas, Ricardo 180, 197, 248, 261, 263
Romero, Antonio A. 185
Romero, José L. 264
Roosevelt, Theodore 212
Rosas, Juan Manuel de 80, 98, 105 f., 111 f., 116
Rossiter, Peter 170
Rousseau, Jean-Jacques 70 f., 101, 112, 218
Rubiés, Joan-Pau 45
Ruiz de Montoya, Antonio 28, 30
Russell, Bertrand 1, 5, 241, 251

Sack, Alexander P. 87 f.
Sáenz, Antonio 81, 104, 108, 214
Saint-Hilaire, Auguste de 66 f., 69, 112
Saint-Pierre, Bernardin de 138
Saint-Simon, Henri de 5, 112
Sainte-Beuve, Charles-Augustin 280
San Martín, José de 97
Sánchez Labrador, José 25, 27, 30 f., 39, 77
Saraiva, Mateus 39
Sardá y Salvany, Félix 168

Sarmiento, Domingo F. 111–113, 115, 117 f., 133–138, 140, 142, 146 f., 150–153, 159 f., 197, 216, 278 f.
Sarton, George 5
Scalabrini, Pietro (Pedro) J. 160 f.
Scarnicchia, Angelo 167
Schmidt, Martin 36
Schmitz, Hermann 205
Schopenhauer, Arthur 167
Scott, William B. 29, 35, 152, 178
Secchi, Pietro A. 149
Seguens, Jean 59
Segurola, Saturnino 96
Seibold, Jorge 267 f.
Sepp, Anton 31, 36
Seutter, Matthaeus 41
Sigaud de la Font, Joseph-Aignan 78
Simpson, George G. 176, 178
Sisson, Henri 195
Smith, Adam 14, 71, 278
Somellera, Pedro 99
Spencer, Herbert 134, 147, 151–153, 158–160, 162 f., 167, 173, 178, 184, 188, 280 f., 287
Stansel, Valentin 24, 36
Stearns, George A. 160
Stearns, Julia A. 160
Stepan, Nancy 210, 238, 243
Strong, John D. 217
Strümpell, Adolf von 189
Sturzo, Luigi 259
Suárez, Anastasio Mariano 63
Suárez, Buenaventura 36–40, 45, 48, 58, 97
Sullivan, Pedro José 50 f.

Taine, Hippolyte 171, 280
Tamburini, Michelangelo 43
Taquini, Alberto 254
Taylor, Charles 8, 10–13, 95, 114 f., 221 f.
Teilhard de Chardin, Pierre 181
Tellier, Charles 203
Temkin, Owsei 102
Termeyer, José María 26, 29 f., 45
Tiaragú, José (Sepé) 20
Tonda, Américo 55 f., 82–84, 97
Torino, Inocencio 92, 142 f.

Torquemada, Tomás de 148
Tosca, Tomás V. 58 f.
Tournouër, André 178
Trincado, Joaquín 170
Truxillo, Manuel María 59
Turgot, Anne Robert 5
Tyndall, John 9, 152, 167

Ubach, José 186 f., 200, 202, 218, 225, 239
Uriburu, Félix 196, 223, 227, 257
Urquiza, Justo José de 111

Vacherot, Étienne 122
Valle, Aristóbulo del 127
Varela, Luis Vicente 105, 124
Velasco, Juan de 27
Vélez, Luis 127
Vértiz, Juan José de 57
Vico, Giambattista 112
Vicq d'Azyr, Félix 86
Victor Emmanuel I 92
Vignati, Milcíades 185 f.
Villafañe, Benjamín 121–123
Villemain, Abel-François 112
Virchow, Rudolf 127
Volta, Alessandro 149, 203
Voltaire (François-Marie Arouet) 5, 19, 22, 101, 112, 191

Wargentin, Pehr W. 37
Washington, George 23, 44, 47, 100, 174, 179, 259, 277
Wasmann, Erich 181, 183, 205
Weyenbergh, Hendrik 279 f.
Whiston, William 63
White, Andrew Dixon 4 f., 155
White, Gilbert 78
Whitehead, Alfred 21, 251
Wilde, Eduardo 23–25, 27, 116, 149–151, 166, 277
Williner, Gregorio 205, 268
Willis, Bailey 179
Wiseman, Nicholas 129 f., 154
Woodward, John 63

Yriberry, Arturo 268
Yrigoyen, Hipólito 196, 206, 223, 228

Zahm, John A. 212
Zavaleta, Diego E. 83
Zavaleta, Mariano 84
Zeballos, Estanislao 118
Ziegler, H. E. 210
Zola, Émile 172
Zorrilla de San Martín, Juan 175

Index of Places

Ambato 37
Amberg 37
Athens 40, 81
Augsburg 36, 41
Azcoitía 78

Bahia 36
Banda Oriental 64, 96
Barcelona 21, 37, 145, 154, 168 f.
Bavaria 189, 205
Beijing 37
Berlin 9, 180, 183, 188, 210, 214, 228, 267
Boston 9, 134
Brussels 138, 264
Buenos Aires 2 f., 5, 10 – 12, 17, 20 f., 24, 27, 31 – 36, 38 – 42, 48, 50, 52 – 64, 67, 71 – 101, 103 – 106, 108, 110 – 119, 121 – 126, 128 – 143, 145 – 156, 158 – 160, 162 – 177, 179 – 185, 187, 189 f., 193 – 210, 212 – 219, 221, 223 – 225, 227 – 231, 234 – 236, 238 – 240, 247 – 260, 262 – 273, 279 f., 282 – 287, 291

Cantabria 103
Cervera 101
Chaco 21, 26, 29 f., 40, 153
Chamonix 249
Charcas 20, 56, 58
Córdoba 20, 32, 34 – 36, 38 f., 50 – 52, 55 – 57, 59, 63, 74, 77, 82, 90, 96, 100, 116 – 118, 126 f., 134, 137, 147, 165, 195, 211, 217, 228 – 231, 273, 279, 285
Corrientes 159 – 161
Courland 209

East Sussex 184
Entre Ríos 111, 117, 121, 159 f.

Faenza 21

Galicia (Spain) 32, 71, 76, 180
Genoa 180, 188, 283
Guatemala 90

Hendaye 214

La Plata 3, 23, 42, 118, 125, 143, 153, 160, 173 f., 176 – 179, 182, 184, 186, 189, 193, 195 f., 204 f., 214, 216, 228, 231, 267, 283
Lima 37, 54, 90, 229
Lisbon 37 f., 76
London 2 f., 10, 13, 15, 23, 28, 32, 35, 38, 40, 53, 56, 68, 71, 80 f., 85, 87, 89, 93 – 96, 110, 112, 116, 125, 129 f., 132, 136, 152, 159, 164, 179, 211, 221, 235, 238, 240, 248 f., 254, 264, 277, 279, 289
Luján 84, 176, 180, 182, 188, 282 f.

Madrid 20 – 22, 26, 28 f., 32, 37, 43 f., 54, 56, 59, 70, 76 f., 80, 108, 119, 129 f., 151, 155, 162, 207, 212, 225, 239
Mar del Plata 236
Mendoza 55, 57 f., 61, 83, 90, 148, 159, 166, 225, 248, 255, 268
Mercedes 161, 189
Milan 26, 94, 243, 288
Miramar 184 – 188, 192
Moneglia 182
Montevideo 52, 64 f., 67, 70, 76, 78, 97 f., 112 f., 181

New York 4 f., 9 f., 14, 16, 22, 24, 40, 53 f., 66, 87, 89, 95, 110, 114, 119, 140, 150, 152, 162, 165, 169, 172, 176, 178, 209 f., 212, 228 f., 242, 248 f., 252, 262, 283

Pampas 35, 54, 116, 135, 139, 176, 181, 186, 213, 228, 279, 283
Paraná River 32, 65
Paris 9, 11, 19, 29, 47 f., 53, 55, 66 f., 78, 80, 83, 86, 92 – 94, 98 – 100, 102, 113, 119 – 122, 130, 138, 141 f., 144 f., 154, 169, 171 f., 174, 179, 186, 195, 203 f., 209 f., 214, 222, 226, 236, 257, 280, 285, 288

Index of Places

Patagonia 26, 35f., 40, 55, 125f., 129, 135, 146, 178, 198, 253, 269
Piacenza 160
Piedmont 92
Pittsburgh 4f., 27, 47, 165, 176f., 217
Potosí 52, 76
Punta Arenas 253

Quito 27, 219

Rio de Janeiro 10, 39, 64
Río de la Plata 3, 11, 16, 19–21, 23, 25–27, 31f., 34f., 40, 43f., 47–58, 60–62, 64–66, 68, 70–83, 85–87, 89, 91–93, 95–98, 103, 107f., 116, 128f., 132, 181, 187, 275f., 279, 282, 284
Rio Grande do Sul 20
Rome 21f., 27, 36, 41, 44, 48, 55, 82, 120, 128f., 153, 156, 179, 198f., 213, 215, 218, 264, 283, 291
Rosario 28, 55, 82, 118, 199, 229, 248–251, 253f., 256, 262, 266, 271, 282

Salta 55, 82

San Andrés de Giles 169f.
San Cosme 36, 40
San Miguel 42, 205, 217, 262, 267f., 285f.
Santa Cruz de la Sierra 42
Santa Fe 82–84, 121, 140, 164, 200f., 205, 250, 266
Santiago de Chile 10, 54, 103, 196, 219, 267
Siam 40
St. Petersburg 37

Tortosa 180, 200, 202, 205, 284
Tucumán 26f., 29, 31f., 121, 134, 136, 213, 261f.
Tyrol 31, 205

Uruguay River 20, 32, 65, 159

Valkenburg 205, 267
Vatican 4, 120, 125, 171f., 199, 225, 265, 278

Warsaw 40
Worcestershire 35

Index of Subjects

Academy   71, 100, 132, 136, 140, 206, 232, 256, 263
- Académie des sciences 214
- Académie française 171
- Academy of Mathematics   57, 72, 78
- Academy of Medicine   81
- Academy of Sciences   118, 134, 279
- Nautical Academy 72, 78
Acclimation Garden   86–88, 204
Action Française   207
Adam   29, 65, 76, 120, 126 f., 179, 218
agnostic   117, 138, 187, 233
almanac   37, 97
anarchism   158, 165, 195, 198, 282, 284
anti-Catholic   161, 190, 225, 264, 274, 277
anticlerical   5, 48, 74, 82, 95, 115 f., 160, 170, 187, 193, 198, 209, 229, 253, 274, 279, 282, 285, 289
anti-Darwinian   120, 140, 142, 209
anti-evolutionist   134, 140, 142 f., 153, 159, 175, 245, 283
anti-Jesuit   20–22, 48, 57, 125
anti-religious   6, 53, 107, 121 f., 128, 148, 159 f., 163, 186, 218, 279, 284, 288
anti-spiritualist   168 f.
Archdiocesan Seminary in Buenos Aires   169, 181, 183, 186, 200, 202, 205, 226, 240
Archdiocese of Buenos Aires   174, 202, 214 f.
Argentine Association for the Progress of Science   255
Argentine Association of Biotypology   243
Argentine Eugenics Society   238
Argentine Hebraic Society   235
Argentine League of Social Prophylaxis   238
Argentine Medical Circle   119, 132, 291
Argentine Rural Society   204
Argentine Scientific Society   118
Argentine Society of Natural History   209, 213
Aristotelian   26, 51, 58 f., 226 f., 230, 251

astronomy   20, 25, 35, 37, 39, 44–46, 58 f., 72, 126, 132, 149, 173, 181, 196, 200, 206, 213, 216, 267 f., 277, 285
atheism   6, 13, 16, 104 f., 124, 136, 151 f., 161, 187, 233, 270
atomic   237, 268 f.
Augustinian   61, 63, 130, 225
Author of nature   51, 68 f., 201

bankruptcy of science   159, 171, 173–175
Baring Brothers   85
baroque   10, 13, 21 f., 24, 36, 44, 55, 107
Basque   64, 78, 202
belle époque   195, 198, 290
Betharramites   202, 214
Bethlemite   85, 89–91
Bible   40, 52, 59 f., 62 f., 66–68, 75 f., 121, 125–127, 138, 151, 158, 172, 186 f., 271
Biblical chronology   67, 127
Biblical geology   16, 52, 62, 64 f.
Bourbon   22 f., 47, 50, 53, 55, 71 f., 75, 82 f., 276
Bourbon reforms   53, 79
Brahmanism   74
brain   99, 101 f., 126, 137, 172 f., 189–193, 279
Bridgewater Treatises   287
British invasions   86, 91, 128
Buddhism   121

cabinet de curiosités   64, 93, 96 f.
cartography   20, 25, 39–41, 43, 45, 160, 277
Casti connubii   241
Catalan 29, 168 f., 205
catastrophism   134
Catholic Church   3 f., 9, 11 f., 16 f., 23, 54, 82, 107, 110, 114 f., 148, 196, 198–200, 203, 207, 221, 227, 241, 243, 247, 265, 278, 281, 285, 290
Catholic Congress   147
Catholic culture   1, 158, 199, 222

Index of Subjects

Catholic Institute of Sciences   248, 260, 285, 291
Catholic nationalism   223, 227, 268, 270
Catholic Popular Party   259
Catholic revival   17, 199, 219, 222, 232, 240, 274, 276, 290
Catholic science   17, 169, 194, 196, 200, 218 f., 222, 244–248, 261, 267, 270, 274, 276, 280, 284–286, 290
Catholic universities   261, 263–265, 271–274, 286
caudillos   80, 105, 113, 278
cemetery   85–88, 134, 195
Centre of Religious Studies for Ladies   199
Chamber of Deputies   124, 148 f., 152, 172 f., 182, 207, 282 f., 286
Charity   33, 64, 89, 147, 250, 261
Christendom   4 f., 11, 155, 223, 227, 284
Christian Democracy   199, 236, 247, 250, 266, 272, 285
Christian Democratic League   202
Christianity   2, 4, 6, 8, 11 f., 14, 16, 23, 25, 33, 44, 46 f., 101, 105, 112, 121 f., 126, 149 f., 194, 200, 218, 261, 288
civic funeral   179, 283
clergy   3, 16, 96, 106 f., 170, 199, 244, 253, 282, 288
– regular clergy   82, 84, 107
– secular clergy   50, 54, 56, 83, 90, 91, 107, 277
College   20, 22, 24, 32, 34 f., 38, 45, 64, 83 f., 129, 213, 249, 254
– Colegio imperial   44
– Colegio Máximo of San Miguel   205
– College of Ecclesiastical Studies   84
– College of El Salvador   114, 140, 181 f., 186, 202, 204, 206, 219, 224
– College of Montserrat   50
– College of San Carlos   57, 60, 69, 71, 73, 83, 97, 103
– Collegium romanum   44
– Downing College, Cambridge   221
– University College, London   249, 254
comet   35, 133, 195, 201, 214
– comet Encke   94
Communist Party   231, 262
Concordat   100, 115, 286

conflict thesis   4 f., 8, 109–111, 116, 119, 123–125, 132, 148 f., 151 f., 154, 156 f., 203, 218
Congregation of Saint Vincent de Paul   242
Congregation of the Index   60, 127
Consortium of Catholic Physicians   221, 239, 291
Constitution   86, 95, 111, 124, 128, 147, 178, 191
– Constitution of 1819   80, 84
– Constitution of 1826   84
– Constitution of 1853   12
convent   57, 62, 84 f., 88, 288
– Convent of Saint Francis   73
– Convent of Santo Domingo   91–95, 106, 109, 277
– Convent of the Recollect Franciscans   86
Copernican   58 f., 62, 120, 219
cosmic evolution   127, 166, 174
Council of Trent   75
Counter-reformation   55
Courses of Catholic Culture   199, 222, 260, 291
Creator   68, 72 f., 78, 95, 112, 137, 141, 183, 233

Darwinism   2, 17, 109, 111, 119, 131 f., 134, 137, 139 f., 142, 156 f., 160, 188, 209, 245, 279 f., 282 f.
Declaration of independence   54 f.
deism   12 f., 38, 51 f., 70, 73 f., 76, 93, 101, 105, 109, 112, 133, 138, 140, 148, 276
De La Salle Brothers   203 f.
Dispute of the New World   27
divine laws   70
Dominican   54, 85, 91, 106, 176, 195

ecclesiastical reform   16, 75, 81–85, 91, 107, 275, 277
eclipse   35, 37, 39, 97, 202, 209
économistes   71, 77
educational reform   50, 146, 150, 152–154, 278
electricity   78, 130, 169, 272
Enlightenment   1, 5 f., 10, 12–16, 25, 45, 47–50, 56, 72–75, 78 f., 89, 95, 99,

Index of Subjects — 361

107f., 112, 129, 133, 190, 229, 275f., 278, 287f.
- Bourbon Enlightenment 72
- Catholic Enlightenment 49, 50, 53, 74–79, 108, 275, 276
- "Enlightened Republic" 80, 95, 288
- Enlightenment science 47f.
- French Enlightenment 19, 22f., 116, 156, 288
- Spanish Enlightenment 53
entomology 183, 200, 204f., 228, 268
epidemic 31, 34, 38
epistemology 144, 226f., 235
Etsi longuissimo 82
eugenics 210, 238–245, 285
evolutionary theory 110, 119, 139, 148, 213, 245, 280, 283
evolutionism 17, 111, 128, 131f., 137, 140f., 153f., 163, 165, 175, 181, 183, 187–189, 194, 212, 218, 232, 245, 280f., 283–285
experimental 32, 81, 118, 127, 144, 189, 192, 210, 225f., 255, 259, 263, 271
- experimental physics 51, 81, 83, 92–94, 206
- experimental science 4, 57, 58, 75, 79, 118, 140

fascism 197, 207f., 224, 230, 235, 238, 243f., 258
Federalist 80, 94, 105f., 111
fin-de siècle 115, 160, 162, 166, 197, 281
First Triumvirate 80, 97
First World War 158, 170, 174f., 190, 194, 197, 242, 284
Flood 63, 65–67, 130
fossil 63–65, 67, 126, 141, 176, 178f., 186
Franciscan 50f., 54, 56, 58f., 61–63, 73, 79, 86, 103, 199, 243
Freemason 5, 74, 106, 116, 121, 133, 158, 160, 194
freethought 9, 158, 164, 275, 282, 284
French Directory 99
French pattern of secularisation 2, 4, 8, 10f., 17, 19, 21–23, 26, 29–31, 37, 41, 48, 66, 69f., 76–78, 86–88, 93f., 98–100, 102, 107f., 112, 114, 116, 118, 120–

122, 129, 133f., 140–142, 146, 154, 156, 159, 161, 163f., 171, 174, 177, 179, 181, 188, 195, 200, 202–205, 210, 212, 218, 222, 224, 251, 257, 259, 281, 283f., 287–290
French Revolution 99, 113, 276, 287
friars 50, 54, 64, 86, 89f., 92, 95, 106f., 229

Gallicanism 22f., 55, 75, 79, 82f., 106f., 115, 199, 219, 222, 277, 288
Generation 55, 116, 124, 127, 130, 134, 137, 156, 159, 218
- Generation of 1837 111f.
- Generation of 1880 113, 161
Genesis 28, 60, 62f., 66f., 103, 120, 125f., 131, 145, 147, 150, 153
geology 62, 66, 134, 152, 218
gravity 56–58, 61, 66, 103
Guaraní 19–24, 28f., 31, 33f., 36, 48

heliocentrism 58–61
herbals 31f., 34, 97
Histoire naturelle 48, 76, 204
history of the earth 62, 66f., 76
Holy Alliance 107, 277, 287
Holy Roman Empire 45
Holy See 11, 54, 82, 114, 286
Hospital 32, 85f., 143, 242, 253
- French Hospital in Buenos Aires 94
- Men's Psychiatric Asylum 189
- Santa Catalina 90f.
- secularisation of hospitals 89–91
- St Thomas' Hospital 221
- Women's Asylum 172, 189, 240
Hullett Brothers 87
human evolution 176, 186, 276
Humani generis 125, 181, 232

Idéologie 17, 80, 93, 99–101, 108
ideology 1, 10, 22, 80, 95, 100, 103, 111, 116, 135, 156, 190, 243, 274f., 277, 281, 284, 288
immortality 101, 127
Indians 21, 23, 105, 139
Inquisition 4, 38, 54, 169, 231, 241
Insect 28, 68, 70, 74, 77, 92f., 193

Institut Catholique 257
Institute of Biology and Experimental Medicine 255
integral Catholicism 208, 222, 234, 238, 244, 246, 257, 285
International Eucharistic Congress 217, 221, 223
Inter Sollicitudines 171
Irish 35, 57, 66, 91, 146, 169, 210, 242

Jansenism 22f., 55f., 82f.
Jesuits 19–29, 31f., 34–36, 38–48, 52f., 56, 58f., 74, 90, 125, 181, 186, 188, 200, 218, 226, 229, 239, 243f., 267f., 276, 284
– Jesuit astronomers 36f., 217, 225, 267
– Jesuit entomologists 183, 204–206
– Jesuit maps 40–43
– Jesuit medicine 31–35
– Jesuit missionaries 21, 28, 37, 43, 47
– Jesuit missions 19, 21, 23–25, 31–35, 37, 39, 42f., 45, 214
– Jesuit natural histories 25–35, 44–47
– Jesuit science 19–49, 200, 206, 275, 277
– Society of Jesus 16, 20–23, 26, 28, 32, 39, 41–43, 47f., 56, 96, 180, 200, 201, 204, 267
Jew 3, 54, 234
Jewish 3, 38, 231, 257
Josephinism 82

Kaiser Wilhelm Gesellschaft 255
Kaiser-Wilhelm-Institut für Anthropologie, menschliche Erblehre und Eugenik 242
Kulturkampf 131

laboratory 154, 170, 180, 189, 201, 248f., 255, 258, 269, 272
– laboratory of chemistry 92
– laboratory of physics 93
– laboratories of physics and chemistry 106, 203
laïcité 9–11, 114, 289

laws 14f., 17, 21, 57, 66, 70, 93f., 112, 126, 149, 164, 198, 205, 211, 239f., 244, 278
lay brothers 32, 91
liberal Catholicism 17, 112f., 229, 249, 276, 285
liberal elite 17, 194, 198, 219, 275, 281, 289
liberalism 71, 82, 100, 106f., 114, 131, 138, 147, 156, 159, 197, 224, 229, 270, 275

madness 143–145, 167, 173, 189f., 240
Marist Brothers 203
materialism 12, 17, 94, 99, 102, 104f., 108, 113, 132, 138, 140f., 145, 154, 159f., 163–165, 177, 187, 190–193, 231, 270, 280–282
materia medica 20, 33, 94
mathematics 38f., 56f., 70–72, 78, 209, 261, 266, 272
medical botany 25, 32, 44–46
megatherium 67, 130, 176
Mendelian 210f., 242f.
metamorphosis 70
metaphysics 71–73, 100, 103, 122, 141, 152, 167, 226f., 233f., 271
Middle Ages 133, 161, 235
Military Medical Institute 101
Modernity 1, 5, 8f., 75, 117, 156, 190, 226, 235, 244, 279, 281
monism 150, 164, 183
Museum 92f., 95–99, 106, 125, 152, 160, 176, 181, 183–185, 213, 228, 279, 288
– American Museum of Natural History 242
– British Museum 178, 204
– Deutsche Entomologische National-Museum 204
– Museum d'Histoire Naturelle in Paris 178, 204
– Museum of Natural History 64, 97, 183, 201, 203f., 236, 242
– Museum of Natural Sciences in Buenos Aires 106, 209
– Public Museum 81, 118, 134f., 209
– San José school museum 203

National Bacteriological Institute 224

Index of Subjects — 363

National Board of Intellectuals 268
National Council of Education 206, 291
nationalism 17, 170, 173, 176f., 196f., 199, 207f., 216f., 219, 223f., 227, 230, 235, 243, 245, 247, 251–253, 255, 257f., 269, 276, 283–285
National Research Council of Science and Technology 270
natural equilibrium 68–70, 78
naturalism 132, 148, 279
natural laws 14, 70, 93
natural philosophy 12, 38, 51f., 57–62, 64, 67, 70f., 75–77, 79, 83f., 95, 103, 206, 219, 227, 268
natural religion 73, 77, 101, 103, 105, 109, 193
natural theology 14, 16, 50–52, 67, 70–73, 75–78, 80, 96, 150, 275f., 287
nature 1, 6, 15, 24–26, 30f., 36, 44, 47–49, 51f., 61, 65–70, 72, 74–76, 78f., 101–104, 106, 109, 112, 117, 131, 135, 152, 162, 177, 193, 201, 206, 210f., 221, 225, 230, 233f., 239f., 255f., 259, 279f.
"neo-Durkheimian" societies 10
neo-Kantian 232
neo-Malthusianism 239, 241
neo-scholastic 123, 224, 269, 276
Neptunism 67, 76
Newtonianism 38, 187
New World 24–31, 33, 36, 38, 41, 46f., 68, 89, 108
Normal School of Paraná 159f.

observatory 92, 94f., 117f., 136f., 195f., 200–202, 204f., 225, 267f., 284–286, 288
– Catholic observatories in Argentina 216f.
– Ebro Observatory 200, 205, 217, 284–285
– Jesuit Observatory in Manila 267
– Mount Wilson 217
– National Astronomical Observatory 132, 135, 137, 201, 216f.
– National Council for Observatories 216
– Observatory of Brera 94
– Observatory of La Plata 195, 214, 267
– Observatory of San Miguel 217, 267, 269, 285f.
– Observatory of the Castle of Abbadia 214
– Observatory of the National College of Buenos Aires 202
– Paris Observatory 214
occultism 30, 58, 164–166, 170, 194
origin of the human being 176, 186, 213, 283

palaeontologist 159f., 175, 177
"paleo-Durkheimian" societies 10f., 221, 275
pantheism 121, 126, 129, 138, 140, 280
parish priest 51, 107, 169
patronato 12, 54, 82, 115, 143, 199
pattern of secularisation 155, 288, 290
Peronism 248, 250, 269f.
pharmacy 32, 92, 94, 209
physica 58f., 61, 71, 83
physico-theology 70, 76f., 101, 201
physiology 93, 99, 102, 104, 141, 144f., 248f., 253f., 261, 263, 266, 268
polygenism 126–128
positivism 1, 110, 116, 118, 122, 132, 134, 153, 155–157, 159f., 162–165, 167, 172, 174, 179, 188, 191, 194, 200, 211, 225, 280–284
private universities 250f., 255–257, 259, 262–264, 286
Protestant 3–5, 10, 43, 54, 68, 76, 85, 87, 107, 115–117, 155, 279, 281, 288
Protestantism 3f., 121, 137, 161
Protomedicato 57, 101
Providence 33, 52, 67–70, 74, 78, 139, 201
pscyhology 190, 243
psychiatry 145, 154, 164, 167, 172f., 188–190, 240
psychology 121, 144, 166, 169, 189
Ptolemaic 58, 219
Public Library 60, 64, 70, 76, 88, 96f.

Radical Civic Union 180, 196, 206
railway 117, 208
Recollect Franciscans 85f.
Reformation 13, 22, 133

regalism   22f., 55, 82f., 199, 222
relationships between science and religion   1, 6, 8f., 16, 23, 52, 75, 131–133, 140, 145, 156, 194, 201, 215, 219f., 226, 244f., 270, 275f., 287, 290
relativity   158, 225f., 228, 230, 285
Religion of Humanity   161f., 188, 280f.
religious beliefs   7f., 145, 151, 153, 203, 212, 250
religious space   276f., 288
Rerum novarum   171
Revelation   72f., 101, 103, 133, 141, 191, 237, 269
Revolution of 1810   55, 60, 78
Rockefeller Foundation   249
Romanisation   115, 117, 199f.
Romantic   104, 112f., 134, 144, 149, 164
Royal Astronomical Society   94
Royal Basque Society of Friends of the Country   78
Royal Society   38f., 211, 248

saint   31f., 71, 73, 121, 173, 179, 234, 257, 283
Saint Bartholomew's night   148
Saint-Simonism   284
Scholasticism   55f., 71, 75, 79
School of Medicine   91, 132, 160, 189, 224, 252–254, 279
– London School of Medicine for Women   248
– School of Medicine in Rosario   253
– School of Medicine of the Central University of Madrid   239
– School of Medicine of the Protomedicato   57, 101
– School of Medicine of the University of Buenos Aires   143, 248, 252, 261
– School of Medicine of the University of Maryland   94
schools   17, 56f., 68, 95, 112, 124, 132, 135, 156, 162, 200–203, 209, 251, 255, 260, 262f., 266, 276, 280, 284–286
– elementary schools   110, 116, 138, 168, 197, 228
– secondary schools   181, 201, 207, 213f.

science   1–9, 11–21, 24f., 29f., 32–34, 36–38, 40, 50, 52, 66–68, 70–81, 85, 89, 92–95, 98–100, 103, 106–110, 112f., 116–122, 124–144, 146, 148–156, 158–168, 170–176, 178–183, 185–188, 190f., 194–196, 200f., 203, 205f., 208–212, 214f., 218–222, 225–230, 232–235, 237, 241f., 244–246, 248–250, 253–255, 257, 260f., 265–282, 284–290
– cult of science 175, 218
– local science 3, 30, 43f.
– missionary science 20, 43–49
– modern science 12–14, 17, 56–62, 73, 117, 124, 130, 134, 166, 226, 244, 275
– patriotic science 176, 187
scientific institutions   85, 107, 110, 117, 119, 135, 196, 205f., 218, 275, 281
scientific instruments   50, 78, 93, 97
scientific theories   3, 16, 128, 245, 269, 289
scientism   119, 157, 171, 227, 237, 239, 243, 280
Second World War   219, 223, 247
secularisation   1f., 7–17, 75, 79, 85, 89, 91, 95, 109f., 114–116, 124, 146f., 156f., 168, 196, 198, 219f., 222, 229, 237, 265, 274–279, 281, 283, 288f.
– theory of secularisation   8, 15
– secularisation thesis   15
secularising discourse   150, 219
secular religion   17f., 159, 163, 180, 187, 284
secular space   43, 276f.
secular state   128, 143, 278, 283
sensibility   102f., 145
separation of Church from state   82, 84, 124, 147
shells   63, 65, 97f.
slaves   24, 53, 90
socialism   17, 112, 114, 121, 158, 163–165, 167, 170, 175f., 179, 181–183, 187f., 194–196, 198, 200, 207, 218f., 223, 229, 231, 244, 248, 274f., 282–284, 289
Socialist Party   168, 282
Sociedad Luz   177, 181

Society of Physico-Mathematical Sciences 81
solar physics 217, 267 f., 286
soul 31, 45, 71, 73, 87, 89, 101–103, 124, 127 f., 130, 134, 170, 173, 187, 189–193, 269
Spanish civil war 103, 236 f., 247, 257
spiritualism 17, 158 f., 164, 166–170, 194, 270, 282, 284
St. Andrew's Scotch Presbyterian Church 85
struggle for life 139, 147, 177
subtraction stories 12
supernatural 13, 15, 158, 161, 164 f., 170, 190, 193
superstition 1, 12, 22, 51, 68, 73, 107, 131, 152, 156, 179, 190, 278, 281, 288
Supreme Being 51, 73, 93
Syllabus 148, 156, 227

telescope 36 f., 51, 97, 106, 195, 201–203, 267
"Tertiary Man" 186 f.
Theatre of nature 77
The Origin of Species 134, 140 f., 177, 233
Theosophy 164–167, 174
Third Republic 11, 171 f., 209, 279, 284
Thomist 219, 227, 230, 232, 235, 237
Tychonic 219

ultramontane 115, 126, 151
unbelief 9, 111, 136, 222, 289
Unitarian 80, 105 f.
United Provinces of Río de la Plata 54, 64, 80, 82, 105
University 1–6, 8 f., 11–17, 20–25, 27–30, 33, 36, 38, 40 f., 43 f., 47 f., 50 f., 53 f., 56 f., 60 f., 66 f., 74, 77 f., 81, 85, 89, 91 f., 94 f., 99 f., 104, 107 f., 111 f., 114, 116, 118, 120, 128, 133 f., 141, 143, 159, 161, 165, 172 f., 177–180, 183 f., 188 f., 197, 204–213, 221, 224, 228–231, 239, 241, 243, 245, 247–261, 263–266, 269, 271–273, 276–280, 282–287, 291
– Catholic University of America 259, 271
– Catholic University of Argentina 205, 248, 250, 260, 265 f., 271–276
– Columbia University
– Gregorian University 213
– Harvard 160, 249, 251
– Illinois Institute of Technology 259, 271
– Polytéchnique Montréal 258
– Sorbonne 122, 171
– Stanford 254
– University of Birmingham 221
– University of Buenos Aires 48, 81, 94 f., 99–105, 128, 165, 172, 180, 189, 206, 208 f., 224, 231, 251, 256 f., 261, 264, 271, 286
– Faculty of Exact and Natural Sciences 118, 260
– Faculty of Philosophy and Letters 189
– School of Agronomy and Veterinary 173
– School of Engineering 263
– University of California 254
– University of Chuquisaca 20, 56, 63, 73
– University of Córdoba 20, 36, 38, 50, 74, 77, 100, 228–231, 285
– University of La Plata 173, 189, 204 f., 283
– University of Michigan 263
– University of Paris 257
– University of Pavia 94
– University of the Sacred Heart in Milan 243
University Reform 143, 228, 231, 251 f.
Unknowable 159, 162, 173, 281
Utilitarianism 17, 80 f., 99

vaccination 64, 96
Viceroyalty of Río de la Plata 53 f., 56
vitalism 141, 144, 210, 228, 230 f.
wisdom 72, 77, 103, 106, 234 f.

www.ingramcontent.com/pod-product-compliance
Lightning Source LLC
Chambersburg PA
CBHW030518230426
43665CB00010B/664